GREAT BASIN LOWER DEVONIAN BRACHIOPODA

The Geological Society of America, Inc.

Memoir 121

Great Basin Lower Devonian Brachiopoda

J. G. JOHNSON
Department of Geology
Oregon State University

Boulder, Colorado

1970

Copyright 1970, The Geological Society of America, Inc.
Library of Congress Catalog Card Number 70-98021
S.B.N 8137-1121-5

Published by
THE GEOLOGICAL SOCIETY OF AMERICA, INC.
P. O. Box 1719
Boulder, Colorado 80302
1970

Printed in the United States of America

The Memoir Series
of
The Geological Society of America, Inc.
is made possible
through the bequest of
Richard Alexander Fullerton Penrose, Jr.
and is partially supported by
a grant from
the National Science Foundation

Contents

Abstract 1

PART I

INTRODUCTION
General remarks 5
Scope and purpose 7
Methods of study 8
Location 8
Acknowledgments 10

PART II. BIOSTRATIGRAPHY

Previous Work
    Lower Nevada zones 15
    Pre-Nevada zones 17

The Faunal Sequence
    Sulphur Spring Range 19
    Northern Roberts Mountains 21
    Southern Roberts Mountains 24
    Lone Mountain 24
    East slope of the Monitor Range 28
    Toquima Range 30
    Coal Canyon, northern Simpson Park Range 32
    Cortez Range 36
    Lynn Window, Tuscarora Mountains 39
    Antelope Peak 39

AGE AND CORRELATION

Siegenian Stage
    Discussion 41
    Quadrithyris Zone 43
    Spinoplasia Zone 48
    Trematospira Zone 52

Emsian Stage
    Acrospirifer kobehana Zone 55
    Eurekaspirifer pinyonensis Zone 58

Summary 63

PART III. SYSTEMATIC PALEONTOLOGY
PHYLUM BRACHIOPODA
CLASS ARTICULATA

Order Orthida 69
  Suborder Orthoidea 69
    Superfamily Orthacea 69
      Family Hesperorthidae 69
        Subfamily Dolerorthinae 69
          Genus *Dolerorthis* 69
      Family Orthidae 71
        Subfamily Orthinae 71
          Genus *Orthostrophella* 71
  Suborder Dalmanelloidea 72
    Superfamily Dalmanellacea 74
      Family Dalmanellidae 74
        Subfamily Isorthinae 74
          Genus *Levenea* 74
        Subfamily Cortezorthinae 78
          Genus *Protocortezorthis* 78
          Genus *Cortezorthis* 79
      Family Rhipidomellidae 80
        Subfamily Rhipidomellinae 80
          Genus *Dalejina* 80
          Genus *Discomyorthis* n. gen. 84
          Genus *Rhipidomella* 85
    Superfamily Enteletacea 86
      Family Schizophoriidae 86
        Subfamily Schizophoriinae 86
          Genus *Schizophoria* 86
        Subfamily Draboviinae 89
          Genus *Muriferella* 90
Order Pentamerida 91
  Suborder Syntrophioidea 92
    Superfamily Camerellacea 92
      Family Camerellidae 92
          Genus *Anastrophia* 92
          Genus *"Camerella"* 93
  Suborder Pentameroidea 94
    Superfamily Pentameracea 94
      Family Gypidulidae 94
        Subfamily Gypidulinae 94
          Genus *Sieberella* 94
          Genus *Gypidula* 96
Order Strophomenida 101
  Suborder Strophomenoidea 101
    Superfamily Strophomenacea 102
      Family Strophomenidae 102
        Subfamily Leptaenoideinae 102
          Genus *Leptaenisca* 102
        Subfamily Leptaeninae 102
          Genus *Leptaena* 102
    Superfamily Davidsoniacea 106
      Family Schuchertellidae 106

Subfamily Schuchertellinae 106
   Genus *Schuchertella* 106
   Genus *Aesopomum* 111
Superfamily Stropheodontacea 112
  Family Strophonellidae 112
   Genus *Strophonella* 112
  Family Stropheodontidae 115
   Genus *Brachyprion* 115
   Genus *Mesodouvillina* 117
   Genus *McLearnites* 118
   Genus *Cymostrophia* 120
   Genus *Megastrophia* 120
   Genus *Stropheodonta* 123
  Family Leptostrophiidae 125
   Genus *Leptostrophia* 125
  Family Pholidostrophiidae 129
   Genus *Pholidostrophia* 129
   Genus *Phragmostrophia* 130
Suborder Chonetoidea 132
 Superfamily Chonetacea 132
  Family Chonetidae 132
   Subfamily Strophochonetinae 133
    Genus *Strophochonetes* 133
   Subfamily Chonetinae 135
    Genus *Chonetes* 135
   Subfamily Parachonetinae n. subfam. 135
    Genus *Parachonetes* 135
  Family Anopliidae 137
   Subfamily Anopliinae 137
    Genus *Anoplia* 137
Suborder Productoidea 138
 Superfamily Productacea 138
  Family Productellidae 138
   Subfamily Productellinae 138
    Genus *Spinulicosta* 138
Order Rhynchonellida 141
 Suborder Rhynchonelloidea 141
  Superfamily Camarotoechiacea 141
   Family Rhynchotrematidae 141
    Subfamily Orthorhynchulinae 141
     Genus *Machaeraria* 141
   Family Hebetoechiidae 143
    Genus *Pleiopleurina* 143
   Family Trigonirhynchiidae 144
    Genus *Ancillotoechia* 144
    Genus *Trigonirhynchia* 147
    Genus *Astutorhyncha* 148
   Family Pugnacidae 148
    Genus *Corvinopugnax* 148
   Family Leiorhynchidae 149
    Genus *Leiorhynchus* 150
Order Spiriferida 153
 Suborder Atrypoidea 153
  Superfamily Atrypacea 154

Family Atrypidae 154
    Subfamily Atrypinae 154
        Genus Atrypa 154
    Subfamily Carinatininae 157
        Genus *Spirigerina* 157
    Subfamily Atrypininae 159
        Genus *Atrypina* 159
Family Palaferellidae 160
    Subfamily Karpinskiinae 160
        Genus *Toquimaella* 160
Superfamily Dayiacea 160
    Family Anoplothecidae 160
        Subfamily Anoplothecinae 160
            Genus *Bifida* 160
        Subfamily Coelospirinae 162
            Genus *Coelospira* 162
    Family Leptocoeliidae 165
        Genus *Leptocoelia* 165
        Genus *Leptocoelina* n. gen. 170
Suborder Athyridoidea 173
Superfamily Athyridacea 173
    Family Meristellidae 173
        Subfamily Meristellinae 173
            Genus *Meristina* 173
            Genus *Meristella* 175
    Family Nucleospiridae 177
        Genus *Nucleospira* 177
Suborder Retzioidea 179
Superfamily Retziacea 179
    Family Retziidae 179
        Genus *Trematospira* 179
        Genus *Pseudoparazyga* n. gen 181
Suborder Spiriferoidea 182
Superfamily Delthyridacea 184
    Family Delthyrididae 184
        Subfamily Delthyridinae 184
            Genus *Howellella* 184
            Genus *Acrospirifer* 188
            Genus *Dyticospirifer* 193
        Subfamily Costispiriferinae 193
            Genus *Costispirifer* 194
        Subfamily Hysterolitinae 195
            Genus *Hysterolites* 195
            Genus *Brachyspirifer* 198
        Subfamily Kozlowskiellininae 201
            Genus *Kozlowskiellina* 201
                Subgenus *Megakozlowskiella* 201
        Subfamily *Spinellinae* n. subfam. 205
            Genus *Spinella* 205
            Genus *Eurekaspirifer* 207
    Family Reticulariidae 208
        Genus *Reticulariopsis* 208
        Genus *Quadrithyris* 209
        Genus *Elythyna* 210

Family Ambocoeliidae 212
    Genus *Ambocoelia* 212
    Genus *Spinoplasia* 213
    Genus *Metaplasia* 214
    Genus *Plicoplasia* 215
Superfamily *Cyrtinacea* 216
Family *Cyrtinidae* 216
    Genus *Cyrtinaella* 217
    Genus *Cyrtina* 218
Order Terebratulida 223
Suborder Terebratuloidea 223
    Superfamily Terebratulacea 224
    Family Dielasmatidae 224
        Subfamily Mutationellinae 224
            Genus *Mutationella* 224
    Family Centronellidae 224
        Subfamily Rensselaeriinae 224
            Genus *Rensselaerina* 224
            Genus *Rensselaeria* 225

PART IV

Plate section 227
Appendix of localities 379
References cited 393
Index of genera and species 409

FIGURES

1. Index map of Nevada and its counties. The shaded area is shown in Figure 2 4
2. Eureka County and vicinity, central Nevada, showing the principal marine upper Lower Devonian collecting localities 9
3. Lower Devonian cross-section between Coal Canyon, Simpson Park Range and the northern Roberts Mountains 37
4. Correlation of Great Basin Lower Devonian brachiopod zones with European stages 62
5. Serial sections x 10 of *Ancillotoechia aptata* Johnson, n. sp. 145
6. Serial sections x 6 of *Toquimaella kayi* Johnson, 1967 161

TABLES

1. Brachiopods of the Rabbit Hill Limestone in the vicinity of Rabbit Hill 30
2. Brachiopods of the lower McMonnigal Limestone 31
3. Cortez Range, *Eurekaspirifer pinyonensis* Zone brachiopods 38
4. Fauna of the *Quadrithyris* Zone 44
5. Fauna of the *Spinoplasia* Zone 48
6. Fauna of the *Trematospira* Zone 53
7. Fauna of the *Acrospirifer kobehana* Zone 56
8. Fauna of the *Eurekaspirifer pinyonensis* Zone 59
9. Stratigraphic occurrence of Siegenian and Emsian brachiopod species in central Nevada 63

Abstract

Continuous sections of fossiliferous Lower Devonian (Siegenian and Emsian) strata comprise five major faunal assemblage zones based primarily on brachiopods. The faunal succession which furnishes the basis for zonation is documented by description and illustration of the brachiopod species obtained from relatively large collections made in sequence at critical localities.

The Siegenian begins with the *Quadrithyris* Zone which has its type occurrence in the upper Windmill Limestone of Coal Canyon, from 10 to 36 feet above beds with *Monograptus hercynicus nevadensis*. The *Quadrithyris* Zone brachiopod fauna most closely resembles faunas from western and Arctic Canada and from the USSR (Old World Province) and is less like faunas of the Appalachian Province of eastern North America.

The *Spinoplasia* Zone succeeds the *Quadrithyris* Zone and carries a large brachiopod fauna whose affinity is almost exclusively with the Appalachian Province of eastern North America. The *Spinoplasia* Zone brachiopod fauna indicates a late Helderbergian age and a position within the Siegenian.

The *Trematospira* Zone succeeds the *Spinoplasia* Zone and is characterized by an Appalachian Province fauna of Oriskany age and a position at or near the top of the Siegenian.

The *Acrospirifer kobehana* Zone succeeds the *Trematospira* Zone and contains a few newly-appearing brachiopod genera that indicate Old World faunal affinity together with a few Appalachian Province genera that carry over from the *Trematospira* Zone below. By position, the *A. kobehana* Zone appears to be correlative with the Esopus Formation (*Etymothyris* Zone) of eastern North America and is assigned to the Lower Emsian.

The *Eurekaspirifer pinyonensis* Zone succeeds the *A. kobehana* Zone and carries a rich and well-preserved brachiopod fauna of Old World aspect. The zone is assigned to the Emsian and probably includes beds correlative with parts of both the Lower and Upper

Emsian. The oldest productids and the oldest goniatites in the western United States occur in this zone.

The *E. pinyonensis* Zone is succeeded by beds formerly included in the zone, but which have been separately designated *Elythyna* beds. The latter are probably of Late Emsian age and appear to represent the highest fossiliferous Lower Devonian strata in the Great Basin. The brachiopods of the *Elythyna* beds are most closely related to those of the *E. pinyonensis* Zone below and not to faunas from succeeding beds that can be confidently identified with the Middle Devonian.

The Siegenian and Emsian zonal sequence of central Nevada serves as a standard for the western United States and in some respects for all of western North America because Siegenian-Emsian brachiopod faunas from western Canada are wholly of Old World aspect and, considering the megafauna, can only be related to Appalachian Province faunas by way of an intermediate and transitional faunal succession.

All brachiopods known from the five major assemblage zones are listed according to their stratigraphic occurrence; 111 are described and illustrated. New subfamilies are Parachonetinae and Spinellinae. New genera are *Leptocoelina, Pseudoparazyga,* and *Discomyorthis.* New species are *Schizophoria parafragilis,* "*Brachyprion*" *mirabilis, Aesopomum varistriatus, Atrypina simpsoni, Cyrtinaella causa, Orthostrophella monitorensis, Leptostrophia inequicostella, Leptocoelina squamosa, Spinoplasia roeni, Leptocoelia murphyi, Pleiopleurina anticlastica, Megakozlowskiella magnapleura, Levenea navicula, Stropheodonta filicosta, Stropheodonta magnacosta, Megastrophia transitians, Sieberella pyriforma, Ancillotoechia aptata, Coelospira pseudocamilla, Gypidula praeloweryi, Mclearnites invasor, Anoplia elongata, Nucleospira subsphaerica, Levenea fagerholmi, Brachyspirifer pinyonoides,* and *Spinella talenti.*

PART I. INTRODUCTION

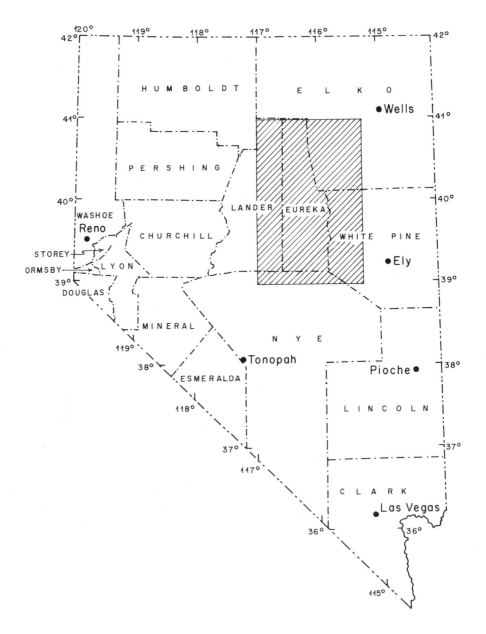

Figure 1. Index map of Nevada and its counties. The shaded area is shown in Figure 2.

Introduction

GENERAL REMARKS

For many years Lower Devonian strata went unrecognized in the Great Basin. The early surveys of Clarence King and the Eureka district study of Arnold Hague yielded fossils that are now recognized as Lower Devonian forms, but at that time the age range of these species had not been determined. F. B. Meek (1870, 1877) was the first to describe a Lower Devonian brachiopod species from the Great Basin (*Eurekaspirifer pinyonensis*). However, it was C. D. Walcott's unique monograph (1884) on the paleontology of the Eureka district that first produced illustrations and descriptions of a number of Lower Devonian species. Walcott described four new species of Lower Devonian brachiopods (*Parachonetes macrostriatus*, "*Strophochonetes*" *filistriatus*, *Leptocoelia infrequens*, and *Trigonirhynchia occidens*). He illustrated at least one species that might have been recognized as Lower Devonian (*Acrospirifer kobehana*, listed as *Spirifera raricosta*, 1884, Pl. 4, Fig. 2). H. S. Williams (Williams and Breger, 1916, p. 95) identified Walcott's specimen as the Lower Devonian species *Spirifer murchisoni*, but as late as 1927 Edwin Kirk stated that no Lower Devonian beds were known in the Great Basin.

In a series of papers W. L. Bryant (1932, 1933, 1934, 1935) and Erling Dorf (1934) described an Early Devonian biota, comprising fish and plant remains from an isolated outcrop at Beartooth Butte, Wyoming. A similar fish fauna was recovered from the Water Canyon Formation of northern Utah (Branson and Mehl, 1931; J. Stewart Williams, 1948). Subsequently, the fauna was studied by R. H. Denison (1952, 1953, 1958) who correlated the Beartooth Butte-Water Canyon assemblage with the European stages (1956, 1958). More recently, these Lower Devonian fish-bearing strata have been shown to be widespread geographically (Sandberg, 1961). Correlation with part of the southern Nevada section was suggested by J. G. Johnson and Anthony Reso (1964).

It remained for Charles Merriam (1940) to align western Devonian marine faunas in a zonal arrangement and to demonstrate conclusively the presence of Lower Devonian fossiliferous beds in central Nevada. Merriam did not discuss Lower Devonian correlations in detail, but placed the Lower Devonian-Middle Devonian boundary between the *Acrospirifer kobehana* and *Eurekaspirifer pinyonensis* Zones. Later G. A. Cooper correlated the *pinyonensis* Zone with the Coblenzian (Cooper and others, 1942). Restudy of the *Eurekaspirifer* Zone brachiopods (Johnson, 1962a) and its goniatites (House, 1962) confirmed Cooper's assignment and suggested an Emsian age. The writer has discussed the position of the Lower Devonian-Middle Devonian boundary eastward from the fossiliferous localities (Johnson, 1962b).

Merriam's discovery of the fauna of the Rabbit Hill Limestone and his recognition of its Helderbergian age (1954, 1963) was a major step toward the elucidation of the total Lower Devonian assemblage. His work was complemented by Johnson's recognition (1965) of the Rabbit Hill fauna in a structurally simple sequence in the northern Simpson Park Range and of its relation to the older Windmill Limestone fauna. The Simpson Park Range sequence, studied in conjunction with the Lower Devonian of the Monitor and Toquima Ranges, allowed recognition of the *Spinoplasia* and *Quadrithyris* Zones and proved to be a key sequence in central Nevada. An important revison of the lower part of the succession exposed at Coal Canyon resulted from detailed mapping by M. A. Murphy in 1967. Murphy discovered a fault within the Windmill Limestone and found *Monograptus microdon* and *M. praehercynicus* interbedded with a pre-*Quadrithyris* Zone shelly fauna in the lower Windmill beds which are judged to represent the highest Gedinnian. A still older Lower Devonian (Gedinnian) brachiopod assemblage, characterized by *Gypidula pelagica,* was noted from the upper Roberts Mountains Formation in the Roberts Mountains (Johnson, 1965), making it clear that the Siluro-Devonian boundary lies within the Roberts Mountains Formation.

It is now known that two diverse and richly fossiliferous brachiopod assemblages succeed the Ludlovian faunas in the northern Roberts Mountains (Johnson and others, 1967) and that they correspond to the Pridoli and Gedinnian, respectively. Each is fully as distinctive in its faunal content as the five younger Early Devonian zones (Johnson, 1965, Fig. 4). Because no one stratigraphic section exposes the complete Lower Devonian fossiliferous sequence from Gedinnian to Emsian, the brachiopods are conveniently studied in two parts. This paper deals with the brachiopods of five zones that span the Siegenian and most of the Emsian. A second paper in preparation by the writer, together with A. J. Boucot and M. A. Murphy, deals with Ludlow to Gedinnian stratigraphy and brachiopods of the Roberts Mountains Formation.

INTRODUCTION 7

The mutually dependent stratigraphic framework and faunal succession worked out by C. W. Merriam, M. A. Murphy, and the writer and based largely on studies of contained shelly megafauna have been tested through investigation of the conodont sequence by Gilbert Klapper (Klapper, 1968; Klapper and Ormiston, 1969). These conodont studies have confirmed the zone by zone arrangement defined by megafossils, both by direct age assignment, wherever that is possible, and by demonstration that the zonal sequence coincides with the conodont sequences known in other regions.

In Part 2 of this paper, the unusually complete Siegenian-Emsian faunal sequence and zonation based on brachiopods, as well as the age and correlation of the zones, are fully discussed. Part 3 deals with the systematic paleontology of all the post-Gedinnian Lower Devonian brachiopod species known and well represented in collections available to the writer as of January, 1967. The upper Lower Devonian brachiopods of central Nevada compose a large assemblage of which 127 species distributed among 83 genera are presently known. Almost all are known from silicified specimens that allow a better understanding of a number of genera for the first time.

SCOPE AND PURPOSE

The present work is an attempt to discover the make-up of brachiopod faunas of Siegenian and Emsian age in the Great Basin, particularly as they occur in the more fossiliferous localities in central Nevada. The writer's brief encounter with these faunas in the field during the late nineteen fifties, together with the laboratory study of many large and small collections over a 10-year period beginning in 1957, convinced him of the need for a preliminary but comprehensive account of the brachiopods and their succession according to a zonal scheme. The present work fills these requirements with the fullest possible description and illustration of the common brachiopods and of many of the rare ones. Nevertheless, it is regarded as a preliminary effort affording a basis for future investigations. Much remains to be done because even the better known species are understood with respect to only a few common occurrences, making it impossible to adequately judge morphologic variation in space and time. Almost without exception when good silicified collections are available, a single brachiopod assemblage consists of 15 or more species, but in an assemblage of this size only a few species are really common, and most can be satisfactorily characterized only at the generic level. Many more large collections need to be made before the rarer species can be described.

The faunal sequence, expressed in terms of assemblage zones, is neither the result of nor supported by numerous collections made in measured section, nor by field identification of the fossils. Without exception, the brachiopod species listed and discussed in this paper have been identified in the laboratory. The assemblage zones are discrete faunal units recognizable by association of the more abundant species. Since not every stratum in the field is abundantly fossiliferous, and since no detailed bed-by-bed sampling program has ever been undertaken, the faunal succession at the several localities is known in terms of a number of collections in sequence, from as few as two to ten or more in a few cases. Almost without exception such collections are easily assigned to zones and the zonal distribution of species expressed in Table 9 became known in this way. Small collections, which furnished a number of the better preserved specimens used for illustration, were not utilized in determining zonal ranges. The sequence of assemblage zones from *Quadrithyris* Zone at the base of the Siegenian through *Spinoplasia, Trematospira, Acrospirifer kobehana,* to *Eurekaspirifer pinyonensis* Zone at the top is well tested, but future investigation, including detailed sampling of measured sections, is required before the transitions from zone to zone can be fully understood.

METHODS OF STUDY

Most collections consist of silicified single or articulated valves enclosed in a limestone matrix. These were etched free in a strong solution of hydrochloric acid in order to process the most rock in the shortest possible time. The insoluble residues were dried and picked and were hardened by dipping in a solution of Alvar dissolved in acetone. Nonsilicified specimens were either broken out of the enclosing rock, or were found weathered-free on talus slopes. The latter kind of preservation is typical of *E. pinyonensis* Zone collections. The internal structures of the nonsilicified specimens were investigated by means of internal molds which were prepared by first roasting the shell over a Bunsen burner flame and then removing the shell with the aid of a bevel-edged needle. Some internal features, such as muscle scars, are best studied and illustrated by internal molds and must be prepared in this way. In a few cases, rubber internal molds were prepared from silicified free specimens to facilitate illustration and comparison.

LOCATION

The Great Basin centers in the state of Nevada and includes parts of California, Oregon, Idaho, Wyoming, and Utah. The brachiopod-rich sections of marine Lower Devonian rocks are most abundantly represented in central Nevada (Fig. 1) and particularly in Eureka County and vicinity (Fig. 2), the area investigated in this paper.

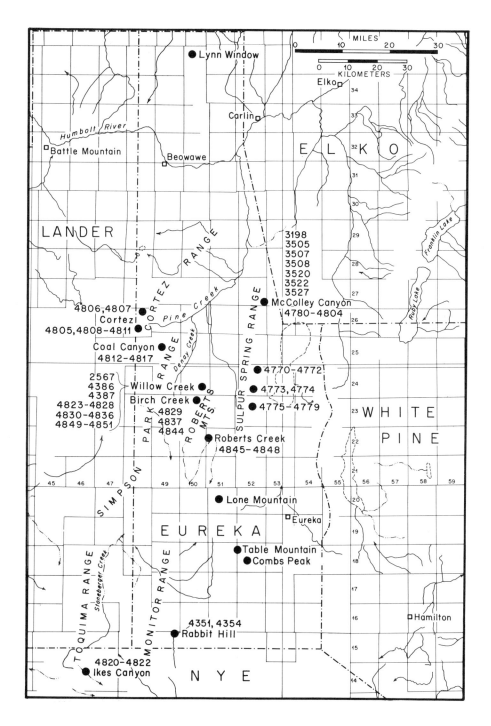

Figure 2. Eureka County and vicinity, central Nevada, showing the principal marine upper Lower Devonian collecting localities. Locality numbers are Univ. California, Los Angeles, Department of Geology, locality catalog numbers.

ACKNOWLEDGMENTS

The writer is greatly indebted to a number of paleontologists and their institutions who have loaned specimens for comparison or who have furnished topotype material as gifts. Without their help the present paper would have been impossible. The writer is most indebted to Professor A. J. Boucot for many discussions as well as the free use of his library and his extensive brachiopod collection while the writer was employed at the California Institute of Technology. In addition, the cost of the majority of photographs of fossil specimens illustrated in this paper was underwritten by funds from a National Science Foundation grant to Boucot for the study of Silurian and Lower Devonian Brachiopoda. The remainder of the cost was covered by a grant from the Research Committee of the Academic Senate, University of California, Los Angeles.

Dr. M. A. Murphy, University of California, Riverside, made many valuable collections available and furnished much vital information on field relations of collections and lithologic units. Dr. G. A. Cooper, Department of Paleobiology of the U.S. National Museum, initially loaned the types of the described Lower Devonian species of Meek, Walcott, and Merriam, and furnished the photographs of these species that are used in this paper. Dr. Cooper also responded numerous times to additional requests for smaller loans which the writer found necessary. Professor Marshall Kay of Columbia University provided collections from the McMonnigal Limestone in the Toquima Range. Professor Norman D. Newell of the American Museum of Natural History loaned Hall's types of *Trematospira perforata, T. multistriata, Coelospira concava, Howellella cycloptera,* and *H. tribulis.* Professor Harry B. Whittington, then of the Museum of Comparative Zoology, Harvard University, very kindly made available loans from the Schary Collection, allowing comparison of some of the Nevada material with Barrande's species from the Devonian of Bohemia. Additional material was received from Dr. V. Havlícek who sent several examples of Bohemian species not otherwise available to the writer. Professor Tor Ørvig and Dr. H. Mutvei of the Swedish Museum of Natural History, Stockholm, provided specimens of *Strophochonetes cingulatus* from Gotland. Dr. N. L. Bublitschenko, Akademia Nauk, Ust-Kamenogorsk, Kazakh S.S.R., gave specimens of his species *Leptocoelia biconvexa* to the writer.

Canadian material was put at the writer's disposal through the courtesy of Dr. D. J. McLaren, Geological Survey of Canada, allowing recognition of faunal resemblance between Arctic brachiopod species and some from central Nevada. Further material, of Siegenian age, was sent by Dr. Alfred Lenz, then of the California Standard Company, Calgary, Alberta. Loans of material from the Inyo Range and from northern California were made by Joseph E. Peck and

W.B.N. Berry from the Museum of Paleontology, University of California, Berkeley. Special thanks are due to Professor Berry for identification of the graptolites listed in this paper. Drs. Jean M. Berdan and W. A. Oliver, Jr., U.S. Geological Survey, Washington, D. C., identified ostracodes and corals listed herein. Dr. Gilbert Klapper, of the University of Iowa, identified most of the conodonts discussed in this paper and made the writer aware of the details of Lower Devonian conodont biostratigraphy through discussions, correspondence, and exchange of manuscripts. Several conodont collections were processed and identified by Professor O. H. Walliser, Göttingen. The writer also thanks Dr. C. W. Merriam, U.S. Geological Survey, Menlo Park, California, for a long and valued correspondence dealing with problems of Devonian paleontology and stratigraphy in the Great Basin.

The collections on which the present paper is based, in addition to those made by the writer, were gathered from many sources beginning in 1949. Many of these were obtained by the staff and students of the University of California, Los Angeles, during annual summer camps in the Cortez Mountains, the Sulphur Spring Range, and the Roberts Mountains under the guidance of Donald Carlisle, J. C. Crowell, James Gilluly, C. A. Nelson and E. L. Winterer. New collections, made since the spring of 1963 by M. A. Murphy and by A. J. Boucot, are the largest and most comprehensive to date and have been of immeasurable value. Harold Masursky and J. B. Roen, both of the U.S. Geological Survey, sent collections from the Cortez Range and Simpson Park Range and from the Tuscarora Mountains of northern Eureka County. R. H. Waines kindly sent Lower Devonian spirifers from the Piute Formation in the Arrow Canyon Range. Gilbert Klapper and M. A. Murphy read Parts 1 and 2 of the manuscript. A. J. Boucot read Part 3. Their reviews are appreciated. Serial sections in Figures 5 and 6 were prepared by Mr. Michael Stephens. The plates were prepared by the writer's assistant, Miss Vera Rose. Drafting was done by Mrs. Ruth Talovich of the California Institute of Technology. The manuscript was transcribed from tape and seen through several typescript copies by the writer's wife, Miriam Johnson. Mr. Nelson W. Shupe prepared most of the photographs of fossil specimens.

To all those mentioned above, the writer wishes to express his heartfelt thanks for their cooperation and assistance.

PART II. BIOSTRATIGRAPHY

Previous Work

LOWER NEVADA ZONES

The first study of a zonal nature dealing with Lower Devonian faunas in the Great Basin was by Merriam (1940). Merriam's principal investigations of Lower Devonian strata were made in the lower Nevada Group at Lone Mountain and in the Roberts Mountains. He defined the *Acrospirifer kobehana* Zone (1940, p. 52) and the *Eurekaspirifer pinyonensis* Zone (p. 53) and showed that the *pinyonensis* Zone overlies the *kobehana* Zone at Lone Mountain and at several places in the Roberts Mountains. Merriam was able also to demonstrate a twofold subdivision of the upper or *Eurekaspirifer pinyonensis* Zone on the basis of corals, dividing it into a lower *Radiastraea* Subzone and an upper *Billingsastraea* Subzone. Both were recognized at Lone Mountain and in the Roberts Mountains. In addition to the two zones formally recognized, Merriam also collected large retzioid brachiopods in an assemblage that he named the *Trematospira* fauna. Merriam (1940, p. 52, Table 3) listed the assemblage of two collections under the title *Trematospira* fauna and noted that one, lacking the name-bearer, might be slightly younger. The principle listed elements of the type faunule are *Costispirifer* sp. and *Pseudoparazyga cooperi*. The assemblage from the second-named locality which was thought to be younger was reported to contain *Acrospirifer kobehana* and *Gypidula praeloweryi*?, elements which are now known to be strictly assignable to the *Acrospirifer kobehana* Zone and to beds definitely younger than those bearing the *Trematospira* fauna.

Merriam later (Merriam and Anderson, 1942, p. 1692) reported that the original assemblage from the Roberts Mountains, called the *Trematospira* fauna, had been found in the Sulphur Spring Mountains below the range of several fossil forms characteristic of the upper *Acrospirifer kobehana* Zone and in 1956, Nolan and others (p. 46) regarded the *Trematospira* fauna as a formal zone.

In the early nineteen fifties, a mapping program, centered principally around the Mineral Hill quadrangle in the Sulphur Spring Range, was initiated by members of the staff of the Department of

Geology, University of California, Los Angeles. Work on the Devonian culminated in a paper (Carlisle and others, 1957) that afforded considerable additional data on the Lower Devonian faunal sequence. The *Trematospira* Zone, the *kobehana* Zone, and the *pinyonensis* Zone were shown to be present in ascending order at Williams Canyon and at McColley Canyon. Four additional sections were shown to contain the *Trematospira* fauna beneath the *Eurekaspirifer pinyonensis* Zone, with strata barren of fossils intervening. With three exceptions, these zones were found to be limited to the McColley Canyon Formation (then the McColley Canyon Member of the Nevada Formation). The exceptional occurrences reported the *pinyonensis* Zone fauna continuing well up into the overlying Union Mountain Sandstone Member. The fossils which served as a basis for these reports were studied by Johnson (1962b, p. 543) and were found, however, not to be assignable to the *Eurekaspirifer pinyonensis* Zone.

The collections of fossils on which the Sulphur Spring zonal assignments were based have been re-examined by the writer, and some are described and figured in this report. The writer is in complete agreement with the sequence of brachiopod assemblages set out in that paper although some named species were found to be more widely reported than a critical interpretation of species now allows. Specimens listed as *Gypidula* cf. *G. coeymanensis* and *Spirifer kobehana* from the *Trematospira* Zone correspond to *Sieberella pyriforma* and *Acrospirifer* aff. *murchisoni* of this report. The misidentification of *kobehana* with the earlier *Acrospirifer* aff. *murchisoni* led also to the report of *Costispirifer* in the *Spirifer kobehana* Zone (Carlisle and others, 1957, p. 2186). Of these, *Costispirifer* is correctly identified and, "*Spirifer kobehana*" is instead *Acrospirifer* aff. *murchisoni*. The beds are assignable to the *Costispirifer* Subzone within the upper part of the *Trematospira* Zone.

Using Merriam's work at Lone Mountain and in the Roberts Mountains and the work of Carlisle and others (1957) in the Sulphur Spring Range as a base, the writer undertook a re-examination of the Lower Devonian brachiopod collections which had been gathered at the University of California, Los Angeles, from the several regions. With the data from these collections and those made by the writer from the Roberts Mountains and from the northern Simpson Park Range, some preliminary revisions and additions were made (Johnson, 1962a). The *Trematospira* Zone was renamed *Costispirifer* Zone because of relative abundance of the latter genus in contrast to the exceedingly rare occurrence of *Pseudoparazyga* ("*Trematospira*") *cooperi*. A new subzone, the *Leptocoelia infrequens* Subzone, was recognized on the basis of brachiopods in the Roberts Mountains and in the northern Simpson Park Range. In addition, a number of genera were recorded for the first time from the western United States.

However, up to 1960 when Johnson's paper (1962a) was submitted for publication, all the accumulated data substantiated and amplified Merriam's original work, and no new zones had been recognized.

An important stratigraphic study by Murphy and Gronberg (1970) has given us a much better framework for understanding the lithologic succession in the post-Rabbit Hill Lower Devonian. Murphy and Gronberg's work was available too late to be used throughout the text of the present work, but their named members, Kobeh, Bartine, and Coils Creek, and the paleontological data that date them, are well known to the writer.

PRE-NEVADA ZONES

In 1954, Merriam focused attention briefly on a new Lower Devonian fauna (p. 1284, 1285) to which he assigned a Helderbergian age, and which was not known to be present in the previously investigated fossiliferous regions in central Eureka County. This is the fauna of the Rabbit Hill Limestone from the east slope of the Monitor Range (Merriam, 1963, p. 43). Although structurally complex, the Rabbit Hill, at its type locality, was recognized to rest upon graptolite-bearing limy shale and siltstone, but no beds were found that overlie the formation, and it was assigned to Helderbergian age according to an objective appraisal of its contained fauna. There was no satisfactory appraisal of the zonal relations of the Rabbit Hill fauna to the previously described faunas constituting the *Trematospira, Acrospirifer kobehana,* and *Eurekaspirifer pinyonensis* Zones until 1965, when the Rabbit Hill Formation was found in a structurally simple sequence in the northern Simpson Park Range (Johnson, 1965). Proposal of the *Spinoplasia* Zone for the Rabbit Hill fauna and the *Quadrithyris* Zone for the underlying Windmill Limestone fauna completed initial zonation of the Siegenian and Emsian in central Nevada on the basis of brachiopods.

The Faunal Sequence

SULPHUR SPRING RANGE

The work of Carlisle and others (1957) has been recently supplemented by large collections made by Murphy at measured intervals in McColley Canyon. The lower 105 feet of beds (USNM locs. 10726, 27, 30-37, 41-49) carries a prolific and well-preserved brachiopod assemblage that represents the *Trematospira* Zone and consists of the following species:

Dalejina sp. B
Discomyorthis musculosa
Levenea navicula
Sieberella pyriforma
Leptaena cf. *acuticuspidata*
"Schuchertella" sp. B
Strophonella cf. *punctulifera*
Stropheodonta filicosta
Stropheodonta magnacosta
Megastrophia transitans
Leptostrophia cf. *beckii*
Anoplia sp. A
Pleiopleurina anticlastica
Ancillotoechia aptata
Coelospira pseudocamilla

Leptocoelia murphyi
Acrospirifer aff. *murchisoni*
Dyticospirifer mccolleyensis
Costispirifer sp.
Megakozlowskiella magnapleura
Reticulariopsis sp. B
Elytha sp.
Cyrtina cf. *varia*
Metaplasia cf. *paucicostata*
Nucleospira sp.
Meristella cf. *robertsensis*
Pseudoparazyga cooperi
Trematospira perforata
Rensselaeria sp.

These collections for the first time allow recognition of the true position of *Costispirifer* beds within the lower part of the McColley Canyon Formation. *Costispirifer* is found to appear in the upper 85 feet of beds of the zone. The lower 20 feet of beds, which do not yield *Costispirifer*, carry *Pleiopleurina anticlastica* and *Pseudoparazyga cooperi*, which are not known to ascend into the overlying *Costispirifer* beds. Thus *"Costispirifer* zone" as used by Johnson (1962a, p. 165) is revised to the rank of an upper subzone within the *Trematospira* Zone as originally defined by Merriam. At present, the two subzones can be recognized only in the Sulphur Spring Range and at two localities in Willow Creek on the north side of the Roberts Mountains. The occurrence of true *Trematospira* (that is *T. perforata*) allows preservation of the name for the zone, although the original name-giver has [now] been renamed *Pseudoparazyga*. This is convenient because *Trematospira perforata* is slightly more abundant than *Pseudoparazyga* in the *Trematospira* Zone at the examined sections. One large collection (USNM loc. 12777) from the *Costispirifer* Subzone yielded the following articulate brachiopods; number of specimens shows relative abundance of species:

Dalejina sp. B, 336
Leptaena cf. *acuticuspidata*, 14
"*Schuchertella*" sp., 23
Leptostrophia? sp., 1
Megastrophia transitans, 15
indet., rhynchonellid, 1
Coelospira pseudocamilla, 6
Leptocoelia murphyi, 3
Nucleospira sp., 2

Meristella cf. *robertsensis*, 903
Trematospira perforata, 32
Acrospirifer sp., 10
Costispirifer sp., 30
Megakozlowskiella magnapleura, 7
Elytha sp., 15
Metaplasia cf. *paucicostata*, 81
Cyrtina cf. *varia*, 66.

In the interval from 185 to 205 feet above the base of the formation, three separate horizons (USNM locs. 10738-10740) have yielded abundant brachiopods diagnostic of the *Acrospirifer kobehana* Zone. These collections contain the following forms:

Dalejina sp. C
Leptaena sp.
"*Schuchertella*" sp. C
Leptostrophia? sp.
Strophonella cf. *punctulifera*
Megastrophia transitans
Mclearnites invasor
Stropheodonta filicosta

Anoplia elongata
Atrypa sp.
Acrospirifer kobehana
Megakozlowskiella cf. *raricosta*
Metaplasia cf. *paucicostata*
Cyrtina sp. C
Meristella robertsensis.

A large collection (USNM loc. 12778), probably from the same horizon but away from the measured section, contains:

Dalejina sp. C, 785
Salopina sp., 18
Leptaena sp. C, 73
"*Schuchertella*" sp. C, 127
Strophonella cf. *punctulifera*, 28
Leptostrophia sp. D, 19
Mclearnites invasor, 103
Megastrophia transitans, 55
Stropheodonta filicosta, 22
Phragmostrophia cf. *merriami*, 6
Pholidostrophia (*Pholidostrophia*) sp., 22
Chonetes? sp., 4

Anoplia elongata, 1974
Katunia? sp., 18
Atrypa nevadana, 16
Coelospira sp., 143
Leptocoelia aff. *murphyi*, 38
Nucleospira cf. *subsphaerica*, 81
Meristella robertsensis, 349
Acrospirifer kobehana, 418
Megakozlowskiella cf. *raricosta*, 18
Metaplasia sp., 64
Cyrtina sp. C, 56.

Another collection made away from the measured section (USNM loc. 10710) contains:

Levenea navicula
Gypidula praeloweryi
Leptaena sp. C
"Schuchertella" sp. C

Acrospirifer kobehana
Megakozlowskiella cf. raricosta
Meristella robertsensis.

In the strata overlying the *Acrospirifer kobehana* Zone, several large collections have been made from the *Eurekaspirifer pinyonensis* Zone, but no clear subdivision within that zone or understanding of the inter-relations of the separate collections has been possible. One horizon yields very abundant well-preserved specimens of *Parachonetes macrostriatus*, almost to the exclusion of other species (a notable exception is *Spinulicosta?* sp.). A second horizon, which is easily traceable for a distance of over a mile, is filled with silicified specimens of *Eurekaspirifer pinyonensis*. This horizon (USNM locs. 10704, 10729) yields the following forms:

Rhipidomella sp.
Leptaena sp.
Phragmostrophia merriami
"Strophochonetes" filistriata
Trigonirhynchia occidens

Atrypa nevadana
Eurekaspirifer pinyonensis
Hysterolites sp.
Mutationella? sp.

Specimens of *Eurekaspirifer pinyonensis* from this horizon of abundant silicified shells have been compared and agree in all respects with the type lot of specimens upon which Meek originally based his species, and this may be the topotypical horizon for *Eurekaspirifer pinyonensis*.

NORTHERN ROBERTS MOUNTAINS

The lowest beds assignable to the Devonian belong to the Roberts Mountains Formation, although at the time of the naming of the formation it was considered to be entirely Silurian (Merriam, 1940, p. 11). Johnson (1965, p. 368) first established the presence of Lower Devonian beds in the Roberts Mountains Formation. The sequence, including graptolitic and shelly faunas of Late Llandovery, Wenlock, Ludlow, Pridoli, and Gedinnian ages, has been briefly described by Johnson and others (1967).

Overlying the fossiliferous limestones of the upper Roberts Mountains Formation, there is at least 1000 feet of barren Lone Mountain Dolomite (Winterer, oral commun., 1961) which appears to be equivalent to the Windmill Limestone (Johnson, 1965, Fig. 3). Scaling the map outcrop at Willow Creek suggests a thickness of about 1675 feet. Lying unconformably above the Lone Mountain Dolomite and with convergence along the contact (Winterer and Murphy,

1960, p. 133), there is a sequence of fossiliferous limestones that represents the McColley Canyon Formation of the Nevada Group. Two small collections made within 25 feet of the base of the McColley Canyon Formation (USNM locs. 10714, 10715) are assignable to the *Trematospira* Zone and probably to its basal *Trematospira* Subzone. Combined, they contain the following brachiopods:

Dalejina sp. B
Levenea navicula
Leptaena cf. *acuticuspidata*
"*Schuchertella*" sp.
Megastrophia transitans
Ancillotoechia aptata

Leptocoelia murphyi
Nucleospira sp. B
Trematospira perforata
Acrospirifer aff. *murchisoni*
Megakozlowskiella magnapleura.

About 20 feet above this faunule, abundant silicified remains of *Costispirifer* sp. appear in the section. One collection from this horizon (USNM loc. 10778) has yielded the following brachiopods:

Dalejina sp. B
Levenea navicula
Leptocoelia murphyi
Coelospira pseudocamilla
Meristella sp.

Acrospirifer aff. *murchisoni*
Costispirifer sp.
Megakozlowskiella magnapleura
Cyrtina cf. *varia*.

A large collection (USNM loc. 13268) yielded the following species; numbers of specimens are included to show relative abundance:

Dalejina cf. sp. B, 457
Levenea navicula, 115
Leptaena sp., 35
"*Schuchertella*" sp. (coarse ribs), 7
Strophonella? sp., 1
Leptostrophia sp., 10
Stropheodonta filicosta, 1
Megastrophia transitans, 4
rhynchonellids? indet., 2

Coelospira pseudocamilla, 707
Leptocoelia cf. *murphyi*, 290
Nucleospira sp., 1
Meristella cf. *robertsensis*, 458
Acrospirifer aff. *murchisoni*, 620
Costispirifer sp., 476
Elytha sp., 2
Metaplasia cf. *paucicostata*, 38
Cyrtina cf. *varia*, 55.

This is the *Costispirifer* Subzone of the upper part of the *Trematospira* Zone and its assemblage was the basis for erecting the subzone (Johnson, 1962a, p. 165). The assemblage listed above, from the east flank of Willow Creek, is an easily recognizable one which has been traced southwest across the crest of the ridge separating Willow Creek and Birch Creek and into the bottom of Birch Creek. At several points along the course of the occurrence of the *Costispirifer* Subzone, *Acrospirifer kobehana* was found a few feet above the occurrence of *Costispirifer* sp., and the zones are easily recognized on the occurrence of these two forms in the absence of a large assemblage. A number of small collections from the *Acrospirifer kobehana* Zone have been studied. The largest collection and the one from which some of the described and illustrated specimens in this paper are derived (USNM loc. 10777) contains the following brachiopod species:

Dalejina sp. C
"Schuchertella" sp. C
Leptocoelia aff. murphyi
Acrospirifer kobehana

Megakozlowskiella cf. raricosta
Cyrtina sp. C
Meristella robertsensis
Nucleospira sp.

The forms from the three zones or subzones listed above in the lower part of the McColley Canyon Formation were collected from a relatively nonargillaceous limestone facies which yields silicified specimens. The next main canyon east of Willow Creek is Dry Creek, and at that locality the *Acrospirifer kobehana* Zone is represented by an argillaceous limestone facies which bears unsilicified articulated specimens and a slightly different assemblage (USNM locs. 10705, 10772) that contains the following brachiopods:

Dalejina sp. C
Gypidula praeloweryi
Leptaena sp. C
Strophonella cf. punctulifera

Mclearnites invasor
Parachonetes macrostriatus
Atrypa sp. A
Acrospirifer kobehana?

At present there is no evidence to suggest whether the latter faunule is exactly correlative with the occurrences immediately to the west, or whether the two represent slightly different horizons within the zone.

Directly overlying fossiliferous beds of the *Acrospirifer kobehana* Zone at Willow Creek and at Birch Creek, the *Eurekaspirifer pinyonensis* Zone fauna is abundantly represented in the argillaceous limestone facies. Its brachiopod fauna is an abundant and easily recognizable one which commonly (USNM locs. 10774-10776, 10781-10789) is composed of the following species:

Dalejina sp. C
Schizophoria nevadaensis
Gypidula loweryi
Megastrophia iddingsi
Phragmostrophia merriami

"Strophochonetes" filistriata
Parachonetes macrostriatus
Atrypa nevadana
Brachyspirifer pinyonoides.

This is the typical assemblage of the zone and is found at many localities throughout the Roberts Mountains. At Willow Creek and Birch Creek where the writer has most carefully examined the fossiliferous sections the argillaceous limestone facies and its contained *Eurekaspirifer pinyonensis* fauna as listed above, are overlain by less argillaceous gray limestone, about 140 feet thick, with *Elythyna* sp. in its upper part. These beds are overlain in turn by gray, platy, commonly siliceous limestones bearing the wholly new fauna of the *Leptathyris circula* Zone and *Pentamerella* Subzone (Johnson, 1962a, 1966a) which are assigned to the Eifelian.

SOUTHERN ROBERTS MOUNTAINS

Several of the significant localities that were discussed by Merriam (1940) are in the southern part of the Roberts Mountains, and some of the best brachiopod collections studied by him came from these localities. Several collections made by Winterer and his students from the same region have been studied by the writer. The *Acrospirifer kobehana* Zone fauna is particularly well developed. The occurrence of *Anoplia elongata* with *Acrospirifer kobehana* is the most notable feature of these collections. The *Eurekaspirifer pinyonensis* Zone is also well represented by an assemblage closely comparable to that at Lone Mountain which is reported below. The fauna of the *Trematospira* Zone has not yet been found to occur in the southern Roberts Mountains. Merriam's locality 17 (1940, p. 51, 52) is properly associated with the *Acrospirifer kobehana* Zone rather than the *Trematospira* Zone.

LONE MOUNTAIN

Merriam (1940, p. 52) reported the fauna of the *Acrospirifer kobehana* Zone to be present between 100 and 150 feet above the base of the Nevada Formation at Lone Mountain and that it is overlain by beds carrying the typical fauna of the *Eurekaspirifer pinyonensis* Zone. A number of the localities listed by Merriam (1940, Table 5, p. 54, 55) from the *Eurekaspirifer pinyonensis* Zone were collected at Lone Mountain. Several collections from Lone Mountain were studied by the writer. A collection made from float and combined from several levels within the *pinyonensis* Zone at Lone Mountain (USNM loc. 12775) includes the following brachiopods:

Schizophoria nevadaensis
Gypidula loweryi
Leptaena sp.
Strophonella cf. *punctulifera*
Megastrophia iddingsi
Chonetes sp.
"*Strophochonetes*" *filistriata*
Parachonetes macrostriatus
Spinulicosta? sp.

Trigonirhynchia occidens
"*Leiorhynchus*" sp.
Leptocoelia infrequens infrequens
Atrypa nevadana
Hysterolites sp. A
Brachyspirifer pinyonoides
Spinella talenti.

Another collection, obtained from the lower part of the *pinyonensis* Zone (USNM loc. 11712), contains the following brachiopods:

Dalejina? sp.
Gypidula loweryi
Schuchertella nevadaensis
Strophonella cf. *punctulifera*
Stropheodonta sp.
Megastrophia iddingsi
Phragmostrophia merriami
"*Strophochonetes*" *filistriata*
Parachonetes macrostriatus

Anoplia elongata
Trigonirhynchia occidens
Atrypa nevadana
Nucleospira sp.
Meristina sp.
Spinella talenti
Ambocoelia sp.
Cyrtina sp.

Especially valuable have been collections made in two measured sections by E. C. Gronberg. Fossils obtained from Gronberg's collections are listed below. Brachiopods and other megafauna were identified by the writer; conodonts were identified by Gilbert Klapper.

Section EF, Lone Mountain

Upsection to NE. from a point at the base of the McColley Canyon Formation 600 ft N., 2100 ft W. of VABM 7936, Bartine Ranch quad., Eureka Co., Nevada, (*see* Gronberg, 1967).

EF-1, 5-10 ft, UCR 4496
   Meristella sp.
   large indet. corals
EF-2, 35-40 ft, UCR 4497
   Leptaena sp.
   "*Schuchertella*"? sp.
   Meristella sp.
   Acrospirifer cf. *kobehana*
   indet. corals
EF-3, scree UCR 4498
   Leptaena sp.
   "*Schuchertella*" sp.
   Megastrophia? sp.
   Meristella sp.
   indet. spirifer, medium-coarse ribs
   indet. corals
   Favosites sp.
EF-4, 100 ft, UCR 4499
   Dalejina sp.
   "*Schuchertella*"? sp.
   Mclearnites? sp.
   Anoplia elongata
   Coelospira sp.
   Meristella sp.
   Acrospirifer kobehana
   Favosites sp.
   indet. corals
   small gastropods, indet.
   Phacops sp.
Age: *kobehana* Zone
EF-5, 120 ft, UCR 4500
   "*Schuchertella*" sp.
   Meristella? sp.
   Acrospirifer kobehana
Age: *kobehana* Zone
EF-6, 143 ft, UCR 4501
   Dalejina sp.
   "*Schuchertella*" sp.
   Meristella robertsensis
   Acrospirifer cf. *kobehana*
   corals, indet.
   gastropods, indet.
Age: *kobehana* Zone
EF-7, 183 ft, UCR 4502
   Acrospirifer sp.
   corals, indet.
Age: *kobehana* Zone
EF-8, 190 ft, UCR 4503
   Leptaena sp.
   "*Strophochonetes*" cf. *filistriata*
   Atrypa sp.
   Loxonema sp.
   dalmanitid, indet.
Age: probably *pinyonensis* Zone
EF-9, 195-200 ft, UCR 4504
   Gypidula loweryi?
   Leptaena sp.
   Strophonella cf. *punctulifera*
   indet. stropheodontid
   Megastrophia? sp.
   "*Strophochonetes*" *filistriata*
   Atrypa sp.
   Meristella sp.
   Phacops sp. (pygidium)
   Orthoceras sp.
Age: *pinyonensis* Zone
EF-10, 200-290 ft, UCR 4505
   Dalejina sp.
   Leptaena sp.
   Strophonella cf. *punctulifera*
   Stropheodonta sp. (= sp. A of C.W.M.)
   Parachonetes macrostriatus
   Atrypa nevadana
   Spinella sp.
   Orthoceras sp.
   Loxonema sp.

bellerophontid?, indet.
Phacops sp.
Dalmanites sp.
Age: *pinyonensis* Zone
EF-11, 290-315 ft, UCR 4506
   Parachonetes macrostriatus
   Atrypa nevadana
   Spinella talenti
   Proetus sp.
   Loxonema sp.
   Syringopora sp.
   corals, indet.
   bivalves, indet.
Age: *pinyonensis* Zone
EF-12, 315-340 ft, UCR 4507
   Gypidula loweryi
   Megastrophia iddingsi?
   "*Strophochonetes*" *filistriata*
   Parachonetes macrostriatus
   Atrypa nevadana
   Hysterolites? sp.
   Spinella? sp.
   Orthoceras sp.
   gastropod, indet.
   Favosites sp. (with small coralla)
   Syringopora sp.
   Dalmanites sp.
Age: *pinyonensis* Zone
EF-13, 400-450 ft, scree UCR 4508
   Dalejina sp.
   Schizophoria nevadaensis
   Phragmostrophia merriami
   Atrypa nevadana
   Brachyspirifer pinyonoides
   Hysterolites sp.
   Favosites sp.
   bivalves, indet.
   Loxonema sp.
   Phacops sp.
Age: *pinyonensis* Zone
EF-14, 465-500 ft, UCR 4509
   Schizophoria nevadaensis
   Chonetes sp.
   Trigonirhynchia occidens
   Atrypa nevadana
   Brachyspirifer pinyonoides
   bivalve, indet.
   Favosites sp.
   Phacops sp.
   Proetus? sp.
Age: *pinyonensis* Zone
EF-15, 520-530 ft, UCR 4510
   Schizophoria nevadaensis
   Leptaena sp.
   Megastrophia iddingsi
   Phragmostrophia merriami
   Chonetes sp.
   Trigonirhynchia occidens
   Atrypa nevadana
   Brachyspirifer? sp.
   Dalmanites? sp.
Age: *pinyonensis* Zone
EF-16, 540-580 ft, UCR 4511
   Schizophoria nevadaensis
   Gypidula loweryi
   Megastrophia iddingsi
   Chonetes sp.
   Atrypa nevadana
   Trigonirhynchia occidens
   Brachyspirifer pinyonoides
   Orthoceras sp.
   corals, indet.
Age: *pinyonensis* Zone
EF-17, 580-630 ft, UCR 4512
   Cortezorthis cortezensis
   Gypidula loweryi
   indet. schuchertelloid
   Megastrophia iddingsi
   Phragmostrophia merriami
   Atrypa nevadana
   Brachyspirifer pinyonoides
   Hysterolites sp. (coarse ribbed)
   planospiral gastropod, indet.
Age: *pinyonensis* Zone

Section FF, Lone Mountain

Upsection to NE. from a point at the base of the McColley Canyon Formation, 1600 ft S., 1900 ft W. of VABM 7936, Bartine Ranch quad., Eureka Co., Nevada, (*see* Gronberg, 1967).

FF-1, 0-85 ft, UCR 4524
   Meristella sp.
   Acrospirifer cf. *kobehana*
   corals, indet.
FF-2, 85-135 ft, UCR 4525
   Leptaena sp.
   Strophonella? sp.
   deeply convex chonetid aff.
      "*Strophochonetes*" *filistriata*
   Atrypa sp.
   Meristella sp.
   Acrospirifer kobehana
   Phacops sp.
   indet. corals
Age: *kobehana* Zone
FF-3, 135-185 ft, UCR 4526
   Leptaena sp.
   Strophonella cf. *punctulifera*
   Megastrophia transitans?
   deeply convex chonetid aff.
      "*Strophochonetes*" *filistriata*
   Parachonetes macrostriatus
   Atrypa sp.
   Meristella sp.
   Acrospirifer cf. *kobehana*
   Conocardium sp.
   conical gastropod, indet.
Age: *kobehana* Zone
FF-4, 185-235 ft, USNM loc. 13271, UCR 4527
   Dalejina sp.
   Gypidula loweryi
   Leptaena sp.
   Parachonetes macrostriatus
   Atrypa sp.
   Meristella sp. (sulcate)
   Howellella cf. *textilis*
   Orthoceras sp.
   Phacops sp.
Age: *pinyonensis* Zone
FF-5 235-285 ft, UCR 4528
   Dalejina sp.
   Leptaena sp.
   "*Strophochonetes*" *filistriata?*
   Parachonetes macrostriatus
   Atrypa nevadana
   Spinella sp.
   Platyceras sp.
   Favosites sp. (large coralla)
Age: *pinyonensis* Zone

FF-6, 285-335 ft, UCR 4529
   "*Strophochonetes*" *filistriata*
   Atrypa sp.
   Spinella or *Brachyspirifer* sp.
   bellerophontid, indet.
Age: *pinyonensis* Zone
FF-7, 335-385 ft, UCR 4530
   Atrypa sp.
   Leptocoelia infrequens
   coarse-ribbed spirifer (possibly
      Howellella sp.)
   Hysterolites? sp.
   platycerids, indet.
   Favosites sp.
   Phacops sp.
Age: *pinyonensis* Zone
FF-8, 385-435 ft, UCR 4531
   Dalejina sp. C
   Schizophoria nevadaensis
   Leptaena sp.
   Megastrophia iddingsi
   Phragmostrophia merriami
   Trigonirhynchia occidens
   Atrypa nevadana
   Leptocoelia infrequens
   Meristina? sp.
   Nucleospira sp.
   Brachyspirifer pinyonoides
   Phacops sp.
   bivalves, indet.
   corals, indet.
Age: *pinyonensis* Zone
FF-9, 435-485 ft, USNM loc. 13270, UCR 4532
   Schizophoria nevadaensis
   Cortezorthis cortezensis
   Gypidula loweryi
   Leptaena sp.
   Megastrophia iddingsi
   Phragmostrophia merriami
   Chonetes sp.
   indet. rhynchonellid
   Atrypa nevadana
   Howellella cf. *textilis*
   Brachyspirifer pinyonoides
   Hysterolites sp.
   Favosites sp.
   Platyceras sp.
   bivalves, indet.
   bryozoa, indet.

trilobite, indet.
Age: *pinyonensis* Zone
FF-10, 485-535 ft, USNM loc. 13269,
  UCR 4533
  Schizophoria nevadaensis
  Gypidula loweryi
  Megastrophia iddingsi
  Chonetes sp.
  Trigonirhynchia occidens
  Atrypa nevadana
  Brachyspirifer pinyonoides
  Spinulicosta? sp.
  bivalves, indet.
  corals, indet.
Age: *pinyonensis* Zone
FF-11, 535-585 ft, UCR 4534
  Schizophoria nevadaensis
  Gypidula loweryi
  Megastrophia iddingsi
  Phragmostrophia merriami
  Atrypa nevadana
  Hysterolites sp. (wide,
    multiplicate)
  coral, indet.
  Phacops sp.
Age: *pinyonensis* Zone
FF-12, 585-625? ft, UCR 4535
  Schizophoria nevadaensis
  Gypidula loweryi
  "Schuchertella"? sp.
  Megastrophia iddingsi
  Trigonirhynchia occidens
  Atrypa nevadana
  Brachyspirifer pinyonoides
  coral, indet.
FF-12 620 ft, UCR 4536
  Spathognathodus exiguus
  Polygnathus foveolatus

Icriodus latericrescens huddlei
Age: *pinyonensis* Zone
FF-16.2, 660-665 ft, UCR 4538
  Spathognathodus exiguus
  Polygnathus foveolatus
  Icriodus latericrescens huddlei
  Ozarkodina denckmanni
FF-12a, 665-670 ft, UCR 4539
  Bifida sp.
  Styliolina sp.
  Spathognathodus steinhornensis
  Polygnathus foveolatus
Age: *pinyonensis* Zone
FF-13, 685 ft, UCR 4540
  Bifida sp.
  Styliolina sp.
  Spathognathodus steinhornensis
  Polygnathus foveolatus
  Icriodus sp. indet.
Age: *pinyonensis* Zone
FF-13a, 685-690 ft, UCR 4541
  "Leiorhynchus" sp.
  Spathognathodus steinhornensis
  Polygnathus foveolatus
  Icriodus latericrescens huddlei
  Ozarkodina denckmanni
Age: *pinyonensis* Zone
FF-16.4, 685-690 ft, UCR 4542
  Icriodus latericrescens huddlei
FF-16.6, 775-780 ft, UCR 4543
  Spathognathodus exiguus
  Polygnathus foveolatus (very
    late examples)
FF 805 ft, USNM loc. 13633
  Gypidula sp. (lacking median
    septum)
  fish plate, indet.
Age: *Elythyna* beds

EAST SLOPE OF THE MONITOR RANGE

Merriam (1963) mapped a significant Cambrian to Lower Devonian sedimentary section adjacent to Antelope Valley and along the east slope of the Monitor Range. He showed that the Silurian is represented by graptolite-bearing platy calcareous siltstone and shale of the Roberts Mountains Formation.

Merriam and Anderson (1942, p. 1687) reported *Monograptus acus* Lapworth and *Monograptus pandus* Lapworth a short distance above the cherty member at the base of the formation. These species were identified by Ruedemann who reported the horizon to be "of the approximate age of the Lower and Middle Gala beds of Great Britain, or of Clinton (and younger age) of New York." Graptolites collected by the writer from the lower 200 feet of the Roberts Mountains Formation (UCLA locality 4353) were determined by W.B.N. Berry as *Monograptus flemingii* (Salter) and *M*. cf. *M. praedubius* (Bouček) which he regards (written commun., 1962) as probably of Late Wenlock age. Berry has also reported (oral and written commun., 1963) the presence of a probable *Monograptus ultimus* graptolite assemblage in beds lithologically similar to those of the Roberts Mountains Formation.

Berry has also identified *Monograptus hercynicus nevadensis* from beds of uncertain formation assignment in the vicinity of Rabbit Hill (in a collection made by D. L. Clark). Finally, Berry has identified *Monograptus* sp. in a collection made by M. A. Murphy from a position in Clark and Ethington's (1966, p. 664, Text-Fig. 4) section between their horizons 13 and 14, but Murphy noted that no shelly fossils have been seen by him below Clark and Ethington's horizon 11 (oral commun., 1967).

The presence of a probable *Monograptus ultimus* age graptolite assemblage indicates the presence of latest Silurian (Pridoli) age beds. The Gedinnian is represented by a brachiopod faunule noted by Johnson and others (1967). The presence of a younger fauna, that of the *Quadrithyris* Zone, is indicated by the discovery of *Monograptus hercynicus nevadensis*.

Merriam (1963, p. 33) suggested the possibility of an unconformity between the Roberts Mountains and the Rabbit Hill Limestone, but the evidence that Ludlow, Pridoli, Gedinnian, and lower Siegenian (*Quadrithyris* Zone) beds are present in the Rabbit Hill area largely negates his assumption that the upper part of the Roberts Mountains Formation is missing. Merriam's scheme (1963, p. 28, Fig. 7), which recognizes the Rabbit Hill as the initial representative of an eastwardly transgressing sequence that spreads more widely during the time of the *Trematospira* and *Acrospirifer kobehana* Zones, is essentially in agreement with the writer's interpretation of facies relations across Denay Valley (Johnson, 1965, Fig. 3).

Merriam was first to list the fauna of the Rabbit Hill, and he concluded that it is of Helderbergian age (1963, p. 44). The writer has collected from both the type locality at Rabbit Hill (USNM loc. 10713) and from the Rabbit Hill Limestone at a well-exposed section approximately a mile north of the type section (USNM loc. 10712). These collections were supplemented by those made by E. L. Winterer and M. A. Murphy during their study of the Silurian rocks in central Nevada (1960). More recently, large collections were made by A. J.

Boucot from the type locality. These have been etched, and the silicified residues were studied by the writer. The brachiopod collection from the Rabbit Hill Limestone at the type locality is a relatively large and varied one containing thousands of free, silicified, individual valves. The species now known from the type locality compose a more extensive list than the various other localities combined. The brachiopod species listed in Table 1 are present at the type area (USNM locs. 10712, 10713, and 12324); actual numbers of specimens are listed in order to show relative abundance of the species.

The Rabbit Hill Limestone is structurally disturbed in the vicinity of Rabbit Hill where it is overlain by a thrust-plate sequence (Merriam, 1963, p. 42, 43).

TOQUIMA RANGE

Fossiliferous Lower Devonian beds are represented by the dark-gray bioclastic McMonnigal Limestone of Kay (1960; Kay and Crawford, 1964, p. 440). The McMonnigal is underlain by a platy carbonate facies of the Roberts Mountains Formation called Masket by Kay and Crawford. The lithology and sequence of the units is much like that

TABLE 1. BRACHIOPODS OF THE RABBIT HILL LIMESTONE IN THE VICINITY OF RABBIT HILL

	10713	12324	10712
Orthostrophella monitorensis	123	..	..
Dalejina sp. B	191	60	5
Levenea navicula	1011	234	39
"Camerella" sp.	17	..	1
Leptaena cf. acuticuspidata	87	46	2
"Schuchertella" sp. B	75	23	2
Strophonella cf. punctulifera	1	..	..
Leptostrophia inequicostella	46	11	..
Stropheodonta filicosta	3	..	..
Stropheodonta magnacosta	3	1	1
Megastrophia transitans	..	8	..
Pholidostrophia? sp.	3	..	..
"Strophochonetes" sp.	1	..	..
Anoplia sp. A	33	12	..
Pleiopleurina anticlastica	124	..	..
Atrypa sp.	..	..	1
Leptocoelia murphyi	172	18	..
Leptocoelina squamosa	531	925	324
Coelospira concava	403	118	20
Nucleospira sp. B	2	..	1
Meristella cf. robertsensis	220	83	68
Trematospira perforata	1	1	..
Pseudoparazyga cooperi	28	..	..
Howellella cycloptera	372	410	99
Megakozlowskiella magnapleura	223	14	..
Spinoplasia roeni	628	200	118
Plicoplasia cooperi	14	..	..
Cyrtina cf. varia	10	..	..
Rensselaerina? sp.	34	..	..

found in the Windmill Window (Johnson, 1965). Probably the upper part of the Roberts Mountains Formation includes beds of early Devonian age as elsewhere, but the section has not yet been carefully collected. The Tor Limestone (Kay and Crawford, 1964, p. 440) is a probable lateral equivalent of part of the Roberts Mountains Formation and probably is about Pridoli age (Johnson and Boucot, 1970).

As noted by the writer (in Kay and Crawford, 1964, p. 440, Table 4), the McMonnigal Limestone carries two brachiopod faunas, one belonging to the *Quadrithyris* Zone and another to the *Spinoplasia* Zone. Two collections were made from beds that represent the *Quadrithyris* Zone; the brachiopods of these collections are listed in Table 2. In addition, Gilbert Klapper briefly revisited the collection site of USNM loc. 10765 and found that it yielded *Icriodus pesavis* (oral commun., 1967). The conodonts from USNM loc. 10808, identified by O. H. Walliser, were listed by the writer (Johnson, 1967, p. 877).

A third small, but diagnostic collection (USNM loc. 10766) from the McMonnigal yielded brachiopods indicating assignment to the *Spinoplasia* Zone and correlation with the Rabbit Hill Limestone in the Monitor Range to the east. The following species occur:

Orthostrophella monitorensis
Dalejina sp. B
Levenea navicula
Leptaena cf. *acuticuspidata*

Pseudoparazyga cooperi
Trematospira cf. *perforata*
Meristella sp.

TABLE 2. BRACHIOPODS OF THE LOWER MCMONNIGAL LIMESTONE

	USNM 10808	USNM 10765
Dalejina sp. A	X	..
Schizophoria parafragilis	X	..
Levenea sp. A	X	..
Anastrophia cf. *magnifica*	X	X
Gypidula cf. *pseudogaleata*	X	X
Leptaena sp. A	X	X
Leptaenisca sp.	X	..
"*Schuchertella*" sp. A	X	X
Strophonella cf. *bohemica*	X	X
Mesodouvillina sp.	X	..
"*Brachyprion*" sp.	X	..
Atrypa "*reticularis*"	X	X
Spinatrypa sp.	X	..
Toquimaella kayi	X	X
Dubaria sp.	X	..
Howellella sp.	X	X
Megakozlowskiella sp. A	X	..
Quadrithyris cf. *minuens*	X	..
Ambocoelia sp.	X	..
Cyrtina sp. A	X	X
Cyrtinaella causa	X	..
Meristina sp. A	X	..
Nucleospira sp.	X	..

This is an especially significant collection because it comes from beds identical in lithofacies to beds that contain the *Quadrithyris* Zone fauna. At Coal Canyon in the northern Simpson Park Range, the *Quadrithyris* Zone and *Spinoplasia* Zone faunas are found in sequence, but their occurrence is restricted to relatively different lithofacies, so it might be postulated that major differences of the brachiopod faunas of the two zones might be attributed to ecologic factors. The occurrence of brachiopods characteristic of the two zones in the McMonnigal Limestone allows more confidence to be placed in the faunal discontinuity than would be appropriate were the fossils confined to the Windmill and Rabbit Hill Limestones which differ markedly in their lithofacies.

COAL CANYON, NORTHERN SIMPSON PARK RANGE

The stratigraphic section at this locality, called Windmill Window by Roberts and others (1958, Fig. 4), was described by Johnson (1965). The part of the section with Devonian graptolites was described in detail by Johnson and Murphy (1969). The base of the section, beginning in a small canyon about three-fifths of a mile west of the mouth of Coal Canyon, lies in the Hanson Creek Formation. It is overlain by a 1- to 2-foot black-chert member at the base of the Roberts Mountains Formation. The chert member is succeeded by graptolite-bearing platy limestone with late Llandovery, Wenlock, and early Ludlow age graptolites occurring in the lower 400 feet (Johnson, 1965, p. 368).

In the summer of 1967, the Silurian and lower Lower Devonian part of the section exposed in the vicinity of Coal Canyon was remapped and measured by M. A. Murphy. Graptolite collections, made in measured sections by Murphy and examined by W.B.N. Berry, confirm the ages stated above and support a Ludlow age as high as about 480 feet above the base of the Roberts Mountains Formation. Murphy's remapping convinced him that the small northeasterly fault shown by Johnson (1965, p. 367, Fig. 2) cutting across the lower part of the formation does not exist. Accordingly, Murphy has measured 1612 feet of Roberts Mountains Formation compared to the estimated 1400 feet cited by Johnson. A single collection with shelly fossils at a horizon about 570 feet above the base of the formation yielded fossils suggesting a position high in the Ludlow compared to the faunal sequence being studied in the Roberts Mountains by the writer, A. J. Boucot, and M. A. Murphy. In the Hanson Creek area, a similar fauna occurs between 800 and 900 feet above the base of the formation.

At 1255 feet above formation base Murphy collected graptolites determined by W.B.N. Berry as *Monograptus* cf. *praehercynicus* and *M.* aff. *praehercynicus* (thin form). In the highest beds of the

Roberts Mountains Formation, as defined here, at a level 1610 to 1611 feet above the base of the formation, Murphy collected *Monograptus* cf. *praehercynicus* and *M. microdon* identified by W.B.N. Berry (oral commun., February, 1968). These species are indicative of an assignment to the *Monograptus praehercynicus* Zone (Berry, 1968).

Murphy's remapping showed the presence of a north-south fault through the Windmill Limestone on the west flank of Coal Canyon. This fault proves to repeat section so that most of the Windmill Limestone crops out on both flanks of Coal Canyon; the so-called lower breccia of Johnson has been determined to be a down-faulted remnant of the upper breccia. Murphy has redefined the base of the Windmill Limestone at a level 1612 feet above the base of the Roberts Mountains Formation. The lowest beds of the Windmill in the measured section yielded a silicified brachiopod fauna at 1617 feet. The following brachiopods have been determined (USNM loc. 13659):

Dicaelosia cf. *varica*
Dalejina sp.
Isorthis? sp.
Protocortezorthis sp.
Salopina aff. *crassiformis*
Gypidula sp. (ribbed)
"Schuchertella"? sp.
Strophonella sp.
Mesodouvillina sp.
"Chonetes" sp.
Atrypa sp. (large)
"Atrypa" sp.
indet. atrypid (small, flat, smooth)
Coelospira cf. *virginia*
Ambocoelia sp.

The same fauna occurs about 10 feet higher in the section. At 1666 feet there is an occurrence of *Monograptus microdon* and *Linograptus* sp., identified by W.B.N. Berry, and at 1716 feet a new shelly fauna, somewhat resembling that of the *Quadrithyris* Zone, but not exactly like it, occurs. Brachiopods identified include (USNM loc. 13670):

Skenidioides sp.
Dalejina sp.
Salopina cf. *crassiformis*
Schizophoria sp.
Levenea? sp.
Gypidula sp. (smooth)
schuchertellid indet.
Atrypa spp. (small)
Atrypina simpsoni
Nucleospira? sp.
spiriferid indet.
Cyrtina sp.

Highest *praehercynicus* Zone graptolites, including *M. praehercynicus*, *M. microdon*, and *Linograptus posthumus* occur at 1800. Ten feet higher, *M.* sp. of the *hercynicus* type occurs.

The measured section west of Coal Canyon continues to a level of approximately 2000 feet where the "lower breccia" occurs at a position estimated by Murphy to be about 30 stratigraphic feet above the type occurrence of *Monograptus hercynicus nevadensis* Berry (1967).

Apparently most of the Windmill Limestone is repeated on the east flank of Coal Canyon where a new section has been measured by Murphy and by Gilbert Klapper. It includes approximately 295 feet of beds below the upper Windmill breccia and approximately 60 feet of breccia. In this section the lowest bioclastic beds have not been collected yet. A horizon 210 feet below the upper Windmill breccia yielded *Monograptus microdon* and *Linograptus posthumus,* identified by W.B.N. Berry (oral commun., February, 1968; *Monograptus microdon* was first identified in the field by Professor Bedřich Bouček, October 4, 1967). Conodont faunas obtained from about 12 feet above the *microdon* occurrence are part of the same fauna recovered by Klapper from the 1617 horizon on the west side, thus bracketing the *microdon*-bearing beds within a single fauna whose important brachiopods are *Dicaelosia* cf. *varica, Coelospira* cf. *virginia,* and the ribbed *Gypidula.* Between 50 and 100 feet above the *microdon* occurrence, a few brachiopods are found that apparently belong to the same faunule located above *microdon* at 1716 on the west slope of the canyon. These are succeeded at a level 36 feet below the base of the upper breccia by *Monograptus hercynicus nevadensis,* identified by Berry. The fauna of the upper Windmill breccia, including its downfaulted remnant, yields a relatively rich fauna that has been discussed before (Johnson, 1965, p. 370, 371) and which has been named *Quadrithyris* Zone. The *Quadrithyris* Zone brachiopod fauna from the "upper breccia" (USNM loc. 10758) east of Coal Canyon is as follows:

"*Dolerorthis*" sp.
Dalejina sp. A
Protocortezorthis windmillensis
Levenea sp. A
Schizophoria parafragilis
Anastrophia cf. *magnifica*
"Camerella" sp.
Gypidula cf. *pseudogaleata*
Sieberella cf. *problematica*
Leptaena sp. A
Leptaenisca sp.
"*Schuchertella*" sp. A
Aesopomum cf. *varistriatus*
Strophonella cf. *bohemica*
Leptostrophia sp. A
"*Brachyprion*" *mirabilis*
Mesodouvillina cf. *varistriata*

Machaeraria sp.
"Camarotoechia" cf. *modica*
Atrypina simpsoni
Spirigerina supramarginalis
Atrypa "reticularis"
**Toquimaella kayi*
**Lissatrypa?* sp.
**Coelospira* sp.
Meristina sp. A
**Nucleospira* sp. A
Howellella sp.
**Megakozlowskiella* sp. A
Reticulariopsis? sp. A
Quadrithyris cf. *minuens*
Ambocoelia sp.
Cyrtina sp. A
**Cyrtinaella causa.*

The "lower breccia" collection (USNM loc. 10757) from west of Coal Canyon is the same, except for species with an asterisk which are absent from the collection. Above the upper breccia, the Windmill Limestone is composed of dark, thin-bedded, calcareous shale and calcarenite estimated to be about 140 feet thick. This interval includes beds with abundant tentaculitids at several horizons. A few brachiopods have been collected from this unit (*Anastrophia?* sp., *Leptaena* sp., and *Cyrtina* from USNM loc. 12344), plus a single specimen of *Leonaspis tuberculata favonia* Haas (1969).

Above the tentaculite-bearing beds at the top of the Windmill Limestone, there is a gradational contact with the Rabbit Hill Limestone. The Rabbit Hill fauna is not so abundantly developed as at the type locality, but it is the characteristic one of the *Spinoplasia* Zone (USNM locs. 10759, 10760) as represented by the following brachiopods:

Lingula spp.
Dalejina sp. B
Levenea navicula
Leptaena cf. *acuticuspidata*
"*Schuchertella*"? sp.
Megastrophia transitans
"*Strophochonetes*" sp.

Leptocoelina squamosa
Leptocoelia murphyi
Coelospira concava
Acrospirifer aff. *murchisoni* (in uppermost beds)
Spinoplasia roeni.

Float from low on the east slope of Coal Canyon (USNM loc. 13660) yielded the following:

Dalejina sp. B
Levenea navicula
Leptaena sp.
Leptaenisca sp.
Strophonella? cf. *punctulifera*
Leptostrophia inequicostella
Megastrophia sp.
Stropheodonta? cf. *magnacosta*
Anoplia sp. A

Pegmarhynchia sp.
indet. small atrypid sp.
Coelospira concava
Leptocoelia murphyi
Leptocoelina squamosa
Meristella sp.
Megakozlowskiella magnapleura
Cyrtina sp.

Howellella cycloptera is common in some beds (in USNM loc. 12726) and in a collection made by Allen Ormiston. In addition to brachiopods, there are leonaspid and phacopid trilobites, conulariids, orthocerids, platycerids, and a single small species of the solitary rugose coral *Syringaxon* (Johnson, 1965, p. 374). The Rabbit Hill Limestone is overlain conformably by nearly barren flaggy limestone representing the lower part of the McColley Canyon Formation. No fine fossil debris characterizes these beds as it does the underlying beds of the Rabbit Hill Limestone and the few contained fossils are recovered one or two at a time at various places along strike. Typically, it bears platycerid and *Loxonema*-type gastropods and in addition *Ancillotoechia aptata*, the characteristic rhynchonellid species of the *Trematospira* Zone. No other fossils were collected from the several hundred feet of limestone overlying these basal beds of the McColley Canyon Formation, but at about 400 feet above the base, argillaceous beds enter the section bringing with them a rich brachiopod assemblage. The lowest collections within the argillaceous beds are those characteristic of the *Leptocoelia infrequens* Subzone of the *Eurekaspirifer pinyonensis* Zone. The faunule representing the subzone (USNM loc. 10761, 12393) contains the following species:

Dalejina sp. C	*Leptocoelia infrequens*
Levenea fagerholmi	*infrequens*
Leptaena sp. C	*Howellella* cf. *textilis*
	Elythyna sp. A.

Several hundred feet of yellow argillaceous limestone (USNM loc. 10762, 12396) overlying the basal *Leptocoelia* beds contains a brachiopod assemblage typical of the *Eurekaspirifer pinyonensis* Zone from which the following brachiopod species have been identified:

Levenea fagerholmi	*Phragmostrophia merriami*
Cortezorthis cortezensis	*Parachonetes macrostriatus*
Schizophoria nevadaensis	*Trigonirhynchia occidens*
Gypidula loweryi	*Atrypa nevadana*
Leptaena sp.	*Brachyspirifer pinyonoides*
leptostrophiid, gen. indet.	*Meristina* sp. B.
Megastrophia iddingsi	

A few corals determined by Oliver and reported by Johnson (1965, p. 377) are also present. Figure 3 depicts the Lower Devonian section in the northern Simpson Park Mountains and in the northern Roberts Mountains.

At the top of the section exposed along the front of the range east of Coal Canyon the argillaceous beds of the *Eurekaspirifer pinyonensis* Zone are succeeded by fossiliferous thin-bedded, platy, gray and pink-weathering beds of the Denay Limestone carrying abundant silicified shells of the *Leptathyris circula* Zone fauna (Johnson, 1962a, 1962b, 1966a). The contact relations have not been observed.

CORTEZ RANGE

The Cortez Mountains contain a Cambrian to Devonian sedimentary section below the Roberts Mountains thrust (Roberts and others, 1958, Fig. 2, column 5). The Silurian part of the section is graptolite-bearing and the Lower Devonian bears a shelly fauna. No detailed stratigraphic work has been done, however, to show the actual sequence of the various collections which are representative of the Silurian and Devonian; therefore, the collections listed below are intended only to show the presence of beds of various ages as indicated by an appraisal of the fossils in the separate collections.

A particularly fine collection of monograptids from a black limestone horizon within the Roberts Mountains Formation was

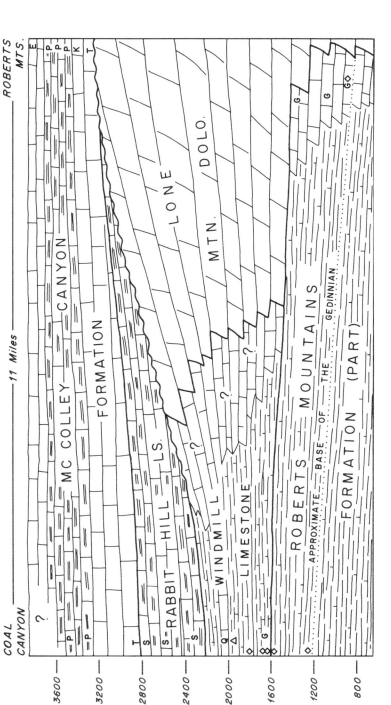

Figure 3. Lower Devonian cross section between Coal Canyon, Simpson Park Range and the northern Roberts Mountains. Note that the vertical scale on the left expresses feet above the base of the Roberts Mountains Formation, west of Coal Canyon. Considerable interbedding occurs between the platy and the thick-bedded facies of the Windmill Limestone and of the Roberts Mountains Formation—the simple lateral equivalence of these rock facies is schematic. Diamonds show position of Gedinnian graptolites. The triangle shows the position of *Monograptus hercynicus nevadensis*. G=Gedinnian age brachiopods. Q=*Quadrithyris* Zone brachiopods. S=*Spinoplasia* Zone brachiopods. T=*Trematospira* Zone brachiopods. K=*kobehana* Zone brachiopods. P=*pinyonensis* Zone brachiopods. E=*Elythyna* beds brachiopods.

referred to W.B.N. Berry by the writer. This collection contains the following species:

Cyrtograptus symmetricus Elles
Dendrograptus sp.
Monograptus cf. M.
   praedubius (Bouček)
Monograptus parapriodon
   (Bouček)?
Monograptus priodon (Brown)

Monograptus priodon n. var.?
Monograptus aff. M.
   triangulatus (Harkness)
Monograptus vomerinus n. var.
   (aff. var. gracilis)
Retiolites geinitzianus
   (Barrande).

Berry regards the assemblage as probably of the Zone *Cyrtograptus symmetricus* (Zone 28) of Elles and Wood, that is, about middle Wenlock age. The writer has examined a single collection from the Wenban Limestone which contains *Anastrophia?* sp., *Sieberella* cf. *problematica*, *Leptaena* sp. A, *Machaeraria* sp., *Toquimaella kayi*, *Atrypa* sp., and *Howellella* sp., indicating *Quadrithyris* Zone. Other collections from the Wenban contain *Leptocoelina squamosa* and *Spinoplasia roeni?*, species that are restricted to the *Spinoplasia* Zone. A third collection yielded *Costispirifer* sp., *Trematospira perforata*, and *Dalejina* sp. This assemblage belongs to the *Costispirifer* Subzone.

The brachiopod collections mentioned above are small and obviously give only a brief indication of the Lower Devonian faunas present in the Cortez Range. However, large collections were made by A. J. Boucot from the *Eurekaspirifer pinyonensis* Zone on both sides of the range. The fossiliferous horizons are developed in a black terrigenous limestone facies and the shells are commonly silicified. The brachiopod assemblage collected at the two most prolific localities is listed in Table 3. Also present in addition to the brachiopods are *Syring-*

TABLE 3. CORTEZ RANGE, EUREKASPIRIFER PINYONENSIS ZONE BRACHIOPODS

	USNM 10754	10752
Dalejina sp. C	X	..
Schizophoria nevadaensis	X	X
Muriferella masurskyi	X	X
Cortezorthis cortezensis	X	X
Gypidula loweryi	X	X
Leptaena sp.	X	..
Mclearnites sp. B	X	X
Phragmostrophia merriami	X	X
Trigonirhynchia occidens	X	..
"*Leiorhynchus*" sp.	X	..
Atrypa nevadana	X	X
Bifida sp.	X	..
Leptocoelia infrequens globosa	X	X
Hysterolites sp. A	X	X
Elythyna sp.	X	X
Cyrtina sp. D	X	X
Nucleospira sp.	..	X

axon sp., *Pleurodictyum* sp., and *Leonaspis* sp. The brachiopod assemblage is an interesting one to which the following species are geographically restricted in central Nevada: *Muriferella masurskyi*, *Mclearnites* sp. B, and *Leptocoelia infrequens globosa*.

A single collection containing two specimens of goniatites was obtained, but no other fossils were present. The specimens, sent to Dr. M. R. House for identification, have been made the types of a new species *Erbenoceras erbeni* House (1965). He regards the species to be indicative of the lower Emsian since it is closest to a form from the lower Emsian of the Harz Mountains in central Germany.

LYNN WINDOW, TUSCARORA MOUNTAINS

The sedimentary section below the Roberts Mountains thrust ranges from the Hamburg Dolomite of Cambrian age into unnamed Devonian limestone strata (Berry and Roen, 1963, p. 1123). These writers show that the lowest fossiliferous beds of the Roberts Mountains Formation, approximately 15 feet above its base, bear a graptolite assemblage diagnostic of the *Monograptus riccartonensis* Zone of early Wenlock age. The lithology is typical of exposures in the northern Simpson Park Mountains and in the Monitor Range. Well over a thousand feet higher in the section (Roen, oral commun., 1959), limestones with a silicified shelly fauna appear. A small collection made from this horizon by Roen was studied by the writer. It contains *Skenidioides* sp., *Schizophoria* sp., *Aesopomum* sp., and either *Megakozlowskiella* or *Cyrtina* together with other fragmentary brachiopods. The identifiable specimens are certainly Devonian, but zonal assignment is uncertain.

ANTELOPE PEAK

North of Wells, Elko County, Nevada, there is a sequence of Silurian and Lower Devonian rocks, studied by Douglas Gardner and Ben L. Peterson, much like the sequence developed in the Toquima Range and in the northern Simpson Park Range. According to Peterson (written commun., June 4, 1968) the sequence of the Silurian and Lower Devonian begins with a basal black, bedded chert overlain by approximately 950 feet of thin-bedded to thinly laminated graptolite-bearing limestone, such as is typical of the Roberts Mountains Formation. This is overlain by approximately 500 feet of dark-gray limestone. The latter unit is suggestive of the McMonnigal Limestone of Kay and Crawford (1964), and has yielded a good silicified brachiopod fauna of the *Spinoplasia* Zone from one locality (USNM loc. 17272). The brachiopods from this collection, studied by the writer, are as follows:

Orthostrophella monitorensis *Levenea navicula*
Dalejina sp. *Schizophoria* sp.

Sieberella cf. *pyriforma*
Leptaena sp.
Leptaenisca sp.
"*Schuchertella*" sp.
indet. stropheodontid sp.
Meristella sp.
Nucleospira sp.
Trematospira cf. *perforata*
Howellella cf. *cycloptera*
Megakozlowskiella magnapleura
Spinoplasia roeni
Cyrtina cf. *varia*
Rensselaerina? sp.

This collection, clearly assignable to the *Spinoplasia* Zone, is suggestive of the assemblage from the upper McMonnigal Limestone (USNM loc. 10766), although the latter is smaller and still must be regarded as poorly known because only a small sample, of nonsilicified material, was examined. With three exceptions, all of the brachiopod species from the Antelope Peak locality occur in the type Rabbit Hill Limestone. One of these is *Leptaenisca* sp. which, however, has been found in typical Rabbit Hill lithology and in a *Spinoplasia* Zone brachiopod assemblage from the east flank of Coal Canyon (USNM loc. 13660). *Sieberella* cf. *pyriforma* is known elsewhere only in the *Trematospira* Subzone of the *Trematospira* Zone, but its occurrence in a *Spinoplasia* Zone assemblage is not surprising because the two faunas share a number of species in common. *Schizophoria* does not occur in other *Spinoplasia* Zone assemblages and probably is a holdover from the *Quadrithyris* Zone.

The Antelope Peak locality is the northeasternmost occurrence of the *Spinoplasia* Zone.

Age and Correlation
SIEGENIAN STAGE
Discussion

The fossiliferous graptolite and shelly sequence of the Roberts Mountains Formation (Johnson and others, 1967) is laterally equivalent to lower parts of the Lone Mountain Dolomite as shown by the detailed mapping of Winterer and Murphy (1960, p. 120, 121), but fossiliferous lateral equivalents of the upper beds of the Lone Mountain Dolomite that overlie the Gedinnian are not exposed in the Roberts Mountains. Winterer and Murphy (1960, p. 133) pointed out, however, that equivalents of the uppermost Lone Mountain were to be found in the Simpson Park Range, and the subsequent faunal and stratigraphic study of the Coal Canyon section at Windmill Window in the northern Simpson Park Range by Johnson (1965) showed this suggestion to be well based. At Coal Canyon the Roberts Mountains Formation is represented for the most part by platy graptolite-bearing carbonates with *Monograptus* cf. *praehercynicus* in the upper part of the formation. The uppermost few feet of the formation, as defined on the west flank of Coal Canyon, yields *Monograptus praehercynicus* and *M. microdon* indicating a position equivalent to the *M. praehercynicus* Zone of central Europe. These occurrences are followed above, in the lower part of the Windmill Limestone by *Monograptus microdon, M. praehercynicus*, and *Linograptus posthumus* and higher in the Windmill Limestone by *Monograptus hercynicus nevadensis*. The principal source of *Quadrithyris* Zone brachiopods is the breccia of the upper Windmill Limestone beginning 30 to 36 feet above the lower occurrence of *M. hercynicus nevadensis*. Since *Monograptus hercynicus* appears to occur in a position equivalent to some part of the Siegenian of Europe (Solle, 1963; Jaeger, 1965, chart), the evidence of the graptolites suggests that the *Quadrithyris* Zone is no older than Siegenian. No evidence has been discovered that firmly establishes the position of the Gedinnian-Siegenian boundary in the graptolite zonal scheme.

However, as a working hypothesis, the writer equates the Lower Gedinnian, Upper Gedinnian, and an undetermined lower part of the Siegenian with the *M. uniformis*, *M. praehercynicus*, and *M. hercynicus* Zones, respectively.

Factors of faunal provinciality and rock facies greatly influence our ability to correlate distant faunas with beds in the Ardennes and in the Rhineland where the standard Lower Devonian stage nomenclature has been established. The terrestrial and terrigenous facies of the type region for the most part lack a sequence of faunas that includes such important groups as graptolites, conodonts, or goniatites. Instead, brachiopods have been utilized as important guides to the Siegenian and Emsian. However, the brachiopods also leave many questions unsolved because of problems of provinciality (see Boucot and others, 1967). The Gedinnian brachiopod fauna of Europe is relatively well known through the work of Barrois and others (1922), Kozlowski (1929), Asselberghs (1946), Dahmer (1951), Boucot (1960), and others. The relative ease with which its brachiopod fauna is recognized at least at some distant localities, such as central Nevada, is notable. However, the Siegenian is a different matter; the highest part of the Gedinnian and the lowest Siegenian are generally in the terrestrial facies in northwestern Europe—a situation responsible for uncertainty of recognition within the shelly fossil sequence. Certainly the Siegenian in general is characterized by a shelly fauna that differs markedly from that of the Gedinnian and bears many new Devonian elements—a fact that has had the effect of excluding Gedinnian faunas from the "Devonian" at some distant localities because age assignment was based on systemic affinity of faunas rather than on correlation.

The important question is, does the new fauna of strong Devonian aspect begin at the base of the Siegenian, or lower, in beds equivalent to the upper Gedinnian, at sections where fossiliferous marine beds are continuous across the boundary? With regard to the brachiopod fauna perhaps the most significant evidence is afforded by Boucot's redescription of the Gedinnian brachiopods of the Ardennes. Boucot annotated the several source horizons of the fossils he studied as Lower Gedinnian because all appear to represent essentially a single fauna, but in fact the major part of the Gedinnian fauna, derived from the Grès de Gdoumont is generally regarded as an equivalent of the lower part of the upper Gedinnian (for example, Schmidt, 1954, as noted *in* Boucot, 1960, p. 285; Asselberghs, 1946, p. 87; Jaeger, 1965, chart). This is sufficient evidence to show that the Gedinnian fauna of the type region lacks significant new elements at least as high as the lower part of the upper Gedinnian, and it is on this point that the writer recommends that a new fauna of Devonian aspect and clearly advanced over the Gedinnian fauna should be assigned to the Siegenian.

Finally, it is possible to obtain a clue that assists direct correlation between Nevada and western Europe on the basis of brachiopods. This involves the series of *Schizophoria* species that occur in the two regions and those that are known at other Gedinnian and Siegenian and younger localities around the world. The earliest *Schizophoria* species that are well represented are those of the Gedinnian (*Schizophoria fragilis* and unnamed relatives). This group is always characterized by brachiophore supporting plates that are not widely divergent. Rather, they make an angle considerably less than 90 degrees and commonly are prolonged anteriorly as ridges that are subparallel to one another. In Siegenian species of *Schizophoria*, from the Rhineland, the brachiophore supporting plates are much more widely divergent, and in the Emsian and higher horizons no species that even remotely resembles the Gedinnian ones with narrowly divergent brachiophore supporting plates was ever seen again. This trend in the development of brachiophore supporting plates of *Schizophoria* should have real value for correlation if it can be accurately enough diagnosed in terms of the species of western Europe. Fortunately, a new species from the Grès à *Orthis monnieri* of Armorica, Brittany, is under study by Dr. J. T. Renouf. The fauna of the Grès à *Orthis monnieri* is one that can be characterized with some confidence as a fauna that is intermediate between the known Gedinnian and Siegenian faunas. It clearly occupies a place rather near the Gedinnian-Siegenian boundary and Renouf (written commun., 1969) believes it more likely that this fauna is of Late Gedinnian age. The important new species of *Schizophoria* in this fauna is characterized by relatively narrowly divergent brachiophore supporting plates; decidedly more narrowly divergent than the brachiophore supporting plates of the *Quadrithyris* Zone species *Schizophoria parafragilis*. This comparison allows an independent direct correlation that indicates a post-Gedinnian age for the *Quadrithyris* Zone as already concluded.

Quadrithyris Zone

At Coal Canyon in the northern Simpson Park Range, the upper Windmill Limestone, with a shelly megafauna predominantly of brachiopods and corals, was named *Quadrithyris* Zone (Johnson, 1965, p. 370). The fauna presently known is a relatively large one consisting of brachiopods, corals, conodonts, and ostracodes, in addition to the occurrence of *Monograptus hercynicus nevadensis*, and is listed below in Table 4.

The "*Dolerorthis*" was originally listed as "*Ptychopleurella*," but it now appears to belong to a new genus most closely related to *Dolerorthis*. Two other occurrences are known; the first is in the *Monograptus yukonensis* Zone at Royal Creek, Yukon Territory, Canada (specimens loaned by A. C. Lenz). The second occurrence is

TABLE 4. FAUNA OF THE QUADRITHYRIS ZONE

	WINDMILL LIMESTONE	LOWER McMONNIGAL LIMESTONE
"Dolerorthis" sp.	X*	..
Dalejina sp. A	X	X
Protocortezorthis windmillensis	X	..
Schizophoria parafragilis n. sp.	X	X
Levenea sp. A	X	X
Anastrophia cf. magnifica	X	X
"Camerella" sp.	X	..
Gypidula cf. peudogaleata	X	X
Sieberella cf. problematica	X	..
Leptaena sp. A	X	X
Leptaenisca sp.	X	X
Leptostrophia sp. A	X	..
Strophonella cf. bohemica	X	X
"Brachyprion" mirabilis n. sp.	X	X
Mesodouvillina cf. varistriata	X	X
"Schuchertella" sp. A	X	X
Aesopomum cf. varistriatus n. sp.	X	..
Machaeraria sp.	X	..
"Camarotoechia" cf. modica	X	..
Coelospira sp.	X	..
Atrypina simpsoni n. sp.	X	..
Atrypa "reticularis"	X	X
Spinatrypa sp.	..	X
Spirigerina supramarginalis	X	..
Toquimaella kayi	X	X
Lissatrypa? sp.	X	..
Dubaria sp.	..	X
Meristina sp. A	X	X
Nucleospira sp. A	X	X
Howellella sp.	X	X
Megakozlowskiella sp. A	X	X
Reticulariopsis? sp. A	X	..
Quadrithyris cf. minuens	X	X
Ambocoelia sp.	X	X
Cyrtina sp. A	X	X
Cyrtinaella causa n. sp.	X	X
Leonaspis tuberculata favonia	WH	..
Calymene sp.	X	..
"Loxonema" sp.	X	..
Striatopora sp.	WAO	..
Favosites sp.	WAO	WAO
Thamnopora ("Pachypora") sp.	WAO	..
Grypophyllum sp.	WAO	..
Endophyllum sp.	WAO	..
Mucophyllum sp.	WAO	..
Rhizophyllum cf. enorme	WAO	..
Spongophyllum? sp.	WAO	..
Enterolasma? sp.	WAO	..
chonophylloid corals, indet.	WAO	..
indet. horn corals	WAO	..
indet. stromatoporoids	WAO	..
Monograptus hercynicus nevadensis	WBNB	..
Spathognathodus johnsoni	GK	..
Spathognathodus transitans	GK	..
Spathognathodus n. sp.	OHW	..
Spathognathodus steinhornensis	OHW, GK	..

TABLE 4. (CONTINUED)

	WINDMILL LIMESTONE	LOWER McMONNIGAL LIMESTONE
Spathognathodus inclinatus	OHW, GK	OHW
Icriodus pesavis	GK	GK
Icriodus latericrescens	GK	GK
Icriodus sp.	OHW	..
Icriodus n. sp.	..	OHW
Ozarkodina typica denckmanni	OHW	..
Hindeodella equidentata	OHW	..
Trichonodella excavata	OHW(?)	OHW
Plectospathodus extensus		OHW
Spathognathodus n. sp. (forerunner of *S. sulcatus*)		GK
†*Eognathodus* sp. aff. *E. Secus*	..	OHW
Welleriopsis? occidentalis	JMB (cf.)	JMB
Apatobolbina? sp.	JMB	..
Treposella? sp.	JMB	JMB
aff. *Dolichoscapha* sp.	JMB	..
aff. *Clintiella* sp.	JMB	JMB
Saccarchites sp.	JMB	..
Myomphalus? sp.	JMB	..
Hollinella? sp.	JMB	JMB
Nezamyslia sp.	JMB	..
Chironiptrum sp.	JMB	..
Poloniella sp.		JMB
aff. *Hypotetragona* sp.	JMB	..
Acanthoscapha sp.		JMB
Condracypris? sp.	JMB	..
Bairdiocypris sp.	JMB	JMB
Bairdia sp.	JMB	..
Tricornina sp.	JMB	..
Libumella? sp.	JMB	..
Tubulibairdia sp.	JMB	JMB
smooth ostracodes, indet.	JMB	..

*X, Identified by the writer; JMB, Jean M. Berdan; WBNB, W.B.N. Berry; GK, Gilbert Klapper; WAO, W. A. Oliver, Jr.; OHW, O. H. Walliser. Conodonts from the lower McMonnigal Ls. identified by Klapper are from USNM loc. 10765; those identified by Walliser are from USNM loc. 10808. WH, Winfried Haas; the single specimen identified by Haas comes from the upper Windmill Limestone above the breccia, probably from about the same horizon as USNM loc. 12344.

†This conodont was listed as *Eognathodus* cf. *sulcatus* by Walliser (*in* Johnson 1967, p. 877), but the present identification results from a re-examination of the specimens (Walliser, written commun., 1968).

in a Siegenian age limestone horizon at Manildra, New South Wales, Australia, being studied by Norman M. Savage (specimens sent to the writer by Savage). The dalmanellid *Protocortezorthis windmillensis,* recently described by Johnson and Talent (1967b) is an important element representing the initial stage of a septate dalmanellid lineage that occurs in the *Monograptus yukonensis* Zone, in the *Eurekaspirifer pinyonensis* Zone, and in beds of Emsian and possibly Eifelian age in Arctic Canada. The species represents a decided advance over *P. fornicatimcurvata* of the Gedinnian of western

Europe. *Anastrophia* cf. *magnifica* occurs in the *Monograptus yukonensis* Zone (specimens loaned by A. C. Lenz), in beds in the U.S.S.R. that the writer believes are correlative, that is, the Solovien Limestone (Kulkov, 1963), and in beds of Siegenian age called Gedinnian by Khodalevich, 1951) below the *Karpinskia conjugula* Zone in the Ural Mountains. These occurrences represent the Krekova Stage of Rzhonsnitskaya (1962). The type occurrence of *Anastrophia magnifica* is in the Borszczow beds of Podolia, so that the species apparently ranges through the Gedinnian and Siegenian.

Gypidula cf. *pseudogaleata* is one of the few brachiopods of Appalachian affinity in the *Quadrithyris* Zone shelly fauna and is significant in indicating a Becraft age for the zone.

Aesopomum was defined from the Pragian by Havlíček (1965); it occurs also in the *Monograptus yukonensis* Zone at Royal Creek (Lenz, 1966, 1967b), but appears first in the Gedinnian beds in the Roberts Mountains. *Spirigerina supramarginalis* (originally reported by Johnson as *Plectatrypa* cf. *P. sibirica*) originates in the Solovien Limestone of the Altai Mountains of the U.S.S.R. (Kulkov, 1963). *Toquimaella kayi*, recently described by Johnson (1967), is closest to *Vagrania* which occurs in the latest Siegenian and is a prominent member of the Emsian faunas in the Ural Mountains. As noted by Johnson, it occurs below horizons with *Monograptus yukonensis* at two places in the Canadian Arctic. *Quadrithyris* cf. *minuens* is an important constituent of the *Quadrithyris* Zone inasmuch as the genus is a very common representative of Siegenian and Emsian brachiopod faunas in Bohemia and in Russia, but is unknown in beds of Gedinnian age. In Bohemia, it does not occur below the base of the Pragian (that is, upper Siegenian and lower Emsian) as noted by Havlíček (1959, p. 28). Several other brachiopods, such as *Schizophoria, Levenea, Megakozlowskiella, Ambocoelia, Cyrtina*, and *Cyrtinaella*, are typical Devonian forms, of which none have been reported in beds as old as Ludlow, but which has no zonal significance in central Nevada because they are (with the exception of *Cyrtinaella*) already present in other, underlying beds of Gedinnian age. The remainder of the brachiopod fauna comprises genera and species that are relatively long-ranging and have little bearing on detailed correlation although some have zonal value. Oliver, who identified the *Quadrithyris* Zone corals from Coal Canyon, has discussed them briefly elsewhere (1964, p. D151), and Berry (1967) has described the *Monograptus hercynicus* specimens from the lower part of the zone as a new subspecies.

Quadrithyris Zone conodonts have been studied by Professor O. H. Walliser from samples sent by the writer. The listed new species of *Spathognathodus* was derived from slabs bearing *Monograptus hercynicus nevadensis* in the lower part of the zone at Coal Canyon. According to Walliser (written commun., 1966) "this species indicates a position in the higher part of the (Lower Devonian) succession; younger than Lower Gedinnian, older than Lower Emsian." The remainder of the forms identified by Walliser from the Windmill Limestone are from one locality in the Windmill breccia at Coal Canyon. Of this collection Walliser stated (written commun., 1966) "younger than Lower Gedinnian, older than Upper Emsian." The species from the lower McMonnigal Limestone are from a single collection (USNM loc. 10808) from the same locality as the *Quadrithyris* Zone brachiopod faunule listed earlier by Johnson (1965, p. 371, column C). Of this fauna, Walliser said (written commun., 1965) "younger than Lower Gedinnian, older than higher Siegenian. The *Icriodus* n. sp. is a new species or subspecies within the transition from *Icriodus woschmidti* (lowermost Gedinnian) to *Icriodus latericrescens* (sensu Ziegler; beginning within the Siegenian)." *Eognathodus secus* occurs in the Siegenian age Coopers Creek Formation of Victoria, Australia (Philip, 1965). It must be noted, however, that Walliser qualified the identification of the McMonnigal form as "aff. *secus*," and it may represent an older taxon than true *secus* and *sulcatus* (see below).

The occurrence cited by Klapper (*in* Johnson and others, 1967) of *Icriodus pesavis* Bischoff and Sannemann and *Spathognathodus transitans* Bischoff and Sannemann in the Windmill breccia at Coal Canyon is significant, as it places two important elements of the Bischoff and Sannemann (1958) fauna at a definite level in a known faunal sequence. *Icriodus pesavis* and *Spathognathodus johnsoni* Klapper are present in the *Spirigerina* Unit of Lenz (1966; 1967b), which also contains *Quadrithyris* Zone brachiopods (Johnson, 1967; Johnson and others, 1967). *Spathognathodus* n. sp., reported from the lower McMonnigal (USNM loc. 10765), "is clearly separable from *Spathognathodus sulcatus* and is a morphologic forerunner of that species. In the lower McMonnigal, *Spathognathodus* n. sp. occurs in the same bed with *Icriodus pesavis* and with *Quadrithyris* Zone brachiopods, but it has not yet been found at Coal Canyon" (Klapper, written commun., 1968). The correlation of the Nevada and Royal Creek conodont sequences and position of the *Quadrithyris* Zone conodonts relative to conodonts of known Gedinnian and Emsian age has been demonstrated by Klapper (*in* Klapper and Ormiston, 1969).

The fauna of the *Quadrithyris* Zone is clearly one that is younger than faunas known to be of Gedinnian age. From the introductory discussion of Gedinnian and Siegenian stage assignments, the *Quadrithyris* Zone is considered to be of Siegenian age. Several lines of evidence converge to point to an age older than latest Siegenian. The lineage of *Protocortezorthis-Cortezorthis*, discovered by Johnson and Talent (1967b) shows that the *Quadrithyris* Zone is older than the *Monograptus yukonensis* Zone at Royal Creek. The occurrence of the

Quadrithyris Zone guide, *Toquimaella kayi,* below *Monograptus yukonensis* at two localities in the Canadian Arctic (Johnson, 1967) supports this assignment. As noted by Bouček (1966, p. 165), *Monograptus* cf. *yukonensis* occurs as high as the top of the lower Emsian in Bohemia, but following his Canadian visit in 1967, Bouček (oral commun., 1967) informed the writer that he regards the Canadian occurrences as inclusive of two distinct forms, of which only the higher one (and therefore typical *M. yukonensis*) is actually close to the specimens from the Bohemian Emsian. The writer believes that the range of *M. yukonensis* (in a broad sense) is high in the Siegenian to the top of the lower Emsian because the Canadian occurrences accompanied by brachiopods (*Gypidula* 1 - *Davidsoniatrypa* unit of Lenz, 1967b) appear to be late Siegenian. The *Quadrithyris* Zone also appears to be earlier than latest Siegenian in Nevada because the overlying zones of *Spinoplasia* and *Trematospira* can be independently assigned to the Siegenian. The zone was originally assigned to the upper Gedinnian or Siegenian by Johnson (1965, p. 378, Fig. 4), but the several points discussed above now point to a lower Siegenian assignment.

Spinoplasia Zone

The zone was named by Johnson (1965, p. 374). It coincides with the Rabbit Hill Limestone as restricted in the section at Coal Canyon and also includes beds of the Rabbit Hill Limestone sampled at the type locality. The zone is recognized in the Wenban Limestone of the Cortez Range and in the upper part of the McMonnigal Limestone at Ikes Canyon in the Toquima Range, and at Antelope Peak.

At the type locality of the Rabbit Hill Limestone, the *Spinoplasia* Zone is characterized by a rich and abundant brachiopod fauna comprising 29 articulate species. Corals are relatively common, but are very restricted in number of genera and species; this is particularly true of the Rugosa. Several collections bearing ostracodes have been obtained and those listed from the Coal Canyon section (Table 5) aı

TABLE 5. FAUNA OF THE SPINOPLASIA ZONE

	RABBIT HILL	COAL CANYON	ANTELOPE PEAK
Orthostrophella monitorensis n. sp.	X	..	X
Dalejina sp. B	X	X	sp.
Levenea navicula n. sp.	X	X	X
Schizophoria sp.	..	..	X
"*Camerella*" sp.	X	..	..
Sieberella cf. *pyriforma*	..	..	X
Leptaena cf. *acuticuspidata*	X	X	..
Leptaenisca sp.	..	X	X
"*Schuchertella*" sp. B	X	X	sp.
Strophonella cf. *punctulifera*	X	X	..
Leptostrophia inequicostella n. sp.	X	X	..
Stropheodonta filicosta n. sp.	X	..	..

TABLE 5. (CONTINUED)

	Rabbit Hill	Coal Canyon	Antelope Peak
Stropheodonta magnacosta n. sp.	X	?	..
Megastrophia transitans n. sp.	X	X	..
Pholidostrophia? sp.	X	..	..
"*Strophochonetes*" sp.	X	X	..
Anoplia sp. A	X	..	..
Pleiopleurina anticlastica n. sp.	X	..	..
Pegmarhynchia sp.	..	X	..
Atrypa sp.	X	?	..
Leptocoelia murphyi n. sp.	X	X	..
Leptocoelina squamosa n. sp.	X	X	..
Coelospira concava	X	X	..
Acrospirifer aff. *murchisoni*	..	X	..
Howellella cycloptera	X	X	cf.
Megakozlowskiella magnapleura n. sp.	X	..	X
Spinoplasia roeni n. sp.	X	X	X
Plicoplasia cooperi	X	..	..
Cyrtina cf. *varia*	X	sp.	X
Nucleospira sp. B	X	..	sp.
Meristella cf. *robertsensis*	X	X	sp.
Trematospira perforata	X	..	cf.'
Pseudoparazyga cooperi	X	..	..
Rensselaerina? sp.	X	..	X
Lingula spp.	..	X	..
Conularia sp.	..	X	..
Amphipora? sp.	..	WAO	..
Favosites sp.	CWM	..	..
Cladopora sp.	CWM	..	..
Striatopora cf. *gwenensis* Amsden	CWM	..	..
Michelinia sp.	CWM	..	..
Pleurodictyum cf. *trifoliatum* Dunbar	CWM	..	..
Syringaxon acuminatum (Simpson)	CWM	..	..
Syringaxon sp.	..	WAO	..
Tentaculites sp.	CWM	..	..
Platyceras sp.	CWM	X	X
Orthoceras sp. (small form)	CWM	X	X
hexactinellid sponge spicules, indet.	CWM	..	..
Leonaspis cf. *tuberculatus* (Hall)	CWM	..	..
Phacops sp.	CWM	..	..
Dalmanites sp.	CWM	..	..
Loxonema? sp.	..	X	..
Welleriopsis? cf. *occidentalis* (Walcott)	JMB	..	..
Eukloedenella sp.	..	JMB	..
Libumella? sp.	..	JMB	..
Strepulites sp.	..	JMB	..
Thlipsura aff. *furcoides* Bassler	..	JMB	..
Tubulibairdia sp.	..	JMB	..
Falsipollex? sp.	..	JMB	..
Hollinella sp.	..	JMB	..
Icriodus latericrescens huddlei Klapper & Ziegler	..	GK	..
Icriodus latericrescens n. subsp. B Klapper	..	GK	..
Otarion (*Otarion*) *periergum* Haas	..	WH	..
Phacops (*Phacops*) *claviger* Haas	..	WH	..
Odontochile sp.	..	WH	..
Leonaspis tuberculata favonia Haas	WH	WH	..

*X, Identified by the writer; CWM, C. W. Merriam (1963, p. 43); WAO, W. A. Oliver, Jr., JMB, Jean M. Berdan; GK, Gilbert Klapper; WH, Winfried Haas.

from the upper 50 or 60 feet of the formation. In addition to the groups noted above, conodonts have been described from the type locality of the Rabbit Hill Limestone by Clark and Ethington (1966). It is worth noting, however, that the conodont fauna from bed 26 (Clark and Ethington, 1966, p. 669) is from a part of the section which so far has not yielded brachiopods (M. A. Murphy, oral commun., 1967) and which, therefore, cannot be unequivocably assigned to the *Spinoplasia* Zone. Merriam first studied the megafauna of the *Spinoplasia* Zone at Rabbit Hill and recognized its Helderbergian age (1954, p. 1284, 1963, p. 43). The writer studied the large brachiopod fauna of the *Spinoplasia* Zone and pointed out that it must correlate with the highest part of the Helderbergian as developed in New York State (Johnson, 1965, p. 374).

The *Spinoplasia* Zone is recognized in the Cortez Range, but its brachiopod fauna is rather poorly known there, and for that reason it has not been summarized in Table 5. Both *Leptocoelina squamosa* and *Spinoplasia roeni* are present in the Wenban Limestone near Cortez. In addition, the zone has yielded an important trilobite faunule which was recently described by Haas (1969). The trilobites described by Haas from a single collection of silicified specimens include *Decoroproteus* sp., *Otarion (Otarion) periergum* Haas, *Phacops (Phacops) claviger* Haas, *Phacops (Reedops)* sp., *Leonaspis tuberculata favonia* Haas, and *Koneprusia insolita* Haas.

The affinity of the provincial brachiopod fauna of the *Spinoplasia* Zone is with the Appalachian Lower Devonian brachiopod fauna (Boucot and others, 1967) in contrast to the older *Quadrithyris* Zone, Gedinnian, and Ludlow-Pridoli faunas whose affinities are with Old World faunas. Not only do a large number of typical Appalachian genera and even some species arrive during *Spinoplasia* Zone time, they do so at the expense of the previously established fauna of Old World affinity.

Clearly the specific assemblage of the Rabbit Hill fauna is different enough from the Appalachian assemblage that correlation strictly at that level should not be attempted. On the basis of generic assemblage, there is considerable similarity with the large fauna of the New Scotland Limestone of New York; however, caution is called for in attempting such a correlation. In 1909, J. M. Clarke listed the fauna of the Port Ewen beds summarizing collections and identifications made by himself, Chadwick, Grabau, and others. The assemblage listed includes many forms commonly regarded as typical of the New Scotland, mixed with others equally typical of the Oriskany Sandstone. Among the former are *Dalejina oblata, Platyorthis planoconvexa, Levenea subcarinata, Isorthis (Tyersella) perelegans, Strophonella punctulifera, Leptaenisca concava, Howellella cycloptera, Megakozlowskiella perlamellosa, Hedeina macropleura, Trematospira perforata, Eatonia medialis, Costellirostra singularis, Costellirostra peculiaris, Coelospira concava, Meristella laevis, Meristella bella, Meristella princeps,* and *Anastrophia verneuili?.* A few species typical of the underlying Becraft Limestone fauna were also listed and include *Rensselaeria subglobosa, Schizophoria multistriata,* and *"Spirifer" concinnus.* The typical Oriskany species listed are *Costispirifer arenosus, Cyrtina rostrata, Leptocoelia flabellites, Prionothyris ovalus,* and *Beachia suessana.*

Correlation of Helderbergian faunas away from New York with the rich and varied New Scotland Limestone fauna has been overemphasized because species derived from the New Scotland are better known. But the review by Boucot and Johnson (1967b) showed that many of the important brachiopod species, including those that are usually cited as evidence for New Scotland correlation, persist into higher parts of the Helderbergian, that is, into the Becraft and Port Ewen and even into the Port Jervis Limestone. Higher parts of the Helderbergian can be recognized by the occurrence of species generally regarded as Oriskany forms and by the presence of a few late-appearing Helderbergian species, such as *"Spirifer" concinnus* and its varieties, and by *Gypidula pseudogaleata.* The first estimate of the age of the *Spinoplasia* Zone given by the writer (1965, p. 374) was in the interval Becraft-Port Ewen because of the presence of the Oriskany elements *Pleiopleurina* and *Plicoplasia.* To these may be added *Leptocoelia,* which is abundant, and *Pegmarhynchia,* which is known from a single specimen on the east flank of Coal Canyon. The Oriskany representation is strong enough to require that some attention be given to the possibility of a correlation with some early part of the Deerparkian Stage of Cooper and others (1942), but the presence of *Orthostrophella* and an abundance of spirifers at the *Howellella cycloptera* stage of evolution, which precedes *Acrospirifer,* is good evidence that the *Spinoplasia* Zone is still in the Helderbergian. The presence of *Pseudoparazyga cooperi* might be taken as confirming evidence, but it also occurs in the *Trematospira* Zone assemblages which must be of Oriskany age. It is thus known to range higher in Nevada than in the east.

A correlation in the interval Becraft-Port Ewen can now be modified to a correlation with the Port Ewen interval only and probably that part of the Port Ewen which succeeds the Alsen Limestone because the Becraft guide *Gypidula pseudogaleata* is represented by a closely similar, if not identical, form in the underlying *Quadrithyris* Zone accompanied by the important conodont species *Spathognathodus transitans* (Klapper *in* Johnson and others, 1967). L. V. Rickard found the latter species in the Alsen of New York (written commun., 1967).

A direct assessment of age in terms of the European Lower Devonian stages is difficult, but both the *Quadrithyris* Zone below and the *Trematospira* Zone above can be assigned to the Siegenian and thus bracket the *Spinoplasia* Zone within that stage.

Mainly on the mutual occurrence of *Icriodus latericrescens* n. subsp. B, Klapper (*in* Klapper and Ormiston, 1969, Fig. 4) suggested a correlation of the *Spinoplasia* Zone at Coal Canyon with an interval at Royal Creek that lies above the *Spirigerina* Unit (= *Quadrithyris* Zone) and below the *Gypidula-Davidsoniatrypa* Unit (which has the lowest Royal Creek occurrence of *Monograptus yukonensis*). Consequently, it appears that the *Spinoplasia* Zone occupies a position between the zones of *Monograptus hercynicus nevadensis* and *Monograptus yukonensis* and is thus bracketed to a position within the Siegenian, a conclusion that is in agreement with evidence from the shelly faunas. Earlier, Clark and Ethington (1966, p. 668, 670) suggested a correlation of conodont faunas of the type Rabbit Hill with both the Siegenian ("Transgressions-horizont") fauna described by Bischoff and Sannemann (1958), and the Emsian (Schönauer and Zorgensis) fauna of Ziegler (1956), in an apparent confusion of the two. But "*Icriodus pesavis* Bischoff and Sannemann" (bed 23, Clark and Ethington, 1966, Pl. 83, Fig. 5) was misidentified (*see* Klapper and Ziegler, 1967, p. 70), and crucial affinity with the Schönauer Limestone would require the presence of *Polygnathus foveolatus*, not shown in any collections from the Rabbit Hill. *Polygnathus foveolatus* occurs in the upper part of the *E. pinyonensis* Zone (Klapper, 1968, p. 73).

Trematospira Zone

This zone was originally named to designate a single faunule thought to be closely associated with the *Acrospirifer kobehana* Zone (Merriam, 1940, p. 50), but was accorded zonal status after it was recognized in the Sulphur Spring Range (Nolan and others, 1956, p. 46). Carlisle and others (1957) recognized the zone at a number of localities in the Sulphur Spring Range, and Johnson (1962a) gave additional information on species present in the zone in the northern Roberts Mountains. He also suggested renaming it *Costispirifer* Zone because of the abundance of that genus, but the present work shows that the *Costispirifer* beds represent a discrete interval within the upper part of the *Trematospira* Zone at Willow Creek in the northern Roberts Mountains and at McColley Canyon in the Sulphur Spring Range.

Costispirifer Subzone is defined here to include the beds with *Costispirifer* sp. that compose the upper part of the *Trematospira* Zone. The lower part of the *Trematospira* Zone is here defined as the *Trematospira* Subzone. It includes beds with *Pleiopleurina anticlastica* and *Pseudoparazyga cooperi*, commonly accompanied by an abundance of *Leptocoelia murphyi*, below the lowest *Costispirifer* sp.

The *Trematospira* Zone megafauna is dominated by the brachiopods which are represented by 31 species (Table 6). A number of these, including *Dalejina* sp. B, *Levenea navicula*, *Leptaena* cf. *acuticuspidata*, "*Schuchertella*" sp. B, *Strophonella* cf. *punctulifera*, *Stropheodonta filicosta*, *Stropheodonta magnacosta*, *Megastrophia transitans*, *Anoplia* sp. A, *Pleiopleurina anticlastica*, *Leptocoelia murphyi*, *Megakozlowskiella magnapleura*, *Cyrtina* cf. *varia*, *Nucleospira* sp. B, *Trematospira perforata*, and *Pseudoparazyga cooperi*, are also

TABLE 6. FAUNA OF THE TREMATOSPIRA ZONE

	SULPHUR SPRING RANGE	ROBERTS MOUNTAINS
Dalejina sp. B	X	X
Discomyorthis musculosa	X	..
Levenea navicula n. sp.	X	X
Sieberella pyriforma n. sp.	X	..
Leptaena cf. *acuticuspidata*	X	X
"*Schuchertella*" sp. B	X	cf.
Leptostrophia cf. *beckii*	X	sp.
Strophonella cf. *punctulifera*	X	..
Megastrophia transitans n. sp.	X	X
Stropheodonta filicosta n. sp.	X	X
Stropheodonta magnacosta n. sp.	X	..
Pholidostrophia sp.	X	..
"*Strophochonetes*"? sp.	X	..
Anoplia sp. A	X	..
Ancillotoechia aptata	X	X
Pleiopleurina anticlastica n. sp	X	..
Coelospira pseudocamilla n. sp.	X	X
Leptocoelia murphyi n. sp.	X	cf.
Nucleospira sp. B	X	X
Meristella cf. *robertsensis*	X	X
Trematospira perforata	X	X
Pseudoparazyga cooperi	X	CWM
Acrospirifer aff. *murchisoni*	X	X
Dyticospirifer mccolleyensis	X	..
Costispirifer sp.	X	X
Megakozlowskiella magnapleura n. sp.	X	X
Reticulariopsis sp. B	X	..
Elytha sp.	X	X
Metaplasia cf. *paucicostata*	X	X
Cyrtina cf. *varia*	X	X
Rensselaeria sp.	X	..
Favosites sp.	X	CWM
Pleurodictyum sp.	X	..
streptolasmoid sp. A	..	CWM
horn corals, indet.	X	X
crinoid roots, indet.	X	X
Pterinea? sp.	..	X
Dalmanites cf. *meeki* Walcott	..	CWM
Phacops cf. *rana* Green	..	CWM
Spathognathodus sulcatus (Philip)	GK	GK
Icriodus latericrescens huddlei Klapper & Ziegler	GK	GK

*X, Identified by the writer; CWM, C. W. Merriam (1940); GK, Gilbert Klapper

present in the underlying *Spinoplasia* Zone, but of these *Pleiopleurina anticlastica* and *Pseudoparazyga cooperi* do not occur as high in the zone as the *Costispirifer* beds. A few species are restricted to the zone, for example, *Discomyorthis musculosa, Ancillotoechia aptata* (=*Stegerhynchus* sp. *of* Johnson, 1965, p. 377), *Coelospira pseudocamilla, Dyticospirifer mccolleyensis,* and *Costispirifer* sp. The latter is restricted to the uppermost beds of the zone.

As with the underlying fauna of the *Spinoplasia* Zone, the *Trematospira* Zone megafauna has its provincial affinity with the Appalachian fauna. *Discomyorthis musculosa* is a typical Oriskany fossil and does not occur in the Helderbergian. The zone also contains an abundance of *Acrospirifer* aff. *murchisoni* and of *Costispirifer* sp. and less commonly *Rensselaeria* sp. These are typical Oriskany fossils of the Appalachian Province. The presence of *Pleiopleurina, Trematospira perforata,* and *Pseudoparazyga cooperi* limits the zone to a correlation with beds not younger than the Oriskany Sandstone, or upper Deerpark of the New York terminology.

There is some question as to what part of the Deerparkian the *Trematospira* Zone belongs or *vice versa*. At many localities the upper Deerparkian Oriskany Sandstone lies unconformably upon various older beds (Cooper and others, 1942, chart). Similarly, the correlative Frisco Formation of Oklahoma was found to be unconformable above the Haragan and Bois d'Arc. This appears to be well documented by Amsden (1960, p. 133) from lithic and faunal studies. Until some of the Appalachian sections lacking a hiatus are more thoroughly studied, there remains a very real gap in the understanding of the Appalachian faunal sequence, and precise correlations cannot be confidently made between them and the continuous sequence of zones in central Nevada. In the writer's opinion the *Trematospira* Zone, with its two subzones, probably correlates fairly closely with the whole of the Deerparkian as defined by Cooper and others (1942). The lower part contains a few Helderbergian elements, namely *Pseudoparazyga cooperi* and an *Anastrophia* resembling *A. verneuili* (photos of the latter were shown to the writer by C. W. Merriam), just as does the Port Jervis Limestone which is regarded as lowest Deerparkian in the area of southeastern New York and New Jersey.

As with the *Spinoplasia* Zone, provinciality definitely limits the possibility of detailed intercontinental correlation with brachiopods to the acrospiriferids. The Oriskany species, *Acrospirifer murchisoni* and the comparable Nevada specimens resemble characteristic elements of the Siegenian fauna as developed in the Rhenish sequence. The latter is zoned for the most part on the basis of acrospiriferid genera, and, considering their evolution, the stock proves especially valuable for correlation beyond the Rhenish type area. The Siegenian *Acrospirifer* is a large, strongly plicate derivative of the long-ranging

Howellella, but does not appear as early as the Gedinnian. The Siegenian is characterized by large pauciplicate species with deep U-shaped interspaces. These forms are succeeded in the Emsian by transverse multi-ribbed and flatter-ribbed forms, with the former lineage culminating in the upper Emsian guide *Euryspirifer paradoxus* and with the latter culminating in the Eifelian *Acrospirifer speciosus*. The *Trematospira* Zone includes the highest beds in central Nevada that constitute part of the Siegenian and, in view of their stratigraphic position above two older Siegenian zones and below the *Acrospirifer kobehana* Zone which is regarded as Emsian, a late Siegenian age is probable for this zone.

EMSIAN STAGE

Acrospirifer kobehana Zone

This zone is the Lower Devonian prototype of megafossil assemblage zones in the Great Basin. It was named by Merriam (1940, p. 52) and recognized by him in the Roberts Mountains and at Lone Mountain. The fauna of the zone presently known is only slightly smaller than that of earlier zones in central Nevada and the differences may be due to lack of sufficient collecting. Several silicified collections were studied from the Sulphur Spring Range along with several from the Roberts Mountains. In addition, a few composed of calcareous specimens were examined. The articulate brachiopod fauna consists of 29 species (Table 7). A number of these range up from underlying zones and are the genera of Appalachian affinity; however, a few forms are new including a large, moderately plicate species of *Gypidula* similar in form to *Gypidula loweryi* of the overlying *pinyonensis* beds, *Mclearnites invasor*, *Parachonetes macrostriatus*, and a new elongate species of *Anoplia*. *Mclearnites*, *Pholidostrophia* (*Pholidostrophia*) sp., *Phragmostrophia* cf. *merriami*, and *Parachonetes macrostriatus* appear to be newly introduced elements with affinities to the Old World fauna. Certainly there are no representatives of related forms in either the *Spinoplasia* or *Trematospira* Zones.

Acrospirifer kobehana is a distinctive species although it has been confused with the *Trematospira* Zone guide *A.* aff. *murchisoni* in the past. It may have been derived from the latter, but no form like *Acrospirifer kobehana* is present in the Appalachian Province in the eastern United States. *Anoplia elongata*, characteristic of the *kobehana* Zone and the lower part of the *Eurekaspirifer pinyonensis* Zone, has been seen by the writer in the *Etymothyris* Zone of the Grande Greve Limestone of Gaspé (USNM loc. 17090). Probably the most important Appalachian Province form that occurs in the *kobehana* Zone is *Rensselaeria* sp., found with *Anoplia elongata* in what

TABLE 7. FAUNA OF THE ACROSPIRIFER KOBEHANA ZONE

	SULPHUR SPRING RANGE	ROBERTS MOUNTAINS	LONE MOUNTAIN
Dalejina sp. C	X	X	X
Levenea cf. *navicula* n. sp.	X	..	..
Salopina sp.	X	..	..
Gypidula praeloweryi n. sp.	X	X	..
Leptaena sp. C	X	X	X
"*Schuchertella*" sp. C	X	X	X
Stropheodonta filicosta n. sp.	X	..	..
Stropheodonta magnacosta n. sp.	..	X	..
Megastrophia transitans n. sp.	X	X	..
Mclearnites invasor n. sp.	X	..	..
Strophonella cf. *punctulifera*	X	X	X
Leptostrophia sp. D	X	..	..
Pholidostrophia (*Pholidostrophia*) sp.	X	..	..
Phragmostrophia cf. *merriami*	X	..	..
Chonetes? sp. (deep pedicle valve)	X	..	X
Parachonetes macrostriatus	..	X	..
Anoplia elongata n. sp.	X	X	X
Katunia? sp.	X	..	X
Atrypa sp. A	..	X	..
Atrypa nevadana	X	..	X
Coelospira sp.	X	X	X
Leptocoelia aff. *murphyi* n. sp.	X	X	..
Nucleospira cf. *subsphaerica* n. sp.	X	X	..
Meristella robertsensis	X	X	X
Acrospirifer kobehana	X	X	X
Megakozlowskiella cf. *raricosta*	X	X	..
Metaplasia cf. *paucicostata*	X	X	X
Reticulariopsis sp.	..	..	X
Cyrtina sp. C	X	X	X
Rensselaeria sp.	..	..	X
Loxonema? sp.	X	..	..
Orthoceras sp.	..	X	..
pelecypod indet.	X	..	..
Conocardium sp.	..	..	X
Pterinea? sp.	..	X	..
spiny platycerid, indet.	X	..	..
Platyceras sp.	..	X	..
crinoid roots, indet.	..	X	..
Favosties sp.	..	X	X
horn corals, indet.	X	X	X
Amplexus sp.	..	CWM	CWM
Cyathophyllum sp.	..	CWM	..
streptolasmoid sp. A	..	CWM	?CWM
Papiliophyllum elegantulum Stumm	..	CWM	CWM
Proetus cf. *nevadae* Hall & Clarke	..	CWM	..
Phacops cf. *rana* Green	..	CWM	CWM
Phacops sp.	X	..	X
Ozarkodina denckmanni Ziegler	..	..	C&E
Polygnathus sp.	GK	..	..
Icriodus latericrescens huddlei Klapper & Ziegler	GK	GK	..

X, Identified by the writer; CWM, C. W. Merriam (1940, p. 52, 53); C&E, Clark and Ethington (1966, p. 670); GK, Gilbert Klapper.

may be a lower horizon of the zone at Lone Mountain. In the Appalachian Province, *Rensselaeria* is not known to overlap the range of its successor *Etymothyris*. Nevertheless, it seems to the writer that the *kobehana* Zone is slightly younger than Oriskany, and by position approximately equivalent to the *Etymothyris* Zone in the Appalachian Province. Furthermore, the Emsian age of the *kobehana* Zone suggested by a comparison of *Acrospirifer* species and by the presence of the conodont genus *Polygnathus* is at odds with a Siegenian age that would seem to be required for the Oriskany. As noted below, the degree of faunal affinity between the areas of the Appalachian Province in eastern North America and of the *Acrospirifer kobehana* Zone fauna in central Nevada was considerably decreased compared to the affinities of the older faunas of the *Spinoplasia* and *Trematospira* Zones, and it may well be that the evolutionary processes favoring development of *Etymothyris* from the *Rensselaeria* stock in the east were either nonexistent or at least ineffectual in central Nevada.

Some divergence from a strictly Appalachian character makes direct correlation with the east less than obvious. Except for *Rensselaeria*, the Oriskany or Siegenian elements have dropped out of the fauna and forms that appear new in the *Etymothyris* Zone, which succeeds Oriskany age beds in the Appalachian Province, are not present. It seems probable that unrestricted faunal communication between central Nevada and the westernmost environs of the Appalachian Province (Oklahoma; Texas; Chihuahua, Mexico) was cut off and that a few Old World forms entered the region. Of these, *Mclearnites* is not significant for detailed correlation because it has a long range. *Parachonetes* first appears in the Siegenian (possibly late Siegenian) age Manak beds of Central Asia (Nikiforova, 1937, p. 32), and the genus is common throughout the Old World in beds of Emsian age (Johnson, 1966b). *Mclearnites, Parachonetes macrostriatus*, and *Anoplia elongata* extend upward into the *Eurekaspirifer pinyonensis* Zone, and tend to relate the two zones faunally. Beyond the broad suggestions of post-Oriskany age as indicated by position in the column and the Siegenian or younger age indicated by *Parachonetes*, the most significant element is the nominal species *Acrospirifer kobehana*. It is distinguished from the older *A. murchisoni* and the comparable form in Nevada, in having broader, lower ribs that tend to become asymmetrical, and by a profile in which the flanks adjacent to the dorsal fold are elevated into even, sweeping curves differing from the isolated medial elevation that is the dorsal fold of the older form. *Acrospirifer kobehana* is an advanced species of the genus comparable to Emsian forms, such as *A. ilsae* (Kayser, 1878, Pl. 22, figs. 3, 4) and specimens illustrated as *A.* cf. *fallax* (Drot, 1964, Pl. 2, figs. 7, 8). The indicated correlation is Emsian and probably lower Emsian.

Eurekaspirifer pinyonensis Zone

The *pinyonensis* Zone was named by Merriam (1940, p. 53) and was believed to be of Middle Devonian age, although arguments were not given to support the assignment. Shortly thereafter, Cooper (Cooper and others, 1942, chart) suggested a late Early Devonian assignment. Later work on the brachiopods (Johnson, 1962a), the goniatites (House, 1962, p. 252, 253), and the conodonts (Clark and Ethington, 1966; Walliser *in* Johnson and others, 1967; Klapper, 1968; Klapper and Ormiston, 1969) furnished additional confirmatory evidence of Early Devonian age. The evidence afforded by the brachiopods is discussed below.

The presently known brachiopod fauna of the *E. pinyonensis* Zone comprises 41 species (Table 8). This is a relatively large assemblage for the Devonian of the West and is at least partly due to the broad compass of the zone as presently envisioned. A subzone was proposed by Johnson (1962a, p. 166), but characterization of subzones is badly in need of revision.

In addition to the brachiopods, the shelly megafauna is relatively rich in corals, trilobites, and molluscs, particularly gastropods and pelecypods, and has yielded the only Lower Devonian goniatites in the western United States. Johnson (1962b, p. 544) interpreted the argillaceous limestones of the *pinyonensis* Zone to be shallow water, near-shore deposits of a recessive nature that preceded a time of non-deposition at the beginning of the Middle Devonian. Possibly nondeposition is local. A thorough geographic rearrangement of the sedimentary regime at the beginning of the Middle Devonian may have been the major factor responsible for the marked faunal change exemplified by the appearance of the fauna of the *Leptathyris circula* Zone (*see* Fig. 4).

The *E. pinyonensis* Zone is widely recognized in the Sulphur Spring Range (Carlisle and others, 1957, Fig. 2), in the Roberts Mountains (Merriam, 1940; Johnson, 1962a), in the Cortez Range (*see* Table 3), in the northern Simpson Park Range (Johnson, 1965), at Lone Mountain, and at localities in the Mahogany Hills south of Devils Gate (Merriam, 1940; Nolan and others, 1956, p. 46; Johnson and others, 1968), and in southern Nevada at the Mercury Test Site (Johnson and Hibbard, 1957, p. 354, 355), in the Spotted Range north of Las Vegas (Johnson, 1966c), and in the Desert and Pintwater Ranges.

Several subzones will undoubtedly be defined eventually within the *E. pinyonensis* Zone. The writer proposed a *Leptocoelia infrequens* Subzone (Johnson, 1965, p. 377) for beds within the *pinyonensis* Zone characterized by an abundance of *Leptocoelia infrequens* and *Dalejina* sp. C. This subzone was thought to occupy a basal position within the *pinyonensis* Zone because no underlying fossiliferous beds of the *pinyonensis* Zone were found in the northern Simpson Park Range. However, the *Leptocoelia infrequens* fauna occurs well up in the *pinyonensis* Zone at Lone Mountain and probably should not be regarded as a valid biostratigraphic unit until further studied and better understood.

The brachiopod fauna of the *Eurekaspirifer pinyonensis* Zone is sufficiently well known and diagnostic for a firm Emsian assignment. The genera *Dalejina, Muriferella, Mclearnites, Anoplia, Hysterolites, Brachyspirifer,* and *Mutationella* are all characteristic of the Lower Devonian and restricted to beds of Early Devonian age or older. *Leptocoelia* is similarly restricted to the Lower Devonian; the Onondaga occurrences of *"Leptocoelia" acutiplicata* (Oliver, 1956, p. 1469) do not belong to *Leptocoelia,* but to *Leptocoelina.* In addition, the large chonetid genus *Parachonetes,* recently described by Johnson (1966c), may be restricted to the Lower Devonian. The cited Russian occurrences (Vagran beds of Urals, Salairka beds of Kuznetzk Basin) have been called Middle Devonian, but instead belong to the Upper Emsian according to the views of Rzhonsnitskaya (1960a, p. 130; 1962). The Indochinese occurrences of Mansuy (1908, 1916) and Patte (1926) are very likely Upper Emsian equivalents, and the wide distribution of *Parachonetes* at what appears to be about the same horizon indicates that the relatively high occurrence of the genus in the *Receptaculites* Limestone of New South Wales (Johnson, 1966c, Pl. 63, figs. 9-14) is also of Emsian age as was suggested by Philip (Philip and Pedder, 1964, p. 1323, 1324; Philip and Jackson, 1967). *Bifida* is known to cross the Lower-Middle Devonian boundary; however, it is such a distinctive Old World element that it is worth drawing attention to its presence here. *Bifida* is present in beds on Devon Island which also yield *Polygnathus foveolatus* (Klapper and Ormiston, 1969), and these are the only two occurrences known to the writer in North America. *Howellella* is a long-ranging form. In the

TABLE 8. FAUNA OF THE EUREKASPIRIFER PINYONENSIS ZONE

	CM	SPR	SSR	RM	LM	CP
Dalejina sp. C	X	X	X	X	X	..
Rhipidomella sp.	..	..	X	..	..	..
Levenea fagerholmi n. sp.	..	X	..	..	..	..
Cortezorthis cortezensis	X	X	..	..	X	..
Muriferella masurskyi	X	..	..	..	..	..
Schizophoria nevadaensis	X	X	..	X	X	CDW?
Gypidula loweryi	X	X	X	X	X	CDW
Leptaena sp. C	X	X	X	X	X	..
"Schuchertella" nevadaensis	..	..	..	X	X	..
Leptostrophia? cf. sp. D	..	X	..	X	..	CDW?
Strophonella cf. *punctulifera*	X	..	..	..	X	..
Cymostrophia sp.	..	..	..	X	..	..
Mclearnites sp. B	X	..	..	..	..	..

TABLE 8. (CONTINUED)

	CM	SPR	SSR	RM	LM	CP
Megastrophia iddingsi	..	X	X	X	X	..
Stropheodonta sp.	..	..	..	..	X	CDW?
Phragmostrophia merriami	X	X	X	X	X	CDW
"*Strophochonetes*" *filistriata*	..	..	X	X	X	CDW
Chonetes sp.	..	..	..	X	X	..
Parachonetes macrostriatus	..	X	X	X	X	CDW
Anoplia elongata n. sp.	..	..	..	..	X	..
Spinulicosta? sp.	..	..	X	..	X	..
Trigonirhynchia occidens	X	X	X	X	X	CDW
Astutorhyncha cf. *proserpina*	..	..	..	X	..	..
"*Leiorhynchus*" sp.	X	..	..	..	X	..
Corvinopugnax sp.	..	..	X	..	..	..
Atrypa nevadana	X	X	X	X	X	..
Bifida sp.	X	..	X	..	X	..
Coelospira sp.	..	X	..	..	..	..
Leptocoelia infrequens	X	X	..	X	X	..
Nucleospira subsphearica n. sp.	X	..	X	X	X	..
Meristina sp. B	..	X	X	X	X	..
Howellella cf. *textilis*	..	X	X	X	X	..
Hysterolites sp. A	X	..	..	X	X	..
Hysterolites sp. B	..	..	X	X	..	..
Brachyspirifer pinyonoides n. sp.	..	X	..	X	X	..
Spinella talenti n. sp.	..	..	..	..	X	..
Eurekaspirifer pinyonensis	..	..	X	X	..	..
Elythyna sp. A	X	X	..	..	..	..
Ambocoelia sp.	..	..	..	..	X	..
Cyrtina sp. D	X	..	..	..	X	..
Mutationella? sp.	..	..	X	..	..	..
Lingula lonensis Walcott	..	..	..	..	CDW	..
Discina sp.	..	..	..	..	CDW	..
Pholidops bellula Walcott	..	..	..	..	CDW	...
P. quadrangularis Walcott	..	..	..	..	CDW	..
Hindeodella cf. *priscilla* Stauffer	..	..	..	..	OHW	..
Neoprioniodus bicurvatus Branson & Mehl	..	..	..	..	OHW	..
Ozarkodina typica denckmanni Ziegler	..	..	..	..	OHW	..
Polygnathus sp. ex aff. *P. linguiformis* Hinde	..	..	OHW	OHW	C&E	..
Spathognathodus cf. *primus* (Branson & Mehl) (= *Sp.* cf. *frankenwaldensis* Bischoff & Sannemann)	..	..,	..	OHW	..	..
Spathognathodus spp.	..	..	..	..	C&E	..
Icriodus spp.	..	..	..	OHW	..	..
Icriodus symmetricus Branson & Mehl	..	..	OHW	..	C&E	..
Icriodus latericrescens (sensu Ziegler, 1956, non Branson & Mehl)	..	..	OHW	..	C&E	..
	..	..	OHW	..	C&E	..
Icriodus latericrescens Branson & Mehl	..	..	..	..	C&E	..
Icriodus expansus Branson & Mehl	..	..	..	..	C&E	..
Panderodus unicostatus (Branson & Mehl)	..	..	..	..	C&E	..
Panderodus simplex (Branson & Mehl)	..	..	..	..	C&E	..
Ozarkodina denckmanni Ziegler	..	..	..	..	GK	..
Spathognathodus steinhornensis Ziegler	..	..	..	..	GK	..
Spathognathodus exiguus Philip	..	..	GK	..	GK	..
Icriodus latericrescens huddlei Klapper & Ziegler	..	..	GK	GK	GK	..
Polygnathus lenzi Klapper	..	..	GK	GK	..	..
Polygnathus foveolatus Philip & Jackson	..	..	..	GK	GK	..
Favosites spp.	..	X	..	..	CDW	..
Syringopora sp.	..	X	..	CWM	CWM	..
Alveolites sp.	..	..	..	..	CWM	..
Pleurodictyum sp.	X	..	..	..	..	..
Cystiphyllum sp.	..	..	..	?CWM	..	..

TABLE 8. (CONTINUED)

	CM	SPR	SSR	RM	LM	CP
Cyathophyllum sp.	..	..	..	..	CWM	CWM
Amplexus invaginatus Stumm	..	..	..	..	CWM	..
Radiastrea arachne Stumm	..	..	..	X	..	CWM
Billingsastraea nevadensis Stumm	..	..	..	..	CWM	CWM
Breviphrentis? sp.	..	WAO	..	..	..	..
Heterophrentis cf. nevadensis Stumm	..	WAO	..	..	..	..
Zygobeyrichia sp.	..	..	..	..	KK	..
Beyrichia (Primitia) occidentalis Walcott	..	..	..	..	CDW	..
Hollinella sp.	JMB	..	..	..	..	..
Strepulites sp.	JMB	..	..	..	..	..
Tubulibairdia sp.	JMB	..	..	..	..	..
Camdenidea? sp.	JMB	..	..	..	..	..
Otarion (Otarion) sp.	WH	..	..	..	..	..
Leonaspis sp.	WH	..	..	..	..	..
Proetus cf. nevadae Hall & Clarke	..	..	..	..	CWM	..
Phacops cf. rana Green	..	CF	..	..	CDW	CDW
Dalmanites meeki Walcott	..	CF	..	..	CWM	CDW
Tentaculites sp.	..	X	X	..	CWM	CWM
Styliolina sp.	..	..	..	..	X	..
Conularia sp.	..	..	..	..	..	CDW
Palaeomanon roemeri Walcott	..	..	..	..	CDW	..
Astylospongia sp.	..	..	..	..	CDW	..
Cyrtoceras nevadense Walcott	..	..	..	..	..	CDW
Anetoceras desideratus (Walcott)	..	..	..	..	..	CDW
Teicherticeras nevadense (Miller)	..	..	..	..	CWM	..
Platyceras conradi Walcott	..	..	..	..	..	CDW
Platyceras nodosum Conrad	..	..	..	..	CDW	..
Platystoma lineatum Conrad	..	..	..	..	..	CDW
Euomphalus eurekensis Walcott	..	..	..	..	..	CDW
Callonema occidentalis Walcott	..	..	..	..	..	CDW
"Loxonema" nobile Walcott	..	X	..	..	CWM	CDW
Loxonema? subattenuatum Hall?	..	..	..	..	..	CDW
Scoliostoma cf. americana Walcott	..	..	..	..	CWM	..
Bellerophon combsi Walcott	..	CF	..	..	CF	CDW
Bellerophon perplexa Walcott	..	..	..	..	CWM	CWM CDW
Conocardium nevadensis Walcott	..	..	X	..	CWM	CDW
Actinoptera boydi (Conrad)	..	..	..	..	CDW	..
Mytilarca dubia Walcott	..	..	..	..	CDW	..
M. (Plethomytilus) oviformis (Conrad)	..	..	..	..	CDW	..
Modiomorpha altiforme Walcott	..	..	..	..	CDW	..
M. obtusa Walcott	..	..	..	..	CDW	..
Goniophora perangulata (Hall)	..	..	..	..	CDW	..
Megambonia occidualis Walcott	..	..	..	..	CDW	..
Sanguinolites? gracilis Walcott	..	..	..	..	CDW	..
Sanguinolites? combensis Walcott	..	..	..	..	..	CDW
Paracyclas occidentalis Hall & Whitfield	..	..	..	..	CDW	..
Posidonomya laevis Walcott	..	..	..	..	CDW	..
P. devonica Walcott	..	..	..	..	CDW	..
Microdon (C.) macrostriatus Walcott	..	..	..	..	CDW	..
Anodontopsis amygdalaeformis Walcott	..	..	..	..	CDW	..
Schizodus (Cytherodon) orbicularis Walcott	..	..	..	..	CDW	..
Gennaeocrinus maxwelli Johnson & Lane	..	..	..	X	..	..

   Localities: CM, Cortez Mountains; SPR, N. Simpson Park Range; SSR, Sulphur Spring Range, RM, Roberts Mountains; LM, Lone Mountain; CP, Combs Peak in Mahogany Hills.
   Fossil identifications: X, by the writer; CDW, C. D. Walcott (1884); CWM, C. W. Merriam (1940, p. 54, 55); C&E, Clark and Ethington (1966, p. 670); WAO, W. A. Oliver, Jr.; JMB, Jean M. Berdan; KK, Karl Krömmelbein; OHW, O. H. Walliser; GK, Gilbert Klapper; WH, Winfried Haas.

past some attention was given to the genus while it was thought to be diagnostic of the Silurian, but Boucot (1957) showed that a number of species ranged up into the Lower Devonian. In addition, Havlíček (1959, p. 28, Pl. 18) showed that *Howellella* (*H. koneprusensis*) is present in the Emsian Koneprus Limestone of Bohemia. The writer also shows in this paper that the typically Eifelian species *"Spirifer" aculeatus* Schnur is assignable to *Howellella* as well, thus the genus is of no value in discriminating between Lower and Middle Devonian.

Several *pinyonensis* Zone brachiopod genera, *Rhipidomella*, *Spinulicosta?*, *"Leiorhynchus," Corvinopugnax*, and *Spinella*, are unknown in beds older than Emsian and serve to restrict the zone to that stage. The independent correlation of the underlying *Acrospirifer kobehana* Zone is confirmatory. Four of the five previously mentioned, late-appearing forms (*Rhipidomella, Spinulicosta?*, "*Leiorhynchus*," and *Corvinopugnax*) give an indication that at least a part of the *pinyonensis* Zone is of late Emsian age, as the writer knows of no occurrence of those genera as old as the early Emsian. In the Appalachian Province no *Rhipidomella* is known in beds as old as Emsian, but the Old World affinity of the *pinyonensis* Zone brachiopods makes a conclusive evaluation inadvisable. In their monograph on the productid brachiopods, Muir-Wood and Cooper (1960) indicated that the suborder originated in the Early Devonian, but the oldest productids reported by them are now known to be of Middle Devonian age. In the Old World, no productid (such as *Spinulicosta*) is known below the Upper Emsian.

Just as with the *pinyonensis* Zone goniatites (House, 1962), the conodonts support an Emsian age for the zone (Clark and Ethington, 1966; Walliser *in* Johnson and others, 1967).

The sequence collections of megafossils made by E. C. Gronberg at Lone Mountain (*see* the results of his section FF, from 620 to 780 feet above base of McColley Canyon Formation, listed earlier in this paper) also provided conodont faunas that have a known position

EUROPE		CENTRAL NEV.	
LOWER DEVONIAN	EIFELIAN	*Leptathyris circula* zone	THIS REPORT
		—?—	
	EMSIAN	*Elythyna* beds	
		Eurekaspirifer pinyonensis zone	
		Acrospirifer kobehana zone	
	SIEGENIAN	*Trematospira* zone	
		Spinoplasia zone	
		Quadrithyris zone	
	GEDINNIAN	Beds with *Monograptus praehercynicus* & *M. microdon*	
		Beds with *Gypidula pelagica* & *Icriodus woschmidti*	

Figure 4. Correlation of Great Basin Lower Devonian brachiopod zones with European stages.

within the *pinyonensis* Zone. Evidently the conodont fauna with *Polygnathus foveolatus, Spathognathodus exiguus, S. steinhornensis, Icriodus latericrescens huddlei,* and *Ozarkodina denckmanni* ranges at least to the top of the *pinyonensis* Zone as it is presently defined. According to Klapper (oral commun., 1968) a conodont fauna, intermediate between that of the *pinyonensis* Zone and one clearly of Eifelian (Onondaga) age, approximately corresponds in position to the still poorly known brachiopod fauna with *Elythyna* as delineated by Johnson and others (1967).

SUMMARY

Table 9 lists the Lower Devonian brachiopod species discussed in this study according to their occurrence in the zonal framework.

TABLE 9. STRATIGRAPHIC OCCURRENCE OF SIEGENIAN AND EMSIAN BRACHIOPOD SPECIES IN CENTRAL NEVADA

	Q	S	T	K	P
"Dolerorthis" sp.	X	..	..	..	..
Dalejina sp. A	X	..	..	..	..
Protocortezorthis windmillensis Johnson & Talent	X	..	..	..	..
Schizophoria parafragilis Johnson, n. sp.	X	..	..	..	..
Levenea sp. A	X	..	..	..	..
Anastrophia cf. *magnifica* Kozlowski	X	..	..	..	..
Gypidula cf. *pseudogaleata* (Hall)	X	..	..	..	..
Sieberella cf. *problematica* (Barrande)	X	..	..	..	..
Leptaena sp. A	X	..	..	..	..
Leptostrophia sp. A	X	..	..	..	..
Strophonella cf. *bohemica* (Barrande)	X	..	..	..	..
"Brachyprion" mirabilis Johnson n. sp.	X	..	..	..	..
Mesodouvillina cf. *varistriata* (Conrad)	X	..	..	..	..
Aesopomum cf. *varistriatus* Johnson, n. sp.	X	..	..	..	..
Machaeraria sp.	X	..	..	..	..
"Camarotoechia" cf. *modica* (Barrande)	X	..	..	..	..
Coelospira sp.	X	..	..	..	..
Atrypina simpsoni Johnson, n. sp.	X	..	..	..	..
Atrypa "reticularis"	X	..	..	..	..
Spinatrypa sp.	X	..	..	..	..
Dubaria sp.	X	..	..	..	..
Spirigerina supramarginalis (Khalfin)	X	..	..	..	..
Toquimaella kayi Johnson	X	..	..	..	..
Lissatrypa? sp.	X	..	..	..	..
Meristina sp. A	X	..	..	..	..
Nucleospira sp. A	X	..	..	..	..
Howellella sp.	X	..	..	..	..
Megakozlowskiella sp. A	X	..	..	..	..
Reticulariopsis? sp. A	X	..	..	..	..
Quadrithyris cf. *minuens* (Barrande)	X	..	..	..	..
Ambocoelia sp.	X	..	..	..	X
Cyrtina sp. A	X	..	..	..	..
Cyrtinaella causa Johnson, n. sp.	X	..	..	..	..
Leptaenisca sp.	X	X	..	..	..
"Camerella" sp.	?	X	..	..	..
"Schuchertella" sp. A	X	X	..	..	..
Orthostrophella monitorensis Johnson, n. sp.	..	X	..	..	..
Schizophoria sp.	..	X	..	..	..
Leptostrophia inequicostella Johnson, n. sp.	..	X	..	..	..

TABLE 9. CONTINUED

	Q	S	T	K	P
*Pegmarhynchia sp.	..	X	..	..	..
Atrypa sp.	..	X	..	..	..
Coelospira concava (Hall)	..	X	..	..	..
Leptocoelina squamosa Johnson, n. sp.	..	X	..	..	..
Howellella cycloptera (Hall)	..	X	..	..	..
Spinoplasia roeni Johnson, n. sp.	..	X	..	..	..
Plicoplasia cooperi Boucot	..	X	..	..	..
Rensselaerina? sp.	..	X	..	..	..
Dalejina sp. B	..	X	X	..	..
Leptaena cf. acuticuspidata Amsden	..	X	X	..	..
"Schuchertella" sp. B	..	X	X	..	..
"Strophochonetes" sp.	..	X	X	..	..
Anoplia sp. A	..	X	X	..	..
Leptocoelia murphyi Johnson, n. sp.	..	..	X	X	..
Pleiopleurina anticlastica Johnson, n. sp.	..	..	X	X	..
Acrospirifer aff. murchisoni (Castelneau)	..	..	X	X	..
Megakozlowskiella magnapleura Johnson, n. sp.	..	..	X	X	..
Cyrtina cf. varia Clarke	..	..	X	X	..
Meristella cf. robertsensis Merriam	..	..	X	X	..
Nucleospira sp. B	..	..	X	X	..
Trematospira perforata (Hall)	..	..	X	X	..
Pseudoparazyga cooperi (Merriam)	..	..	X	X	..
Levenea navicula Johnson, n. sp.	..	..	X	X	?	..
Strophonella cf. punctulifera (Conrad)	..	?	X	X	X
Stropheodonta filicosta Johnson, n. sp.	..	X	X	X	
Stropheodonta magnacosta Johnson, n. sp.	..	X	X	X	..
Megastrophia transitans Johnson, n. sp.	..	X	X	X	..
Pholidostrophia sp.	..	?	?	X	..
Metaplasia cf. paucicostata (Schuchert)	..	..	X	X	..
Discomyorthis musculosa (Hall)	..	..	X	..	..
Sieberella pyriforma Johnson, n. sp.	..	cf	X	..	..
Leptostrophia cf. beckii (Hall)	..	..	X	..	..
Ancillotoechia aptata Johnson, n. sp.	..	..	X	..	..
Coelospira pseudocamilla Johnson, n. sp.	..	..	X	..	..
Reticulariopsis sp. B	..	..	X	cf	..
*Elytha sp.	..	..	X	..	..
Dyticospirifer mccolleyensis Johnson	..	..	X	..	..
Costispirifer sp.	..	..	X	..	..
Rensselaeria sp.	..	..	X	X	..
*Salopina sp.	..	..	..	X	..
Gypidula praeloweryi Johnson, n. sp.	..	..	..	X	..
"Schuchertella" sp. C	..	..	..	X	..
McLearnites invasor Johnson, n. sp.	..	..	..	X	..
*Chonetes? sp.	..	..	..	X	..
*Katunia? sp.	..	..	..	X	..
Atrypa sp. A	..	..	..	X	..
*Leptocoelia aff. murphyi	..	..	..	X	..
Acrospirifer kobehana (Merriam)	..	..	..	X	..
Megakozlowskiella cf. raricosta (Conrad)	..	..	..	X	..
Cyrtina sp. C	..	..	..	X	..
Meristella robertsensis Merriam	..	..	..	X	..
Dalejina sp. C	..	..	..	X	..
Leptaena sp. C	..	..	..	X	X
Leptostrophia sp. D	..	..	..	X	X
Phragmostrophia merriami Harper, Johnson, & Boucot	..	..	..	?	X
Anoplia elongata Johnson, n. sp.	..	..	..	X	X
Parachonetes macrostriatus (Walcott)	..	..	..	X	X
Atrypa nevadana Merriam	..	..	..	X	X
*Coelospira sp.	..	..	..	X	X

TABLE 9. CONTINUED

	Q	S	T	K	P
Nucleospira subsphaerica Johnson, n. sp.	..	..	..	?	X
Rhipidomella sp.	..	..	..	..	X
Levenea fagerholmi Johnson, n. sp.	..	..	..	..	X
Cortezorthis cortezensis Johnson & Talent	..	..	..	..	X
Schizophoria nevadaensis Merriam	..	..	..	..	X
Muriferella masurskyi Johnson & Talent	..	..	..	..	X
Gypidula loweryi Merriam	..	..	..	..	X
"*Schuchertella*" *nevadaensis* Merriam	..	..	..	..	X
Cymostrophia sp.	..	..	..	..	X
McLearnites sp. B	..	..	..	..	X
Megastrophia iddingsi (Merriam)	..	..	..	..	X
*Stropheodonta? sp.	..	..	..	..	X
"*Strophochonetes*" *filistriata* (Walcott)	..	..	..	..	X
Chonetes sp.	..	..	..	..	X
Spinulicosta? sp.	..	..	..	..	X
Trigonirhynchia occidens (Walcott)	..	..	..	..	X
Astutorhyncha cf. *proserpina* (Barrande)	..	..	..	..	X
"*Leiorhynchus*" sp.	..	..	..	..	X
Corvinopugnax sp.	..	..	..	..	X
Bifida sp.	..	..	..	..	X
Leptocoelia infrequens (Walcott)	..	..	..	..	X
Howellella cf. *textilis* Talent	..	..	..	..	X
Hysterolites sp. A	..	..	..	..	X
Hysterolites sp. B	..	..	..	..	X
Brachyspirifer pinyonoides Johnson, n. sp.	..	..	..	..	X
Spinella talenti Johnson n. sp.	..	..	..	..	X
Eurekaspirifer pinyonensis (Meek)	..	..	..	..	X
Elythyna sp. A	..	..	..	..	X
Cyrtina sp. D	..	..	..	..	X
Meristina sp. B	..	..	..	..	X
Mutationella? sp.	..	..	..	..	X

Q = *Quadrithyris* Zone.
S = *Spinoplasia* Zone.
T = *Trematospira* Zone.
K = *Acrospirifer kobehana* Zone.
P = *Eurekaspirifer pinyonensis* Zone.

*An asterisk precedes rare or fragmentary forms not illustrated in the present paper.

The sequence of zones and their correlation with the European stages most consistent with the data now available is shown in Fig. 4. The brachiopod assembleage from beds assigned to the Ludlovian and to the Eifelian has been described elsewhere (Johnson and Reso, 1964; Johnson, 1966a; Johnson and others, 1967; Johnson, 1968).

PART III. SYSTEMATIC PALEONTOLOGY

PHYLUM BRACHIOPODA

CLASS ARTICULATA

Order Orthida

SUBORDER ORTHOIDEA

DISCUSSION:

The orthids are represented principally by *Dolerorthis, Ptychopleurella,* and *Skenidioides* in the Silurian of Nevada, but only *Skenidioides* is well represented in the Gedinnian. *Ptychopleurella* evidently did not extend above the basal beds assigned to the Gedinnian in Nevada (Johnson and others, 1967). *"Dolerorthis"* sp. reappears in the Siegenian (*Quadrithyris* Zone of Nevada and the lower part of the *Monograptus yukonensis* Zone in Yukon Territory). Dolerorthids of this type apparently have an important representation in western and Arctic North America and in southeastern Australia (but not in the Appalachian Province Lower Devonian). The evidence of their presence in the Lower Devonian is still largely unpublished. The occurrence of *Dolerorthis persculpta* Philip (1962, p. 194) from the Siegenian Boola Beds of Victoria, Australia, is a noteworthy exception. In addition, Havlíček (1956, p. 623) listed *Ptychopleurella* from the Upper Emsian Zlichov Limestone of Bohemia. Unfortunately, the latter material was not illustrated.

In the Appalachian Province, *Orthostrophia* and *Orthostropella* are well represented in the Helderbergian, and their restriction to beds no younger than that stage is time tested, if not time honored. The only certain occurrence of *Orthostrophella* in Nevada is represented by *O. monitorensis* in the *Spinoplasia* Zone (Rabbit Hill Limestone and upper part of the McMonnigal Limestone) where it serves as one of the better pieces of evidence that the *Spinoplasia* Zone is not as young as early Öriskany.

"Dolerorthis" and *Orthostrophella* are the only two orthid genera in the Siegenian to Emsian Stages of Nevada, and constitute a relatively small part of the extant megafauna.

Superfamily ORTHACEA Woodward, 1854
Family HESPERORTHIDAE Schuchert and Cooper, 1931
Subfamily DOLERORTHINAE Öpik, 1934
Genus *Dolerorthis* Schuchert and Cooper, 1931
TYPE SPECIES: *Orthis interplicata* Foerste, 1909, p. 76, Pl. 3, fig. 44
"Dolerorthis" sp.
(Pl. 1, figs. 1-11)

"*Ptychopleurella*" sp. Johnson, 1965, p. 371.

DISCUSSION: True *Dolerorthis* has a relatively convex, nonsulcate brachial valve.

EXTERIOR: Pedicle valves are pentagonal in outline, and brachial valves are subquadrate. The shells are relatively strongly biconvex in lateral profile with the pedicle valve slightly more convex posteriorly than the brachial valve. The umbo of the pedicle valve is prominent posterior to the hinge line. The ventral interarea is triangular, flat, and commonly gently apsacline or nearly cataclinal and is equal to approximately three-quarters of the maximum width. There is a relatively narrow triangular delthyrium, or the delthyrium may be slightly enlarged apically. The interarea of the brachial valve is long, orthocline, and relatively low triangular. The notothyrium is triangular and well defined, enclosing an angle of about 60 degrees. The beak of the brachial valve is fairly well defined, small, and slightly incurved. The cardinal angles are commonly slightly obtuse, but on some shells they are moderately auriculate and acute. Maximum width of the valves is attained anterior to midlength. On the pedicle valves of large specimens the anterior commissure may be incurved (emarginate). Brachial valves commonly bear a poorly defined sulcus.

The radial ornament consists of coarse angular costae which increase in number anteriorly both by bifurcation and by implantation. Bifurcation occurs most commonly on brachial valves and intercalation is most common on pedicle valves. Arrangement of the costae tends to be symmetrical on the two flanks of a valve unless the valve is irregular in shape, and is characterized by paired medial costae on the brachial valve. A single median costa is characteristic of the pedicle valve. Concentric ornament consists of fine lamellose growth lines which become more prominent anteriorly.

INTERIOR OF PEDICLE VALVE: The hinge teeth are located on the inner edges of the palintrope and do not project medially within the delthyrium and are unsupported by dental lamellae. The tracks of the hinge teeth dorsally overhang the adjustor muscle impressions. The muscle field is broad and flat apically and becomes slightly more impressed and faceted toward its anterior edge which lies about even with the hinge line. The diductor tracks are elongate and flat bottomed and are flanked by parallel, flat-bottomed adjustor scars inclined about 45 degrees to the interior shell surface. Medially, there is a low rectangular ridge separating the diductor scars and extending slightly anterior to the muscle impressions. The lateral parts of the shell are radially ridged and on one specimen there is a high, sharp ridge parallel to the hinge line along the base of the palintrope.

INTERIOR OF BRACHIAL VALVE: The brachiophores are small, elongate, and triangular in cross section. They diverge widely anterolaterally and project ventrally. Small, narrow, shallow sockets are defined between the lateral edges of the brachiophores and the inner edges of the hinge line medially. The notothyrial cavity is partially filled with shell material that is continuous anteriorly with the broad, low, rounded myophragm. The cardinal process consists of a simple wedge-shaped blade which narrows posteriorly. The adductor scars are divided into two suboval pairs. The anterior scars are slightly more divergent than the posterior ones and are bounded along their anteromedial edges by low, short, gently divergent ridges. The interior of the shell is crenulated by the impress of the costae.

OCCURRENCE: *"Dolerorthis"* sp. is present in the *Quadrithyris* Zone at Coal Canyon. Specimens closely resembling the form from the *Quadrithyris* Zone are present in the late Siegenian *Bichonostrophia* Unit of Lenz (1966). N. M. Savage has also sent closely similar specimens from Siegenian limestone at Manildra, New South Wales, Australia. *Dolerorthis* is unknown in higher beds in Nevada and appears to be unrepresented in the Gedinnian in the upper part of the Roberts Mountains Formation. Another species of *Dolerorthis* is present in beds of Ludlow age in Nevada, but it is flatter and more finely costate.

FIGURED SPECIMENS: USNM 156749-156755.

Family ORTHIDAE Woodward, 1854
Subfamily ORTHINAE Woodward, 1854
Genus *Orthostrophella* Amsden, 1968

TYPE SPECIES: *Orthostrophia dartae* Schuchert and Cooper, 1932, p. 71, Pl. 6, figs. 23, 31.

Orthostrophella monitorensis Johnson, n. sp.
(Pl. 1, figs. 12-22)

DIAGNOSIS: *Orthostrophella* with coarse costae posteriorly.

EXTERIOR: Small shells are semicircular to transversely quadrate or shield shaped. Large shells are subpentagonal and transverse. The valves are strongly biconvex with the brachial valve slightly more convex than the pedicle valve. Maximum width is at the hinge line in small shells, but may be anterior to midlength in larger shells. Cardinal angles may be approximately 90 degrees or are gently auriculate in small shells and increase to relatively strongly obtuse in large shells. The ventral interarea is broad, low, triangular, flat, and apsacline. The delthyrium is triangular and unmodified and encloses an angle of about 45 degrees. The dorsal interarea is strongly anacline. It is relatively long, low, and triangular. There is a medial fold on the posterior of pedicle valves, but in large shells the convexity reverses, and the anterior commissure bears a broad, low ventral sulcus.

The radial ornament consists of irregular costae of various sizes that increase in number anteriorly by bifurcation and by implantation. The costae are commonly strong and unbranched posteriorly, but splitting anteriorly tends to produce numerous costae, most of which are of lesser width than those on the umbo. Costae are crossed at irregular intervals by well-developed concentric growth lines. The pattern of splitting and intercalation of costae tends to be symmetrical. Pedicle valves commonly have a strong median costa that gives rise to two smaller costae which split off laterally at equal distances anterior to the umbo. Costae flanking the central costa also give rise to smaller costae laterally at a point slightly anterior to their place of origin. The lateral pair of primary costae also give off new costae on their medial sides somewhat anterior to midlength.

INTERIOR OF PEDICLE VALVE: The teeth are relatively long, although small, and triangular in cross section, with the long sides of the triangle lying parallel to the hinge line. The teeth are supported by short dental lamellae that recurve medially at the base of the valve and enclose the muscle field. The muscle field is relatively smooth and gently rounded. Its anterior edge slopes off abruptly along a transverse line approximately beneath the hinge margin. The interior of the valve is weakly to strongly crenulated by the impress of the costellae.

INTERIOR OF BRACHIAL VALVE: The sockets are set in thick shell material beneath the inner edges of the notothyrium. They are largely covered over anteriorly by the edges of the interarea. The notothyrial cavity is deep and triangular and bears a bladelike cardinal process medially which narrows posteriorly. Shell material is built up at the base of the cardinal process along the anterior edge of the notothyrial cavity forming a low platform which is continuous anteriorly with a broad, low, rounded myophragm that divides the adductor muscle scars. The adductor impressions are divided into two pairs. The posterior impressions are subcircular; the anterior impressions are decidedly elongate with their lateral edges set parallel to the midline. They tend to be moderately elongate slightly lateral to the midline of the valve and then recurve medially at their inner edges. The muscle field is defined by strongly developed muscle-bounding ridges. Anteriorly, the pallial markings consist of a single pair that diverge widely from the midline at the anterior end of the myophragm. They recurve medially and follow a nearly subparallel course to the margin of the valve.

OCCURRENCE: *Orthostrophella monitorensis* is present in the Rabbit Hill Formation in the Monitor Range and in the upper McMonnigal Limestone in the Toquima Range. There are also several specimens of *Orthostrophella* sp. in a collection from the Tor Limestone south of Stoneberger Basin in the Toquima Range, but they are too poorly preserved to compare closely with the specimens described above from the *Spinoplasia* Zone. *Orthostrophella monitorensis* occurs in the *Spinoplasia* Zone at Antelope Peak.

COMPARISON: No other named species with ventral sulcus anteriorly bear costae as coarse as does *O. monitorensis*.

DISCUSSION: Small specimens lack the thickened shell with the strongly impressed muscle scars characteristic of the genus. In the pedicle valve, the muscle scar only becomes strongly anteriorly convex in large specimens. Thus, the small specimens are homeomorphs of *Dolerorthis* and *Schizoramma*, however, none of the specimens, either large or small, bears the accessory ridges in the notothyrial cavity that distinguish *Schizoramma* (Schuchert and Cooper, 1932, Pl. 5, fig. 14).

Amsden (1968) proposed *Orthostrophella* for the species of *Orthostrophia* with a ventral sulcus anteriorly. The former is typically Silurian, but Devonian occurrences are known in the Beck Pond Limestone of Maine and the Pillar Bluff Limestone of Texas, as well as in Nevada.

FIGURED SPECIMENS: The holotype is USNM 156756, illustrated in figs. 20, 21 of Pl. 1. Additional specimens are USNM 156757-156762.

SUBORDER DALMANELLOIDEA

The writer utilizes the dalmanellid hierarchy listed below in organizing the section on Lower Devonian dalmanellids from the Great Basin. Subfamilies with an asterisk are represented in the Siegenian and Emsian of Nevada.

ORDER ORTHIDA

Superfamily DALMANELLACEA
  DALMANELLIDAE
    DALMANELLINAE
    *ISORTHINAE
    *CORTEZORTHINAE
    PROKOPIINAE
  RHIPIDOMELLIDAE
    *RHIPIDOMELLINAE
    PLATYORTHINAE
    HETERORTHINAE
    PROSCHIZOPHORIINAE
  PAURORTHIDAE
  HARKNESSELLIDAE
  DICAELOSIIDAE
Superfamily ENTELETACEA
  ENTELETIDAE
  SCHIZOPHORIIDAE
    *SCHIZOPHORIINAE
    *DRABOVIINAE
  ?MYSTROPHORIDAE
  LINOPORELLIDAE

DISCUSSION: Undoubtedly the ubiquitous rhipidomellid genus *Dalejina* is the most common dalmanellid in the Siegenian and Emsian of Nevada (but no rhipidomellid is known in the Middle Devonian of Nevada). *Dalejina* is common in every zone and in most collections in the Siegenian and Emsian of Nevada. Species from the *Quadrithyris*, *Spinoplasia*, and *Trematospira* Zones are generally small to medium sized and strongly biconvex. Those from the *pinyonensis* Zone are more lenticular. The Appalachian Province lineage, which first became well differentiated with the appearance of *Dalejina oblata* (Hall), makes only a rare appearance in Nevada, represented by the occurrence of a few specimens of *Discomyorthis musculosa*, the well-known Oriskany brachiopod, in the *Trematospira* Zone. It must have migrated westward from the Appalachian Province.

The Isorthinae is relatively well represented by *Levenea*, a genus that is also present in all of the zones of the Siegenian and Emsian of Nevada, but which is common only in the *Spinoplasia* Zone and *Trematospira* Zone, representing, respectively, the highest Helderbergian and the Oriskany. *Levenea* is a typical Appalachian Province genus and evidently also migrated westward because its earliest appearance (in the *Quadrithyris* Zone) post dates the widespread dispersal of *Levenea subcarinata* in the middle part of the Helderbergian of the Appalachian Province. The earliest Nevada representatives (those of the *Quadrithyris* Zone) are hard to evaluate in provincial terms, but *Levenea navicula* of the *Spinoplasia* and *Trematospira* Zones is much more strongly biconvex than any of the Appalachian species.

The third subfamily of the Dalmanellidae, represented in the Siegenian and Emsian in Nevada, is the Cortezorthinae (Johnson and Talent, 1967b). The representatives of this subfamily (*Protocortezorthis windmillensis* and *Cortezorthis cortezensis*) are not common members of the brachiopod fauna, but are of special interest because of their provincial relations. *Protocortezorthis* is widespread in the Upper Silurian and in the Gedinnian Stage of the Lower Devonian and probably in younger beds, but *P. windmillensis* is thought to be the initial divergent member that gave rise to *Cortezorthis*. The latter has a known geographic distribution from central Nevada to the Canadian Arctic to Novaya Zemlya of the U.S.S.R. (and thus has a distribution pattern similar to that of the pholidostrophiid genus *Phragmostrophia*). The evolutionary changes between *Protocortezorthis windmillensis* and *Cortezorthis* of the *cortezensis* and *maclareni* types are well known and probably will prove to be significant guides to correlations between Nevada and the Canadian West and Arctic.

The superfamily Enteletacea is represented principally by the genus *Schizophoria*. The latter is relatively common in the *Quadrithyris* Zone and in the Gedinnian as well. *Schizophoria* is rare in the *Spinoplasia* Zone and no *Schizophoria* has been found in the *Trematospira* or *Acrospirifer kobehana* Zones, but its absence is understandable because of its relative rarity in the upper Helderbergian and Oriskany of the Appalachian Province. The virtual disappearance of the genus at the beginning of the *Spinoplasia* Zone time and its reappearance during *pinyonensis* Zone time evidently can be attributed to migration patterns responding to provincial control. The enteletacean subfamily Draboviinae is represented by the small, septate genus *Muriferella*, which in Nevada is known only from the *pinyonensis* Zone. The genus is known in southeastern Australia and in the Canadian northwest and Arctic but not in the Appalachian Province. *Salopina* has been found in the *Acrospirifer kobehana* Zone, but it is relatively rare.

The dalmanellids as a whole constitute one of the more abundant and common elements of the Siegenian and Emsian brachiopod fauna of Nevada.

Superfamily DALMANELLACEA Schuchert, 1913
Family DALMANELLIDAE Schuchert, 1913
Subfamily ISORTHINAE Schuchert and Cooper, 1931
Genus *Levenea* Schuchert and Cooper, 1931

TYPE SPECIES: *Orthis subcarinata* Hall, 1857, p. 43, figs. 1, 2

DISCUSSION: This genus is most closely related to *Isorthis*, but not to the type species, *I. szajnochai* Kozlowski (1929). The writer considers that there are possibly five isorthoid genera: *Isorthis*, typified by *I. szajnochai* and represented by *I. arcuaria*, is antipodal in a morphologic sense to *Protocortezorthis*. *Protocortezorthis* and its derivatives were removed from the Isorthinae by Johnson and Talent (1967b). Between these two end members are *Tyersella* Philip, including *T. typica* Philip, "*Isorthis*" *perelegans* (Hall), and probably a few others such as *Isorthis amplificata* Walmsley (1965) and "*Schizophoria*" *allani* Shirley (1938). *Isorthis tetragona* (Schnur) and *I. canalicula* (Schnur) have affinities here. The second morphologic group intermediate between *Isorthis* s.s. and *Protocortezorthis* is a *Levenea*-like group including *Isorthis clivosa* Walmsley (1965), *Isorthis mackenziei* Boucot, Johnson, Harper, and Walmsley (1966), *Isorthis uskensis* Walmsley (1065), but not *I. scutiformis* Walmsley, which appears to belong to *Isorthis* s.s.), "*Dalmanella*" *inostranzewi* Peetz (1901), and *Isorthis festiva* Philip, plus "*Levenea*" *taeniolata* and "*L.*" *altaica* Khalfin (1948).

The principal distinction of *Levenea* is its short, pentagonal ventral muscle impression, which is commonly elevated on a platform anteriorly (see Pl. 3, fig. 14). Wright (1965, p. H334) defined *Levenea* as having a trilobate cardinal process. However, most specimens have a bilobate shaft and myophore. Trilobation is the exception.

Levenea sp. A
(Pl. 2, figs. 1-7)

EXTERIOR: Small shells are transversely pentagonal to shield-shaped. Larger shells have width approximately equal to the length. The pedicle valve is subcarinate, and the brachial valve bears a low rounded sulcus. Cardinal angles are acutely rounded on small shells and obtuse on large ones. The ventral interarea is slightly shorter than the maximum width which is attained somewhat posterior to midlength. It is low, triangular, apsacline, and only slightly incurved. The delthyrium encloses an angle of approximately 60 degrees; on some small specimens it is widened apically into a subcircular foramen. The dorsal interarea is relatively well-developed, broad, and anacline.

Ornament consists of strong, rounded, radial costellae which conform to a mild fascicostellate arrangement. Costellae and interspaces are of varying width compared with adjoining costellae or interspaces. There are 12 to 14 costellae in a space of 5 mm, 10 mm anterior to the beak. Indistinct concentric growth lines cross the costellae at widely spaced, irregular intervals.

INTERIOR OF PEDICLE VALVE: Dental lamellae are not present, even in relatively small shells; however, the medial sides of the umbonal cavities are moderately strongly thickened into subpyramidal mounds which adjoin the hinge teeth. The shell material is strongly built up medially forming a transverse platform onto which the muscle field is impressed and which limits it anteriorly. The muscle field is short and broad, and in small specimens it is decidedly wider than long. A barely perceptible myophragm divides widely divergent diductor tracks at the anterior edge of the muscle impressions. The anterior portions of the valves are faintly crenulated by the impress of the costellae.

INTERIOR OF BRACHIAL VALVE: The sockets are short, deep, and widely divergent. They are impressed into shell material built up in the postero-lateral region. Brachiophores adjoin the sockets medially and project almost directly ventrally and only slightly anteriorly. They are trapezoidal in cross section and are attached to the floor of the valve basally. There is an elongate moundlike cardinal process in the notothyrial cavity largely posterior to the brachiophores. A broad low myophragm divides the adductor impressions medially. They are bounded laterally by strong muscle-bounding ridges which are approximately parallel to the midline along their lateral edges and converge smoothly toward the midline anteriorly. The anterior and posterior adductors are separated by a low transverse step dividing them into pairs of approximately equal area. The interior of the shell is crenulated peripherally by the impress of the costellae.

OCCURRENCE: *Levenea* sp. A is present in the *Quadrithyris* Zone at Coal Canyon and in the lower McMonnigal Limestone in the Toquima Range.

COMPARISON: *Levenea* sp. A differs from *L. navicula* principally in its smaller size. Adequate material is still to be collected which should allow a more thorough comparison. *Levenea fagerholmi*, of the *pinyonensis* Zone, is relatively more lenticular.

FIGURED SPECIMENS: USNM 156763-156767.

Levenea navicula Johnson, n. sp.
(Pl. 2, figs. 19-22; Pl. 3, figs. 1-19)

DIAGNOSIS: *Levenea* with strongly biconvex form and naviculate pedicle valve.

EXTERIOR: The shells are transversely suboval to slightly pentagonal in small specimens and elongate suboval in large ones. The valves are unequally biconvex with the pedicle valve considerably the deeper of the two. The cardinal angles are rounded and obtuse; maximum width is posterior to midlength in small shells and anterior to midlength in large shells. The ventral interarea is short, low, and triangular and equals about half the maximum width of the valves. It is apsacline and moderately to strongly incurved. In general, larger shells are more strongly apsacline. The dorsal interarea is poorly defined, nearly linear, and is anacline. The delthyrium of the pedicle valve varies from about 55 to 75 degrees, broader angles being enclosed in large valves. Pedicle valves are strongly incurved at the beak and are moderately to very markedly carinate, particularly in the posterior portion. In an extreme case of a strongly carinate pedicle valve, the flanks are flat. The brachial valve bears a median sulcus.

Ornament consists of fine subangular radial costellae which increase in number anteriorly by bifurcation and implantation. There are 10 to 14 costellae in a space of 5 mm, 10 mm anterior to the beak.

INTERIOR OF PEDICLE VALVE: The hinge teeth are relatively strong and triangular in horizontal section. They are supported by very short, stubby dental lamellae which rest on the inner edges of thickenings of shell material that fill the umbonal cavities. There is commonly a small triangular pedicle callist in the apex of the valve. Shell material in the umbo of the valve is generally built up thick medially as well as laterally so that there is a precipitous slope from the anterior edge of the muscle field. The bases of the diductor tracks tend to be subparallel to the plane of the commissure and rest on shell material filling the strong carinate arch of the valve. However, in a few shells there is even an excavation beneath the anterior edge of the muscle impression leaving a precipitous transverse ledge along its anterior edge. In small shells the muscle field is short and triangular, but in large specimens it is pentagonal and faceted with the diductors divided, in some specimens, by a stout rectangular myophragm, but in others the myophragm is altogether lacking. The central part of the scar may be even deeper medially or may be flat from one diductor scar across to another. The adjustor scars are inclined at strong angles and are triangular and flat, or they may be grooved laterally. Weak to very strong muscle-bounding ridges are developed anterolaterally. The interiors of the valves are smooth and are only slightly crenulated anteriorly.

INTERIOR OF BRACHIAL VALVE: The sockets are deep and widely divergent and are bounded posteriorly by the inner edges of the valves. They may be very slightly covered over by the interarea. The sockets are supported basally by the fulcral plates which are underlain by shell material in a large number of specimens, making them essentially sessile; however, a few have been examined in which fulcral plates are completely free from the base of the valve. The brachiophores diverge anteriorly and toward the base of the valve. They project relatively strongly ventrally on large specimens which also develop thick pads of shell material on their medial sides, forming a constriction approximately at the anterior edge of the notothyrial cavity. The cardinal process is generally an elongate, posteroventrally directed stalk which is essentially triangular on its posterior face, developing two or, less commonly, three lobes. The surface of muscle attachment of large specimens shows well-developed growth lamellae. A high, thick, rounded myophragm is present on large specimens supporting the narrow anterior end of the cardinal process and extending anteriorly and fading toward the anterior edge of the adductor muscle impressions. The adductor impressions are elongate suboval, bounded laterally in large shells by muscle-bounding ridges. The anterior and posterior pairs are separated from one another by deeper impression of the anterior pair. The lines of junction are approximately at right angles to the midline. The interior of the shell is only poorly crenulate anteriorly, as in the pedicle valve.

OCCURRENCE: *Levenea navicula* is very abundant in the Rabbit Hill Limestone in the Monitor Range. It is also present in the upper McMonnigal Limestone in the Toquima Range and in the Rabbit Hill Limestone in the northern Simpson Park Mountains. It is present in the *Spinoplasia* Zone at Antelope Peak. The species is common in the *Trematospira* Zone in the Roberts Mountains and in the Sulphur Spring Range. *Levenea* specimens are generally absent from the *Acrospirifer kobehana* Zone, but a single specimen of *Levenea navicula* has been collected from that zone in the Sulphur Spring Range.

COMPARISON: *Levenea navicula* is a larger form than *L.* sp. A, and it is very much more narrow and biconvex than *Levenea fagerholmi*. *Levenea navicula* differs from all previously named species in the very considerable depth and naviculate shape of its pedicle valve.

DISCUSSION: Bancroft's report (1945, p. 187) of the presence of fulcral plates in immature individuals of certain Dalmanellidae is complemented by identical findings during study of *Isorthis* (Walmsley, 1965) and of *Levenea navicula*. *Levenea* was formerly classified with the Dalmanellinae (Schuchert and Cooper, 1931, p. 246; 1932, p. 123; Williams and Wright, 1963, p. 28), but the elimination of *Levenea* from the Dalmanellinae was suggested by Boucot and others (1965). Because of the discovery of fulcral plates in some small specimens of *Levenea* from the Rabbit Hill Limestone, it appears to the writer that there is no longer any reason to disassociate the genus from the Isorthinae.

FIGURED SPECIMENS: The holotype is USNM 156772. Other specimens are USNM 156773-156780.

Levenea fagerholmi Johnson, n. sp.
(Pl. 2, figs. 8-18)

DIAGNOSIS: Broad, noncarinate *Levenea* with pedicle valve strongly arched in its posterior portion; ventral platform low.

EXTERIOR: The shells are transversely subquadrate to suboval in outline and unequally biconvex in lateral profile. The pedicle valve is about twice as deep as the brachial valve and is strongly arched posteriorly. The ventral interarea is short, low, triangular, and apsacline. It is equal to about half the maximum width. The dorsal interarea is anacline. The delthyrium is rounded at its apex and encloses an angle of about 90 degrees. The pedicle valve is broadly curved from side to side without carination. The brachial valve has a broad, low, poorly defined sulcus. Maximum width of the shells is commonly slightly anterior to midlength.

The ornament consists of fine subangular radial costellae that are crossed at irregular intervals by concentric growth lines. On large specimens the growth lines become numerous and closely spaced at the anterior of the shell. The costellae increase in number anteriorly by bifurcation and implantation. There are commonly 12 to 14 costellae in a space of 5 mm, 10 mm anterior to the beak. Costellae are straight on the midregions of the valve, but posterolaterally they become gently curved and concave posteriorly.

INTERIOR OF PEDICLE VALVE: The hinge teeth are broadly spaced, triangular, and divergent. Dental lamellae are absent. The teeth may bear fossettes medially on large specimens. The muscle field is elongate-pentagonal and faceted, but relatively short. It is deeply impressed laterally, but only moderately impressed anteriorly on a low platform. The adductor track is a broad, low, rectangular median ridge dividing the diductor scars. Broadly spaced, diverging vascula media emanate from the anterior edges of the diductor tracks. The interior of the valve is crenulate peripherally by the impress of the costellae.

INTERIOR OF BRACHIAL VALVE: The brachiophores are blunt, divergent, and rounded, arising from the floor of the valve. Lateral to them, the dental sockets are widely divergent and lack determinable fulcral plates. Medially the cardinal process is an elongate lobe, blunt posteriorly and narrowing anteriorly. It appears to be medially cleft on its ventral surface, but its posterior end was not examined. The adductor muscle impressions are divided into anterior and posterior pairs. The posterior pair is slightly smaller and more deeply impressed laterally. Both pairs are bounded laterally and anterolaterally by muscle-bounding ridges. Medially there is a well-developed, rounded myophragm which originates at the anterior edge of the cardinal process. The interior is crenulate peripherally by the impress of the costellae.

SHELL STRUCTURE: The shell material is finely punctate (endopunctate).

OCCURRENCE: This species of *Levenea* is known only from the northern Simpson Park Mountains where it is present in the *Leptocoelia infrequens* Subzone and in the Zone of *Eurekaspirifer pinyonensis*. The largest shells have been recovered from the *E. pinyonensis* Zone.

COMPARISON: *Levenea fagerholmi* is relatively much more lenticular and less carinate than either of the other two species in Nevada. The broadly arched, posteriorly deep pedicle valve is unlike the flat carinate pedicle valve of *L. subcarinata* (Hall). *Levenea fagerholmi* is also distinct from *L. lenticularis* (Hall) in being larger and relatively wider at the hinge line, and from *L. depressa* Wang (1956), from the Yükiang Limestone, by the absence of a markedly quadrate outline. *L. depressa* also has a more strongly convex brachial valve.

FIGURED SPECIMENS: The holotype is USNM 156768. Other specimens are USNM 156769-156771.

Subfamily CORTEZORTHINAE Johnson and Talent, 1967
Genus *Protocortezorthis* Johnson and Talent, 1967
TYPE SPECIES: *Orthis fornicatimcurvata* Fuchs, 1919, p. 58, Pl. 5, figs. 1-6.
Protocortezorthis windmillensis Johnson and Talent, 1967
(Pl. 4, figs. 3-19)

Protocortezensis windmillensis Johnson and Talent, 1967b, 158, Pl. 21, figs. 1-13.

DISCUSSION: This species was described by Johnson and Talent (1967b). Its distinguishing features are the absence of brachiophore supporting plates and the presence, in some specimens, of an incipient median septum at the anterior of the brachial valve. These features mark *P. windmillensis* as the direct lineal ancestor of the earliest species of *Cortezorthis*, which appears at about the same level as *Monograptus yukonensis* in the Canadian Arctic (Lenz, 1966, 1967b). The *P. windmillensis*—early *Cortezorthis* transition appears to correspond closely with the boundary between the *Monograptus hercynicus* and *M. yukonensis* zones and serves as a significant device for recognition of upper and lower divisions of the Siegenian.

OCCURRENCE: *Protocortezorthis windmillensis* is known only from the Windmill Limestone (*Quadrithyris* Zone) at Coal Canyon.

FIGURED SPECIMENS: USNM 147338-147345, 156785. The holotype is USNM 147339 (here designated). Two specimens of *Protocortezorthis* sp. from the "Tor Limestone" are illustrated on Plate 4 as figures 1, 2 for comparison (USNM 156783, 156784).

Genus *Cortezorthis* Johnson and Talent, 1967[1]

TYPE SPECIES: *Cortezorthis maclareni* Johnson and Talent, 1967b, p. 146, Pl. 19, figs. 1-20; Pl. 20, figs. 28, 29.

Cortezorthis cortezensis Johnson and Talent, 1967
(Pl. 5, figs. 1-12)

Cortezorthis cortezensis Johnson and Talent, 1967b, p. 151, Pl. 20, figs. 14-20.

DISCUSSION: This species was described by Johnson and Talent (1967b). Its principal distinguishing features are the absence of a ventral carina and the presence of peripheral radial septa plus an advanced rhomboidal dorsal adductor muscle field such as is developed in *C. maclareni* (see Johnson and Talent, 1967b, Text-Fig. 3). Since preparation of the definitive paper, three specimens have been collected from the *E. pinyonensis* Zone at Lone Mountain by Mr. Eric Gronberg. One of these shows the external ornament and shell configuration to better advantage than specimens previously available and is described below.

DESCRIPTION: The specimen is 20 mm long, 22.5 mm wide, and 8.6 mm thick. The outline is irregularly suboval to subsemicircular. In lateral profile the valves are unequally biconvex, with the pedicle valve slightly deeper, but a little less than twice as deep as the brachial valve. The ventral beak is short and incurved. The hinge line is curved, and the ventral interarea is flat and apsacline, but slightly incurved apically. It is equal in width to about three-fifths the maximum width and is cleft medially by an open triangular delthyrium, encompassing an angle of approximately 90 degrees. The dorsal interarea is low, triangular, and flat. Its inclination is anacline, approaching the orthocline position. The cardinal angles are obtuse and evenly rounded. Maximum width is attained anterior to the hinge line at about one-quarter of the distance from the ventral beak to the anterior margin. In front of the place of maximum width, the anterior margin

---

[1] *Protophragmapora* Alekseeva (1967, p. 7) is a junior subjective synonym.

curves around evenly in a semicircular fashion. The anterior commissure is rectimarginate. In transverse section the pedicle valve is faintly subcarinate with the flanks less convex outwards than at the middle of the valve. The brachial valve bears a shallow sulcus, developed posteriorly, which flattens out toward the anterior margin. Neither a median carina nor a narrow dorsal furrow is developed.

The ornament consists of rounded to subangular radial costellae, coarser on the midregions than on the posterolateral flanks. In the median forty-five degree sector, the costellae are relatively straight and are disposed in a radial fashion, but costellae toward the cardinal angles are consistently more and more curved, convex toward the midline, so that the costellae closest to the hinge line are concave posteriorly and intersect the posterior margin. The costellae increase in number anteriorly by bifurcation and by implantation. There are a few irregularly disposed, concentric growth lines, developed more commonly in the anterior half of the shell.

OCCURRENCE: This species is known from the *E. pinyonensis* Zone in the Cortez Mountains, in the northern Simpson Park Range and at Lone Mountain. The Lone Mountain occurrences are located in the interval 435-630 feet above the base of the Nevada Group.

FIGURED SPECIMENS: USNM 159066, 159069, 156786; G.S.C. 19614, 19615 (the latter are of *C.* aff. *bathurstensis*). The USNM numbers published for *C. cortezensis* by Johnson and Talent (1967b, Pl. 20) were incorrect because they had been used previously. The incorrect numbers 141450-141453 have been replaced by 159066-159069, respectively.

Family RHIPIDOMELLIDAE Schuchert, 1913
Subfamily RHIPIDOMELLINAE Schuchert, 1913
Genus *Dalejina* Havlíček 1953

TYPE SPECIES: *Dalejina hanusi* Havlíček, 1953, p. 5, Pl. 1, figs. 10, 12-14; Pl. 2, fig. 4.

DISCUSSION: The species of *Dalejina* described here, from central Nevada appears to be closest to the group of species that center around the type, *D. hanusi*, from the Lower Devonian of Bohemia. *Dalejina* sp. A is relatively close to *D. hanusi* in its small size and relatively deeply inflated, biconvex lateral profile. *Dalejina* sp. B, from higher beds, appears to be a more robust derivative. It is deeply biconvex and maintains a relatively small, narrow, ventral muscle scar. *Dalejina* sp. C may be a part of the same lineage, or it may represent a late entry of *Dalejina* stock from outside the region of central Nevada. It is decidedly more lenticular. However, all three forms appear to be distinct from the *Dalejina* lineage that developed in the Appalachian Province and which is exemplified in the Helderbergian of the Appalachian Province by *Dalejina oblata*. The Appalachian lineage is characterized by an increase in size and degree of flabellation of the ventral muscle scar as pointed out by Amsden and Ventress (1963, Text-Fig. 21, p. 64). The end member of this series, as envisioned by Amsden, is *"Dalejina" musculosa* Hall, which is now proposed as type species of a new genus *Discomyorthis*. The appearance of this species in association with the *Dalejina* A, B, C lineage, of Old World aspect, in central Nevada points up the dichotomy between the two *Dalejina* lineages, which is not so obvious in the Appalachian Province where one is forced to deal with forms that include all transitions between the various members of the lineage restricted to that province.

The inclusion of *Dalejina* in the Dalmanellidae by Williams and Wright (1963) and by Wright in the brachiopod Treatise (1965, p. H334) cannot be too firmly rejected (*see* Boucot and others, 1965). *Dalejina* and *Aulacella*, both of which were placed in the Dalmanellidae in the Treatise, have straightforward rhipidomellid form and internal structures and cannot be seriously considered as members of any family but the Rhipidomellidae.

Dalejina sp. A
(Pl. 8, figs. 13-20)

EXTERIOR: The shells are transversely suboval in outline and subequally biconvex in lateral profile. Cardinal angles are strongly rounded and obtuse; maximum width is attained at about midlength or slightly anterior to midlength. There is a low rounded sulcus on the brachial valve. The ventral interarea is narrow, triangular, apsacline, and slightly incurved. It is equal to approximately one-fourth to one-third the maximum width of the shell. The ventral beak is relatively small and projects only slightly beyond the posterior of the brachial valve. The delthyrium is open and encloses an angle of about 60 degrees. The dorsal interarea is narrow, low, triangular, and orthocline. The dorsal beak is small, but prominent.

Radial ornament consists of fine, hollow, subangular costellae which increase in number anteriorly by bifurcation and by intercalation. There are commonly 9 or 10 costellae in a space of 3 mm, 5 mm anterior to the beak. The costellae are crossed at irregular intervals by relatively well-developed concentric growth lines.

INTERIOR OF PEDICLE VALVE: The posteroventral edges of the hinge teeth lie within the delthyrium. The hinge teeth are narrow, elongate, and almost bladelike in horizontal section. They diverge anterolaterally at angles parallel to the inner edges of the delthyrium. A small pedicle callist is present apically. The hinge teeth are supported by very short, thin dental lamellae which enclose the posterior half of the muscle field. The adductor scars are indistinct. The diductor impressions are broad, unfaceted, and very lightly impressed. There may be an indistinct low myophragm near the anterior edge of the diductor impressions. The muscle scars blend imperceptibly anteriorly with the shell interior. The periphery of the shell is crenulated by the impress of the costellae. The ridges corresponding to the external interspaces are flat and are grooved medially.

INTERIOR OF BRACHIAL VALVE: The sockets are relatively closely set medially and diverge broadly anterolaterally. They are set into slight thickenings of shell material along the posterior edge of the shell and bounded by small ridges anterolaterally. The brachiophores are attached directly to the floor of the valve. They are elliptical to nearly circular in cross section near their bases and become relatively more elongate elliptical distally. The brachiophores project relatively strongly ventrally as they diverge slightly anterolaterally. Brachiophore processes are attached to the inner dorsal edges of the brachiophores and consist of small platelike projections only a third or a fourth as thick as the brachiophores themselves. Their planes tend to diverge strongly dorsally. Medially, between the bases of the brachiophores, there is a small erect cardinal process which is bilobate ventrally. There commonly is a poorly defined, low, rounded myophragm dividing the adductor muscle scars. The adductor scars are relatively strongly impressed and are divided into two pairs, the posterior scars are suboval and relatively short in their posterior to anterior dimension. The anterior adductors diverge slightly more broadly laterally and are relatively broader than the posterior adductors. The interior of the valves are crenulate peripherally in the same manner as in the pedicle valves.

OCCURRENCE: The species is common in the *Quadrithyris* Zone at Coal Canyon.

COMPARISON: *Dalejina* sp. A is smaller than *Dalejina* sp. B and when compared with small specimens of the latter species it proves also to have a thinner shell. *Dalejina* sp. A is smaller and relatively much more strongly convex than *Dalejina* sp. C.

FIGURED SPECIMENS: USNM 156806-156809.

Dalejina sp. B
(Pl. 7, figs. 1-22)

EXTERIOR: The shells are subcircular to trapezoidal in outline and relatively strongly biconvex in lateral profile. In large shells the brachial valve may be more convex. Most small shells are transverse, but large ones tend to increase in relative length and are commonly of approximately equal width and length, or they may be slightly elongate. The hinge line is short and is equal to slightly less than half the maximum width which in most shells is attained anterior to midlength. The ventral interarea is very small, triangular, apsacline, and gently incurved. The delthyrium is open, triangular, and encloses an angle of about 60 degrees. The dorsal interarea is orthocline to apsacline, flat, and relatively high. Neither fold nor sulcus are developed on small shells, but on large ones the pedicle valve becomes flattened anteriorly, and the commissure is broadly and gently arched toward the brachial valve.

The ornament consists of fine subangular radial costellae that increase in number anteriorly by bifurcation and by intercalation. There are 9 to 12 costellae in a space of 3 mm, 5 mm anterior to the beak.

INTERIOR OF PEDICLE VALVE: The hinge teeth are relatively strong and triangular, with their long sides adjoining the hinge line. They are supported basally by short, stout dental lamellae which are made obsolescent, in some shells, by deposition of shell material in the umbonal cavities. There is a well-developed pedicle callist apically. The muscle field is subcircular to elongate-suboval in outline, and is unfaceted, commonly only lightly impressed. The anterolateral edges of the diductor scars tend to blend with the interior of the shell, but they may be bounded by low, regularly curving, muscle-bounding ridges. Only in a few instances were shells seen in which the lobes were decidedly impressed and flabellate. The muscle field is relatively small and occupies only one-third to about two-fifths of the length of the valves. For the most part, the diductor scars do not flare widely anterior to the dental lamellae, but the sides in some shells roughly parallel one another, diverging only slightly anteriorly. A low, rounded myophragm is present dividing the muscle field in some specimens, but is absent in others. The periphery is crenulated by the impress of the external costellae; ridges that correspond to the external interspaces are flattened on their inner edges and are deeply grooved medially.

INTERIOR OF BRACHIAL VALVE: The sockets are relatively large and widely divergent. They are bounded posterolaterally by the inner edges of the shell and basally by slight thickenings of shell material at the floor of the valve. Medially, they are defined by the edges of the brachiophores which diverge strongly ventrally and anterolaterally. The brachiophores are relatively ponderous, thick plates. They diverge slightly toward the base of the brachial valve. The anterolateral edges of the dental sockets are defined by ridges elevated above the base of the valve. Medially, between the bases of the brachiophores, there is a stout cardinal process which, on most shells, is erect and thick on its anterior edge and tapers posteriorly. In large shells the cardinal process tends to broaden ventrally and is triangular in horizontal section. A short, stout, rounded myophragm extends from the base of the cardinal process to about the midlength of the adductor muscle field. The adductor scars are relatively strongly impressed posteriorly and laterally, but lack muscle-bounding ridges. The muscle impressions are elongate-oval and blend anteriorly with the interior of the shell. The interior is crenulated peripherally in a manner similar to that of the pedicle valve.

SHELL STRUCTURE: The shell substance is finely punctate (endopunctate).

OCCURRENCE: *Dalejina* sp. B is common in the Rabbit Hill Limestone in the Monitor Range and in the *Trematospira* Zone in the Roberts Mountains and in the Sulphur Spring Range. It is also present in the Rabbit Hill Limestone in the northern Simpson Park Mountains and in the upper McMonnigal Limestone in the Toquima Range.

COMPARISON: *Dalejina* sp. B is relatively more strongly convex than *Dalejina* sp. C. Comparison with *Dalejina* sp. A was discussed under that species.

FIGURED SPECIMENS: USNM 156795-156804.

Dalejina sp. C
(Pl. 6, figs. 1-16)

Rhipidomella spp. Merriam, 1940, Pl. 7, figs. 11, 12.

EXTERIOR: Small shells are transversely oval or slightly quadrate in outline. Large shells may be subpentagonal. The valves are subequally biconvex in lateral profile, but neither is strongly convex. The ventral umbo is low. The beak is small and projects only a short distance posterior to the hinge line. The hinge line is short and commonly is equal to about one-third the maximum width. Cardinal angles are obtuse and rounded and become progressively more so on larger specimens. Maximum width is commonly at midlength on small specimens and anterior to midlength on larger ones. No fold or sulcus is developed; however, the anterior of the pedicle valve may be broadly flattened or the commissure may be very slightly bowed toward the brachial valve. The ventral interarea is low, flat, triangular, and apsacline. The dorsal interarea is orthocline.

The ornament consists of subangular, hollow, radial costellae which increase in number anteriorly by bifurcation and implantation. There are commonly 11 to 13 costellae in a space of 5 mm, 10 mm anterior to the beak.

INTERIOR OF PEDICLE VALVE: The hinge teeth are triangular and broadly divergent and bear fossettes medially. There is a well-developed pedicle callist in the apex. The muscle field is unfaceted, elongate-oval, broadening anteriorly, and is only moderately flabellate. It is bounded laterally and anterolaterally by low, rounded muscle-bounding ridges. There is a relatively well-defined, rounded myophragm beginning at the adductor impressions, posterior to the central part of the muscle impression and continuing anteriorly to the edge of the muscle field.

INTERIOR OF BRACHIAL VALVE: The sockets are small, oval, and deep. They diverge anterolaterally and from the hinge line. The posterior ends of the sockets are covered by the interarea at the hinge line. The bases of the sockets are at the floor of the valve and their medial sides are formed by stout, bladelike brachiophores, elongate suboval in cross section, which basally diverge away from the midline. Short, stout brachiophore plates buttress the brachiophores and the anteromedial edges of the sockets. The notothyrial cavity is thickened by a deposit of shell material supporting a pillar-like cardinal process which stands erect relative to the commissure. The cardinal process narrows posteriorly. A very short, broad, low myophragm is present at the anterior edge of the notothyrial platform. Adductor scars are not well impressed. The interior is crenulate peripherally by the impress of the costellae. Crenulations are commonly somewhat flattened and grooved medially on their interior edges.

SHELL STRUCTURE: The shell substance is finely punctate (endopunctate).

OCCURRENCE: *Dalejina* sp. C is present in the *Leptocoelia infrequens* Subzone in the northern Simpson Park Mountains and the Roberts Mountains. It is also widely distributed in the *Eurekaspirifer pinyonensis* Zone at both localities.

COMPARISON: *Dalejina* sp. C is distinguished from other *Dalejina* species of the Great Basin Lower Devonian by its relatively lenticular form.

FIGURED SPECIMENS: USNM 156788-156794.

Genus *Discomyorthis* Johnson, n. gen.

TYPE SPECIES: *Orthis musculosa* Hall, 1857.

DIAGNOSIS: Large convexiplane rhipidomellids like *Dalejina* internally, but with very large, flabellate ventral diductor scars.

COMPARISON: *Discomyorthis* is easily distinguished from *Dalejina* (=*Rhipidomelloides*) and *Rhipidomella* by its nearly planar pedicle valve and its unusually large ventral diductor scars. *Discomyorthis* arose from the *Dalejina oblata* group by increase of shell size and increase of diductor-scar size. In this transitional plexus, biconvex species are to be excluded from *Discomyorthis*.

Discomyorthis musculosa Hall, 1857
(Pl. 7, figs. 23-26)

Orthis musculosa Hall, 1857, p. 46.
Orthis musculosa Hall, 1859, p. 409, Pl. 91, figs. 1-3; Pl. 95, figs. 1-7.
Orthis musculosa Hall and Clarke, 1892, Pl. 6A, fig. 5.
Rhipidomella musculosa Clarke, 1908, p. 201, Pl. 42, figs. 1-5.
Rhipidomella musculosa Schuchert, *in* Schuchert and others, 1913, p. 305, Pl. 56, figs. 1-4.
Rhipidomella musculosa Stewart, 1923, p. 230, Pl. 61, figs. 6-8.
Rhipidomella musculosa Schuchert and Cooper, 1932, p. 134, Pl. 20, fig. 24.
Rhipidomelloides musculosus Amsden and Ventress, 1963, p. 62, Pl. 1, figs. 18-24.

DISCUSSION: Only two specimens were available from Nevada at the time of writing, and both are pedicle valves. One is poorly preserved, the following description is based on one relatively well-preserved pedicle valve.

DESCRIPTION: The specimen is approximately 31 mm long and, before breakage of one side, must have had a width of approximately 36 mm. The outline is oval, but not far from circular. The lateral profile of the valve is convex, but not deeply inflated, with the principal depth in the posterior half. The valve is almost flat anteriorly in its medial portion. The beak is short and incurved; the interarea is slightly curved, apsacline, and very narrow, equal to one-fourth or slightly less of the maximum width, which is about midlength. The delthyrium is triangular and open, encompassing an angle of about 60 degrees. The exterior is covered with fine radial costellae and only a few growth lines near the anterior margin of the valve.

On the interior the hinge teeth diverge slightly more widely than the inner edges of the delthyrium, and project dorsally. They are nearly blade-like with rounded sides medially. Small flanges extend posteriorly from the medial edges. Dental lamellae are absent, possibly due to obsolescence caused by the presence of thick deposits of shell material in the umbonal cavities. The diductor scar is over-all rhomboidal, nearly semicircular and flabellate anteriorly. It covers about five-sixths of the total length of the valve and about a half of the total area of the interior. The anterior part of the diductor scar is partitioned by small radial ridges that divide the several flabellate lobes. The whole muscle impression is divided by a ridgelike myophragm, stronger in the anterior half. The adductor scars are located posterior to midlength and are relatively elongate-cordate and fixed (Johnson and Talent, 1967b) in position, completely enclosed by the submedian lobes of the diductor scars anteriorly. The internal margin is finely crenulate, but the crenulations are too poorly preserved to describe accurately.

OCCURRENCE: Specimens of *Discomyorthis musculosa* are presently known in Nevada only from the *Trematospira* Subzone of the *Trematospira* Zone at the base of the McColley Canyon Formation at McColley Canyon. They also occur in the upper Hidden Valley Dolomite in the Quartz Spring area, California.

FIGURED SPECIMENS: USNM 156805.

Genus *Rhipidomella* Oehlert, 1890
TYPE SPECIES: *Terebratula michelini* Léviellé, 1835, p. 39.

Rhipidomella sp.
(Pl. 6, figs. 17-20)

DISCUSSION: A single specimen of *Rhipidomella* has been recovered from the *Eurekaspirifer pinyonensis* Zone in the Sulphur Spring Range. It contrasts with *Dalejina* sp. C, which is the typical rhipidomellid of the *pinyonensis* Zone in central Nevada, and must be one of the earliest members of the genus. Fagerstrom (1966) pointed out that *Rhipidomella* first appeared in the Eifelian (Onondaga age beds) in the Appalachian Province. In the Old World, the first appearance of *Rhipidomella* is still to be ascertained. *Dalejina* ranges as high as the Upper Emsian, where it is represented by the type species *D. hanusi* (Havlíček, 1953). The Nevada specimen represents the first evidence for overlap of *Dalejina* and *Rhipidomella*, an overlap which is expected as few brachiopod genera fail to overlap the early ranges of their progeny.

FIGURED SPECIMEN: USNM 156787.

Superfamily ENTELETACEA Waagen, 1884
Family SCHIZOPHORIIDAE Schuchert and LeVene, 1929

DIAGNOSIS: Nonplicate enteletaceans.

DISCUSSION: In the brachiopod Treatise, Wright (1965, p. H329) included *Schizophoria* and *Enteletes* in the same family, the Enteletidae. However, the writer prefers to separate the plicate from the nonplicate shells at family level. Terminological utility is a factor in favor of this arrangement, but equally important is the exotic nature of plications among the dalmanellid brachiopods.

Subfamily SCHIZOPHORIINAE Schuchert and LeVene, 1929
Genus *Schizophoria* King, 1850

TYPE SPECIES: *Conchyliolithus Anomites resupinatus* Martin, 1809, Pl. 49, figs. 13, 14.

Schizophoria parafragilis Johnson, n. sp.
(Pl. 8, figs. 1-12)

DIAGNOSIS: Biconvex *Schizophoria* with elongate ventral diductor scars and wide adductor track; brachial valve with narrowly diverging brachiophore supporting plates.

EXTERIOR: Shells are subcircular in outline and subequally biconvex in lateral profile, or the brachial valve may be slightly more convex. The ventral interarea is triangular and relatively narrow, equal to approximately half the maximum width of the valve. It is apsacline and slightly incurved. The dorsal interarea is relatively long and triangular; anacline to almost orthocline. In transverse section both valves are regularly convex outward on their flanks, without any indication of fold or sulcus medially. Maximum width is commonly at about midlength.

The ornament consists of fine, hollow, radial costellae which are of approximately uniform size over the whole of the valves. On one shell there are about 15 costellae in a space of 5 mm, 10 mm from the beak. Most specimens available are so poorly preserved that an accurate rib count is not possible.

INTERIOR OF PEDICLE VALVE: The hinge teeth are relatively small and triangular in cross section. They are supported by strong dental lamellae which diverge slightly and recurve to project anteriorly as relatively well-marked muscle-bounding ridges. The lateral edges of the muscle field may be almost parallel to the midline. The diductor impressions are separated by a well-defined, low, rectangular myophragm serving as the adductor track. Vascula media are relatively well marked in many shells and diverge only slightly anteriorly. The interior of the shell is crenulated only peripherally by the impress of the costellae.

INTERIOR OF BRACHIAL VALVE: The sockets are short, widely divergent, and are defined basally by thin fulcral plates. On large shells the posterior portions of the sockets may be partially covered by the interarea. The inner edges of the fulcral plates are attached directly to high, narrowly diverging, brachiophore supporting plates. The notothyrial cavity bears a cardinal process of considerable variability in form. Its anterior portion is directed ventrally and it narrows posteriorly. There may be a single plate or lobe, or there may be an additional lateral pair forming a trilobate process. The growth layers of the cardinal process are strongly lamellose in many shells. The cardinal process rests on shell material which is continuous anteriorly with a well-developed rounded myophragm that bisects the adductor muscle field. The posterior adductor scars are triangular and closely set posteriorly, within the brachiophore supporting plates. The anterior adductors are more deeply impressed and are set off from the posterior adductors by bounding shelves which diverge anterolaterally. The anterior adductors tend to blend with the inner surface of the shell, but on the largest specimens may develop fairly well-marked anterior margins. In some specimens the lateral edges of the anterior adductor impressions become moderately wider abruptly at the lateral terminations of the shelf dividing the two sets of impressions.

OCCURRENCE: *Schizophoria parafragilis* is common in the lower McMonnigal Limestone in the Toquima Range. It is present in the *Quadrithyris* Zone at Coal Canyon. *Schizophoria* virtually disappears above this level (*Quadrithyris* Zone) in central Nevada and does not reappear until *E. pinyonensis* Zone time, exemplified by *Schizophoria nevadaensis*.

COMPARISON: The contemporaneous species, *Schizophoria multistriata* Hall (1859-61, Pl. 15) has a more pyriform ventral muscle scar than *S. parafragilis* and tends to be anteriorly bilobate. *Schizophoria parafragilis* has a broader ventral adductor track than does *S. bisinuata* Weller (1903, Pl. 31) or *S. oriskania* Schuchert (*in* Schuchert and others, 1913, Pl. 56). The latter two species are much alike internally and externally. The new species differs from *S. fragilis* Kozlowski (1929) for the same reason. *Schizophoria ferganensis* Nikiforova (1937) and *S? derjawinin* Peetz (1901, Pl. 4, fig. 10) are both poorly known internally. These species, especially in the case of *S? derjawinin*, are more transverse than *S. parafragilis*.

Schizophoria nevadaensis and other Emsian and younger species have more widely divergent brachiophore supporting plates and approach a convexi-plane lateral profile.

FIGURED SPECIMENS: USNM 156810-156820. The holotype is USNM 156820.

Schizophoria nevadaensis Merriam, 1940
(Pl. 9, figs. 1-18)

Orthis impressa Walcott, 1884, p. 115, Pl. 13, fig. 13, *not* Hall, 1843.
Schizophoria nevadaensis Merriam, 1940, p. 79, Pl. 7, figs. 1, 2.
Schizophoria nevadaensis Cooper, 1944, p. 357, Pl. 140, figs. 6, 7.

EXTERIOR: The shells are transversely suboval to subquadrate in outline and unequally biconvex in lateral profile. Small shells are nearly equally convex, but in large specimens the brachial valve is more convex. The pedicle valve is most strongly convex in its posterior half, but the brachial valve curves more uniformly from back to front. Maximum width is at about midlength on small shells and is commonly anterior to midlength in large ones. The pedicle valve bears a broad shallow sulcus beginning at about midlength on large individuals. The brachial valve lacks a fold, but may become slightly emarginate medially. The interareas of both valves are short and approximately equal to about one-third to one-half the maximum width. The delthyrium encloses an angle of about 45 degrees and is enlarged apically into a foramen. The ventral interarea is apsacline to cataeline and may be strongly incurved at the beak. The dorsal interarea is orthocline to apsacline.

The ornament consists of subangular radial costellae and faint concentric growth lines. The growth lamellae are strongly accentuated anteriorly on some large shells. The costellae increase in number anteriorly by both bifurcation and by implantation on each valve. The costellae, 20 mm from the beak on one pedicle valve, number 20 in a distance of 8 mm on the middle part of the valve.

INTERIOR OF PEDICLE VALVE: Short dental lamellae are present supporting stout hinge teeth. Anteriorly the dental lamellae join strong muscle-bounding ridges which enclose the diductor muscle field laterally. The diductor scar is cordate and slightly impressed in small specimens and becomes more elongate and deeply impressed in large shells. The diductor muscle impressions are smoothly U-shaped in transverse section, without any faceting. In the largest shells the diductor impressions become slightly grooved, especially on their lateral lobes. There is a strong, rounded myophragm that medially divides the diductor muscle field and which must have been the site of adductor attachment. The anterolateral portions of large shells are irregularly pitted, except for the medial portion which bears a pair of smooth tracks of the vascula media. Small shells do not show impressions of the vascula media but instead are crenulated by the impress of the costellae.

INTERIOR OF BRACHIAL VALVE: The cardinal process has been observed only on large shells where it is almost linear, narrower at the posterior, and is crossed by transverse growth lamellae. The ventral edge of the cardinal process is rounded in transverse section and is thicker in width than the supporting layers, producing a keyhole-shaped transverse section for the whole structure. In addition, there are small ridges lateral to the anterior part of the cardinal process as illustrated by Schuchert and Cooper (1932, Pl. A, fig. 14) which these writers called accessory cardinal processes. The brachiophore supporting plates are thick and diverge at an angle of about 70 degrees. Fulcral plates serve to define the bases of the sockets. The brachiophore supporting plates merge with muscle-bounding ridges which continue anterolaterally and then recurve to enclose the adductor scars laterally. The adductor scars are divided into posterior and anterior pairs by ridges that diverge anteriorly from one another at angles only slightly greater than 90 degrees. The anterior adductors are separated by an area that widens anteriorly at the place where the pallial markings are first impressed. The latter consist of four ridges and three grooves that originate between the anterior adductors.

COMPARISON: *Schizophoria nevadaensis* is a more transverse and quadrate species than any other known to the writer from the Devonian of the Great Basin. It is readily distinguished from *Schizophoria parafragilis* by shape and by its widely diverging brachiophore supporting plates.

FIGURED SPECIMENS: The holotype, USNM 96375, came from the *Eurekaspirifer pinyonensis* Zone at Lone Mountain. The label states 88 feet above loc. 59 (Merriam, 1940). Measurements made by the writer are as follows: length 29.6 mm, width 35.2 mm, thickness 22.2 mm. Other figured specimens include USNM 96376 and USNM 156821-156826.

Subfamily DRABOVIINAE Havlíček, 1950

DISCUSSION: In the brachiopod Treatise, Wright (1965, p. H330, 332) distinguished Draboviinae from Schizophoriinae according to the disposition of the brachiophore supporting plates (*sic* brachiophore bases). Draboviinae was to include taxa in which the anterior continuations of the brachiophore supporting plates converge onto the myophragm and do not contain the adductor muscle field posteriorly. Schizophoriinae was to include taxa in which brachiophore supporting plates contain the adductor muscle field posterolaterally. However, unless the requirement for draboviinids of convergence onto the dorsal myophragm is dropped, the two categories are not mutually exclusive. *Salopina*, grouped with the Schizophoriinae by Wright, does not fill the principal morphologic requirement because the forward edges of its brachiophore supporting plates lie within the adductor muscle-bounding ridges. *Salopina* does, however, fill the draboviinid requirement that the anterior edges of the brachiophore supporting plates lie within the adductor muscle field, and is, therefore, regarded here as a member of the Draboviinae.

Above and beyond the question of how well *Salopina* fits into one or the other of the subfamilies Schizophoriinae or Draboviinae, in the scheme envisioned by Wright, is the obvious failure of the morphologic requirements of Wright to divide the various Ordovician to Devonian members of the family on a basis consistent with the phylogenetic requirements. The most satisfactory application of Wright's criteria for subfamily division implies that *Hirnantia* is the ancestor of *Schizophoria*. The gross and even detailed morphologies of these two genera are quite close, and *Hirnantia*, alone among the Ordovician members of the Schizophoriidae, is known to extend well into the Silurian, represented by *H. senecta* Hall and Clarke from late Llandovery age beds in the eastern United States, (brought to the writer's attention by C. W. Harper) and in the form of an unnamed species in Wenlock age beds of the Hidden Valley Dolomite of southeastern California. It is possible that the latter is related to

"*Mendacella*" *lenticularis* (Foerste, 1909; Amsden, 1949, Pl. 1, figs. 20, 21) from the Brownsport Formation.

The disadvantage of Wright's scheme is its necessary elimination of *Salopina* as a *Schizophoria* ancestor. The possibility is not yet ruled out, especially as the earliest schizophoriids are smaller and somewhat more closely resemble *Salopina* than do the typical dorsi-biconvex forms, present in Emsian and younger beds. It seems very likely that the Givetian genus *Sphenophragmus* Imbrie is a derivative of *Salopina*. At present no other possibilities are in sight; however, the two fall on either side of the morphologic distinction offered by Wright. This appears destructive to Wright's scheme, attractive as it is morphologically, because it requires polyphyletic origin of the Schizophoriinae.

The classification of *Salopina* is of interest because it is regarded as ancestral to *Muriferella*, described below. Johnson and Talent (1967a) demonstrated the probability that *Muriferella* arose by increase in height of the dorsal median ridge of *Salopina* of the *crassiformis* type to the proportions of a septum. According to A. C. Lenz (written commun., 1966) a complete transitional series exemplifying this evolutionary patterns occurs in the Lower Devonian sequence at Royal Creek, Yukon Territory, Arctic Canada (Lenz, 1966, 1967b).

Genus *Muriferella* Johnson and Talent, 1967

TYPE SPECIES: *Muriferella masurskyi* Johnson and Talent, 1967a.

Muriferella masurskyi Johnson and Talent, 1967

(Pl. 10, figs. 1-17)

Muriferella masurskyi Johnson and Talent, 1967a, p. 46, Pl. 9, figs. 1-14.

DISCUSSION: This species was described by Johnson and Talent (1967a). It is a sulcate, ventribiconvex species with a relatively thick shell and a long, low, triangular median septum that extends to the anterior margin of the brachial valve.

OCCURRENCE: *Muriferella masurskyi* is known only from the *E. pinyonensis* Zone in the Cortez Mountains.

FIGURED SPECIMENS: USNM 156827-156831, 162007, 162009-162014. The USNM numbers originally assigned and reported by Johnson and Talent (1967a, p. 49, 50) had been previously used and are now assigned to the sequence 162007-162014.

Order Pentamerida

DISCUSSION: The typically Silurian family Pentameridae was extinct by the beginning of the Gedinnian and its demise heralded an expansion of the Gypidulidae in some areas. Gypidulidae are the only common pentamerids in the Siegenian and Emsian of Nevada, where each zone appears to be characterized by a particular species. Gypidulids were important members of the Gedinnian fauna and continued prominently in the *Quadrithyris* Zone, but they virtually disappeared with the beginning of the *Spinoplasia* Zone. An interesting parallel can be drawn with the distribution of gypidulids in the Appalachian Province. *Gypidula* is abundant in the Helderbergian and well represented as high as the Becraft by *Gypidula pseudogaleata*. It is not known if the genus persisted to the top of the Port Ewen in the highest Helderbergian, but gypidulids disappeared from the Appalachian Province at this time and are unknown there during the Oriskany and *Etymothyris* Zone. The subfamily is represented in the highest Lower Devonian and in the early Eifelian by *Pentamerella*, but the exact time of re-entry of members of the genus *Gypidula* is uncertain; they are known in the Appalachian Province in the Givetian. However, in Nevada, although the fauna is decidedly of Appalachian aspect in the Oriskany equivalents (*Trematospira* Zone), *Sieberella pyriforma* is a relatively common shell, possibly representing a lone invader from the Old World Province. On the other hand it fairly closely resembles a species of *Sieberella* known in the Helderbergian Beck Pond Limestone of Maine, but possible connections are uncertain. *Gypidula*, exemplified by *G. praeloweryi* and *G. loweryi* in the *kobehana* and *pinyonensis* zones, almost certainly migrated from Old World sources because of the total absence of any similar shell during the late Siegenian and Emsian of the Appalachian Province. Further, their source must have been from the Bohemian Subprovince or the Uralian Subprovince of the Old World Province and not from the Rhenish Community because gypidulids were absent there during the Siegenian and Early Emsian.

The Camerellidae have a small representation exemplified by *Anastrophia* cf. *magnifica* which occurs in the *Quadrithyris* Zone and in older beds in Nevada. The form called *"Camerella"* sp. in the Rabbit Hill Limestone is congeneric with the most abundant form of the Pillar Bluff Limestone brachiopod fauna developed in central Texas. It is not an Appalachian form and its source is unknown, although it is obvious that the *Spinoplasia* Zone fauna and that of the succeeding *Trematospira* Zone had their sources in the Appalachian Province. The occurrence of *"Camerella"* sp. gives a clue to the possible route of migration which seems to have been well to the south of the present southern boundary of the United States rather than across the continental backbone through some marine area whose deposits are now wholly removed by erosion.

SUBORDER SYNTROPHIOIDEA

Superfamily CAMERELLACEA Hall and Clarke, 1895
Family CAMERELLIDAE Hall and Clarke, 1895
Genus *Anastrophia* Hall, 1867

TYPE SPECIES: *Pentamerus verneuili* Hall, 1857, p. 104, Figs. 1, 2.

Anastrophia cf. *magnifica* Kozlowski, 1929
(Pl. 11, figs. 1-8)

Anastrophia magnifica Kozlowski, 1929, p. 140, Text-Fig. 42; Pl. 4, figs. 14-16.
Anastrophia magnifica Khodalevich, 1951, p. 18, Pl. 4, figs. 9A-D.
Anastrophia magnifica Nikiforova, 1954, p. 65, Pl. 4, figs. 1A-D.
Anastrophia magnifica Kulkov, 1963, p. 18, Pl. 1, figs. 3, 4.

EXTERIOR: The shells are transverse in outline, varying from shield-shaped to pentagonal or subquadrate. The valves are relatively strongly biconvex in lateral profile. The valves appear to be of about equal convexity, or the brachial valve may be more convex. The pedicle valve is relatively strongly convex only at the umbo. There is a well-developed sulcus on the pedicle valve and a corresponding fold on the brachial valve; however, the fold is poorly defined. The sulcus is elongated at the anterior commissure as a dorsally projecting tongue. The ventral interarea is flat, apsacline, and of variable length, but on small shells it is considerably shorter than maximum width. It is low and linear rather than triangular. The delthyrium is triangular and encloses an angle of slightly more than 60 degrees dorsal to an apical foramen. Cardinal angles are obtuse and rounded on small shells, but they may be acute or approximately right angles on large specimens.

The ornament consists of strong, subangular, radial costae on the flanks and on the midregions. There are commonly four or five costae in the sulcus, and in some specimens the costae that bound the sulcus bifurcate, simulating the condition of parietal costae. Costae commonly number six or seven on each flank of the pedicle valve. Interspaces are of approximately the same width and shape as the adjoining costae. Low concentric growth lines occur at irregular intervals over the shell and although not strongly imbricating, they appear to disrupt the even continuity of the radial costae at places of intersection.

INTERIOR OF PEDICLE VALVE: Hinge teeth are very small and blunt, approximately triangular in horizontal section. Apically there is a spondylium supported nearly to its anterior edge by a median septum. The shape of the spondylium is typical for the genus, and is very broad dorsally and in its posterior two-thirds, but narrows basally and anteriorly into an elongate trough which projects farther anterior than the main part of the structure. The interior of the valve is strongly crenulated by the impress of the external costae. On one well-preserved pedicle valve, there is an internal flange of shell material which projects as a separate layer parallel to the external surface layer at the posterolateral margin. This flanging mechanism may be related to lengthening of the hinge line in relatively large-sized shells.

INTERIOR OF BRACHIAL VALVE: The brachial valve of this species in Nevada is known only from a single internal mold. It shows divergent sockets and a pair of brachial plates which join the floor of the valve along discrete, subparallel tracks. The interior of the valve is radially ridged and is crenulated by the impress of the external costae.

OCCURRENCE: In addition to Gedinnian occurrences, *Anastrophia* cf. *magnifica* is present in the *Quadrithyris* Zone at Coal Canyon and in the lower McMonnigal Limestone in the Toquima Range. *Anastrophia magnifica* and closely related forms are widespread in Bohemian-Uralian faunas of Siegenian age, but no *Anastrophia* is known to extend into Emsian beds. In addition to occurrences listed in the synonymy above the writer has seen *A.* cf. *magnifica* in Siegenian faunas from Manildra, N.S.W., Australia, sent by N. M. Savage and from Royal Creek in the Canadian Yukon (Lenz, 1967b).

FIGURED SPECIMENS: USNM 156832-156836.

"*Camerella*" sp.
(Pl. 11, figs. 9-18; Pl. 13, figs. 6, 7?)

DISCUSSION: The species described below must be excluded from *Camerella* since that genus is anteriorly plicate. In addition, the specimens at hand do not display alae in the brachial valve, although preservation is less than adequate to be certain that the structures were not developed.

EXTERIOR: Specimens are thin-shelled and are of relatively small size, commonly subcircular in outline and unequally biconvex in lateral profile. Brachial valves are more strongly convex than pedicle valves and are nearly hemispherical in shape. The ventral beak is short and is not strongly incurved. There is a very broad, triangular ventral interarea. The delthyrium is triangular and encloses an angle of about 45 degrees. It is unmodified by a deltidium or by deltidial plates. The dorsal umbo is moderately convex and the beak is slightly more strongly incurved than in the pedicle valve. No dorsal interarea was detected.

The exterior is smooth except for concentric growth lines. A relatively deep sulcus is present in the anterior one-third to one-half of the pedicle valve and there is a corresponding, rounded fold anteriorly in the brachial valve.

INTERIOR OF PEDICLE VALVE: The hinge teeth project relatively strongly anteriorly, beyond the hinge line and are suboval to subtriangular in cross section. The hinge teeth are supported basally by a pair of triangular dental lamellae which converge medially to join, forming a spondylium of subrhomboidal outline. The spondylium is supported on a bladelike median septum along its entire length and in larger specimens the median septum extends a slight distance anterior to the distal edge of the trough of the spondylium. In shape and size the spondylium and supporting septum are closely comparable to those of *Anastrophia* cf. *magnifica* which is present in the underlying *Quadrithyris* Zone.

INTERIOR OF BRACHIAL VALVE: The sockets are narrow, shallow, and diverge moderately anterolaterally. They open anterolaterally into the interior of the shell without a rim of shell material defining the distal edge of the sockets. Medially, along the inner edges of the inner socket ridges there is a pair of lobes that project nearly horizontally, partially covering the notothyrial cavity. Generally there is a narrow slot between the plates, and none of the specimens examined have the plates joined. The brachial plates attach to the inner edges of the socket ridges approximately along the same line as the pair of medial lobes. The brachial plates are nearly subparallel, diverging only slightly anteriorly. They ascend part way toward the floor of the brachial valve, nearly parallel to one another, then recurve medially to join, forming a deep, U-shaped cruralium. In better preserved specimens the cruralium is grooved medially and is bordered along the edges of the median groove by a pair of radial ridges adjacent to it. The cruralium appears to lie almost on the floor of the valve posteriorly, but through most of its length it is strongly elevated on a high, bladelike median septum. The cruralium is relatively long, reaching approximately half the length of the valve in some specimens. In addition, the median septum may extend even further anteriorly, to a length of approximately two-thirds the distance to the anterior commissure.

OCCURRENCE: "*Camerella*" sp. is a rare species, known in Neveda (with certainty) only from the *Spinoplasia* Zone at Rabbit Hill, where 16 specimens have been recovered from a collection of several thousands of individuals. Two specimens from the *Quadrithyris* Zone may belong here.

FIGURED SPECIMENS: USNM 156837-156843, 156855.

SUBORDER PENTAMEROIDEA

Superfamily PENTAMERACEA M'Coy, 1844
Family GYPIDULIDAE Schuchert and LeVene, 1929
Subfamily GYPIDULINAE Schuchert and LeVene, 1929
Genus *Sieberella* Oehlert, 1887

TYPE SPECIES: *Pentamerus sieberi* Von Buch *in* Barrande, 1847, p. 465, Pl. 21, figs. 1, 2.

Sieberella pyriforma Johnson, n. sp.
(Pl. 12, figs. 1-19)

DIAGNOSIS: *S. pyriforma* is characterized by its relatively strongly convex brachial valve, its small, narrow, ventral umbo, its pyriform outline, and its irregular, subangular, bifurcating plications.

EXTERIOR: Small shells are transversely oval, but medium- and larger-sized shells are commonly elongate-pyriform in outline and biconvex in lateral profile. Both valves are strongly convex, with the pedicle valve commonly slightly the greater of the two. The greatest convexity in the pedicle valve is at the umbo, and the beak is strongly incurved so that it touches the umbo of the brachial valve. Brachial valves are commonly most strongly convex slightly posterior to midlength. The palintrope is narrow, and an interarea is only poorly developed in larger shells. The delthyrium encloses an angle of about 60 to 70 degrees. Pedicle valves have a poorly defined, medial fold which first appears at about midlength on large shells. The brachial valve has a corresponding low, broad, and indistinct sulcus. The anterior commissure is only slightly deflected toward the pedicle valve so that the edges of the fold and sulcus are not sharply defined.

The ornament consists of rounded to subangular, radial plications of variable size, which originate on the umbo and increase in number anteriorly by bifurcation. They are crossed by numerous, fine, closely spaced, growth lines. Bifurcation is less common medially than on the flanks. Bifurcation occurs on both valves and is accomplished by the splitting off of a smaller plication from the side of a larger one. No plications have been seen to originate in an interspace by implantation. There are about 16 to 18 plications at the commissure of the medium-sized shells, such as the holotype.

INTERIOR OF PEDICLE VALVE: A short spondylium is present beneath the inner edges of the delthyrium and is supported along its entire length by a thin median septum. The interior is corrugated by the impress of the plications.

INTERIOR OF BRACHIAL VALVE: The sockets are long, narrow, and shallow and are bounded laterally by small thickenings along the inner edge of the posterior of the valve. The bases of the sockets are free, above the floor of the valve and connect medially with the lateral edges of the inner plates. Posteriorly, the inner plates converge toward one another at a high angle, but the angle decreases anteriorly. Basally, the inner plates connect with the outer plates to form a structure of lyre-shaped cross section without noticeable differentiation at the brachial processes. The outer plates converge toward one another to join medially and are supported by a low, but well-defined median septum, as in the type species *Sieberella sieberi*.

OCCURRENCE: The species is common in the *Trematospira* Zone in the Sulphur Spring Range. Poorly preserved specimens that may belong to *S. pyriforma* occur at Antelope Peak.

COMPARISON: *Sieberella pyriforma* is immediately distinguished from *S. siegeri* by its elongate, pyriform shape and its irregular bifurcating plications. *S. lubimovi* Andronov (1961, p. 87, Pl. 13. figs. 34-43) resembles small specimens of the Nevada form, but is plicate further posteriorly. *S. roemeri* Hall and Clarke (1892) is much more strongly convex in the posterior of the pedicle valve than is *S. pyriforma*. The new species has, in addition, a more poorly developed fold and sulcus and more irregular plications. *Sieberella weberi* Khodalevich (1951, Pl. 8) is more finely costate than *S. pyriforma* and is relatively broader posteriorly. *Sieberella? vagranica* Khodalevich (1951, Pl. 8) also is more finely costate and has a different shape than *S. pyriforma*. *Sieberella kakvensis* Khodalevich (1951, Pl. 8) is too poorly illustrated to allow a comparison. *Sieberella pyriforma* closely resembles *S. romingeri* (see Imbrie, 1959, Pl. 50) in outline and type of costae, but has a much more convex brachial valve.

FIGURED SPECIMENS: The specimen figured as nos. 10, 11, 17-19 on Plate 12, USNM 156850, is the holotype. It comes from the *Costispirifer* Subzone of the *Trematospira* Zone, Sulphur Spring Range. Other figured specimens include USNM 156844-156849.

Sieberella cf. *problematica* (Barrande, 1847)
(Pl. 13, figs. 1-5)
Pentamerus problematicus Barrande, 1847, p. 470, Pl. 17, fig. 15.
Pentamerus problematicus Barrande, 1879, Pl. 20, figs. 3-5.

EXTERIOR: The shells are transversely suboval in outline and biconvex in lateral profile with the pedicle valve several times more convex than the brachial valve. The ventral beak is short and not prominent. No interarea is developed. The delthyrium encloses an angle of about 60 degrees. There is a distinct ventral fold which begins in the posterior third of the shell and which is modified by a broad, low, rounded sulcus that effectively divides the fold into a median depression and bounding plications. On the flanks there is a pair of rounded plications close to the lateral edges of the fold. The brachial valve bears a sulcus with a median plication.

INTERIOR OF PEDICLE VALVE: There is a short V-shaped spondylium, beneath the delthyrium, that extends slightly less anteriorly than the height of the delthyrium. There is a short median septum, concave anteriorly, that supports the spondylium along its entire length.

INTERIOR OF BRACHIAL VALVE: The sockets are shallow, narrow grooves between the shell and the posterolateral edges of the inner plates. Inner plates are triangular, diverge anteriorly, and converge toward the midline of the brachial valve. The brachial processes are bladelike with their surfaces approximately normal to the floor of the brachial valve. The outer plates converge toward the midline of the valve and join at approximately their position of intersection with the base of the valve as in *Sieberella sieberi*.

OCCURRENCE: *Sieberella* cf. *problematica* is relatively common in the *Quadrithyris* Zone at Coal Canyon. One specimen has been found in the upper part of the Vaughn Gulch Limestone, USNM loc. 12776.

DISCUSSION: In Nevada the species most like *S. problematica* is *Gypidula? biloba* (Johnson and Reso, 1964, p. 79, Pl. 19, figs. 1-4) from beds of probable Ludlow age at the base of the Sevy Dolomite. If, when the brachial valve structures of *G? biloba* become known and prove to be of the *Sieberella* type, then *biloba* possibly may be regarded as ancestral to the *Quadrithyris* Zone form. Externally, these two species differ significantly only in the presence of a pair of lateral plications and more prominent medial plications in *Sieberella* cf. *problematica*.

Two species in the Siegenian of the USSR closely resemble *Sieberella* cf. *problematica* from Nevada. One is "*Pentamerus*" *kayseri* Peetz (1901, Pl. 3, fig. 8) which Rzhonsnitskaya has listed as a Krekova (Siegen) guide (1962; 1964, p. 1086). The other is *Gypidula? bicostata* Khalfin from the Solovien Limestone of the Altai Mountains (Kulkov, 1963, Pl. 2, fig. 19).

FIGURED SPECIMENS: USNM 156851-156854.

Genus *Gypidula* Hall, 1867
TYPE SPECIES: *G. typicalis* Amsden, 1953, p. 140.

Gypidula cf. *pseudogaleata* (Hall, 1857)
(Pl. 15, figs. 9-11)
Pentamerus pseudogaleatus Hall, 1857, p. 106, figs. 1-6.
Pentamerus pseudogaleatus Hall, 1859, p. 259; 1861, Pl. 48, figs. 2a-l.

Pentamerus pseudogaleatus Hall and Clarke, 1895, Pl. 72, fig. 14.
Sieberella pseudogaleata Clarke, 1909, p. 39, Pl. 7, figs. 24-26.

DISCUSSION: Only a few poorly preserved specimens are available. They appear to be extremely rare in the Windmill Limestone and uncommon in the lower McMonnigal Limestone, at least as represented in the available collections. Only the illustrated specimen, from the lower McMonnigal, is adequate on which to base a discussion; all other available specimens are smooth, featureless shells, not preserving the anterior commissure. The illustrated specimen is a relatively elongate, smooth pedicle valve without a distinctly developed fold. The deflection of the anterior commissure in the ventral direction is very gentle (Pl. 15, fig. 10), as is typical of *Gypidula pseudogaleata*. The anterior and lateral profiles of the Nevada specimen appear to exactly duplicate the form of the valves seen in typical New York specimens illustrated by Hall. Much larger collections of this form from the lower McMonnigal Limestone must be studied before a final conclusion about the specific identity of the Nevada form.

A worthwhile comparison can be made between *Gypidula* cf. *pseudogaleata* from Nevada and *Gypidula pelagica* which constitutes such a prominent part of the preceding fauna in the Gedinnian of the Roberts Mountains Formation. *Gypidula pelagica* (Pl. 15, figs. 12-15) is commonly slightly shorter and broader with a more strongly incurved ventral beak. More elongate specimens of *G. pelagica* are similar in outline, but bear very faint, broad, radial plications in the midregions, and all specimens of *G. pelagica* appear to have a more strongly developed ventral fold with a well-defined trapezoidal re-entrant in the pedicle valve which accommodates the dorsal anterior tongue.

OCCURRENCE: Restricted to the *Quadrithyris* Zone of the Windmill Limestone and lower McMonnigal Limestone.

FIGURED SPECIMEN: USNM 157015.

Gypidula praeloweryi Johnson, n. sp.
(Pl. 15, figs. 1-8)

Gypidula cf. *coeymanensis* Merriam, 1940, Pl. 11, fig. 16; *not* Schuchert, 1913.

DIAGNOSIS: Internal structures essentially as in *Gypidula loweryi*. External size and shape as in *Gypidula loweryi*, but contrasting with that species by the presence of well-developed radial plications developed on the flanks as well as on the midregion.

EXTERIOR: Pedicle valves are elongate pentagonal to subpyriform with the posterior relatively broadly rounded. Brachial valves are transversely suboval to pentagonal. The valves are strongly biconvex with the pedicle valve 2 to 3 times as deep as the brachial valve. The ventral beak is very strongly incurved, touching the dorsal umbo. Curvature of the pedicle valve is greatest posteriorly and decreases regularly toward the anterior. The brachial valve is most strongly convex about one-third its length anterior to the hinge line. A more or less well-developed, slightly curved, bandlike interarea is present on the ventral palintrope. The pedicle valve has a broad low fold. The brachial valve has a corresponding shallow sulcus, but the anterior commissure is sharply deflected, and the brachial valve has a subrectangular tongue which is accommodated by the fold on the pedicle valve.

The ornament consists of low, rounded, radial plications developed over most of the valve, although they appear to be obsolescent in the posterolateral regions. There is no strong demarcation between plications on the fold or sulcus and those immediately adjoining on the flanks. On several large specimens there were found to be about 15 plications at about one-third the distance anterior to the beak. Plications increase in number anteriorly by bifurcation.

INTERIOR OF PEDICLE VALVE: There is a V-shaped spondylium developed beneath the delthyrium and projecting anteriorly. It is supported posteriorly along part of its length by a median septum. The interior is slightly corrugated by the impress of the plications.

INTERIOR OF BRACHIAL VALVE: The brachial plates conform to a structure of lyre-shaped cross section. Outer plates meet the floor of the valve along two long, discrete, slightly divergent tracks. One specimen has transversely corrugated outer plates. The interior of the valve is crenulated by the impress of the plications.

OCCURRENCE: *Gypidula praeloweryi* is known from the *Acrospirifer kobehana* Zone in the Sulphur Spring Range, and on the north side of the Roberts Mountains where it occurs with *Strophonella* cf. *punctulifera* and from locality 17 (Merriam, 1940) on the south side of the Roberts Mountains.

COMPARISON: *G. praeloweryi* is distinguished from *G. loweryi* Merriam as noted in the diagnosis above. It is distinguished from the Helderbergian species *G. coeymanensis* Schuchert and *G. multicostata* Dunbar by development of strong plications on the umbo instead of largely on the anterior portions. In addition, it is much more strongly convex in the posterior of the pedicle valve than is *G. coeymanensis*. No other species, that resembles *G. praeloweryi* closely enough to require comparison, is known to the writer.

FIGURED SPECIMENS: USNM 156864-156867. The holotype is USNM 156867.

Gypidula loweryi Merriam, 1940
(Pl. 13, figs. 8-17; Pl. 14, figs. 1-20)

Pentamerus comis Walcott, 1884, p. 159, Pl. 3, fig. 7; Pl. 14, figs. 15, 15a, 15b; Pl. 15, figs. 5, 5a, 5b; not Owen.
Gypidula loweryi Merriam, 1940, p. 81, Pl. 7, fig. 9.
Gypidula loweryi Cooper, 1944, p. 305, Pl. 115, figs. 4, 5.

EXTERIOR: Pedicle valves are elongate-subpyriform, flaring broadly anterolaterally from the umbo. Brachial valves may be subcircular or more commonly are transversely suboval to somewhat pentagonal in outline. The valves are biconvex in lateral profile with the pedicle valve very deep and strongly curved. Convexity of both valves decreases from posterior to anterior. The umbo may be relatively narrow, or broad with a blunt and strongly incurved beak. Commonly the beak of the pedicle valve touches the dorsal umbo which is incurved beneath it. The ventral interarea is only faintly defined and, when seen, is best developed linearly for a short distance along the hinge line. The delthyrium includes an angle of about 70 degrees.

Most shells are smooth posteriorly. There commonly is a shallow subrectangular sulcus developed in the anterior third of the brachial valve and a corresponding low fold in the pedicle valve. In some shells the dorsal sulcus is accommodated by a slight emargination of the anterior commissure of the pedicle valve. Low, rounded plications are developed anteriorly over the fold and sulcus of some shells. No plications have been seen on the flanks. Near the commissure on some large specimens, the growth lines become prominent and numerous.

INTERIOR OF PEDICLE VALVE: The hinge teeth are small and pointed. The spondylium is attached directly below the teeth and the inner edges of the delthyrium and is about 50 percent longer anterior to the hinge line than beneath the delthyrium. A median septum is present only in the apex of the shell. It supports only a fraction of the length of the spondylium although it may continue along the floor of the valve for as much as 2 cm. The median septum is strongly concave anteriorly. In the bottom of the spondylium there are numerous fine longitudinal grooves of muscle attachment.

INTERIOR OF BRACHIAL VALVE: Sockets are small and are defined laterally by low, rounded thickenings along the posterior margin of the shell. Medially, they are bounded by the lateral edges of the inner plates. The base of the sockets lie on a thickening of shell material on the floor of the valve. The inner plates are triangular and elongate anteriorly. For most of their length they converge toward the midline of the brachial valve, but near their distal ends there commonly is a reversal of slope passing the vertical so that the inner edges converge toward the midline of the pedicle valve. The brachial processes are thin bandlike plates joining the dorsal edges of the inner plates and set approximately normal to the transverse curvature of the shell. Along the anteromedial sides of the inner plates, the ventral edges of the brachial processes project beyond the joint to form a pair of carinae, as the term was employed by Kozlowski (1929, p. 137, Fig. 41E). The outer plates curve medially toward the brachial valve, and each joins the floor of the valve at a low angle.

At the posteromedial edges of the inner plates there are small ridges that converge toward the beak. These ridges are seen on large shells and may have served as the sites of diductor attachment.

OCCURRENCE: The species is common in the *Eurekaspirifer pinyonensis* Zone in the Sulphur Spring Range, Roberts Mountains, Lone Mountain, northern Simpson Park Mountains, and in the Cortez Range.

DISCUSSION: Recently Amsden (1964, p. 232, 235, 236) has advocated separation of the gypidulinid subfamilies Gypidulinae and Clorindinae on the basis of the presence or absence of carinae in the structure of the brachial apparatus. The carinae, or ventral edges of the elongate, ribbonlike brachial processes, were thought by Amsden to characterize the Clorindinae and not *Gypidula* and its allies. The presence of well-defined carinae in the brachial apparatus of *Gypidula loweryi* suggests need for a reappraisal of their value in taxonomy. A perusal of the illustrations of *Gypidula loweryi* in this paper (Pl. 13, figs. 8-17, Pl. 14, figs. 1-20), in the writer's opinion, is adequate to show that the species is a gypidulinid and very different in form from *Antirhynchonella* and genera of similar conformation. The carinae are fairly well illustrated in Plate 13, figure 15, and Plate 14, figures 16-18. These figures compare favorably with those of Amsden, illustrating *Gypidula coeymanensis* (1964, Pl. 40, figs. 5, 6), except that the specimens of *coeymanensis* have somewhat ragged-appearing anterior edges of the inner portions of the brachial plates, suggesting that part of their anterior extremities are missing. In any event, the brachial apparatus of both species is very similar, and the presence of well-developed carinae in *Gypidula loweryi* and their absence or apparent absence in other *Gypidula* species, among specimens that have been studied, indicates to the writer that the feature is an evanescent one. The writer, therefore, recommends its rejection as a structural feature significant for distinctions above the species level.

It is worth noting here that in several of the writer's illustrated specimens (Pl. 13, fig. 17; Pl. 14, fig 18) and in some of Amsden's as well (1964, Pl. 40, figs. 5, 6, 9) the brachial apparatus bears a pair of ridges posteriorly that converge toward the apex of the brachial valve at the place one would expect to find a cardinal process or a similar structure for the attachment of the diductor muscles. Amsden (1964) did not comment on the structures, although he noted (p. 226) the presence of a cardinal process in the pentamerinid genus *Jolvia* Sapelnikov. According to Sapelnikov's serial sections (1960, p. 57, Fig. 1), the structure in *Jolvia* attains the proportions of a long simple stalk and no doubt should be referred to as a cardinal process. Many pentamerinids have such a structure (Amsden and others, 1967). Some additional sections of *Jolvia* Sapelnikov (1960, p. 58, Fig. 3, 5.8 mm) show that its cardinal process developed from a pair of lobes attached apically to the inner sides of the brachial apparatus, somewhat in the manner of the lobes present in *Gypidula loweryi*. It appears to the writer, therefore, that these latter structures may be termed cardinal process lobes.

FIGURED SPECIMENS: The holotype, USNM 96368, is from locality 58 (Merriam, 1940) at Lone Mountain. The collection horizon is low in the *Eurekaspirifer pinyonensis* Zone. The holotype is largely exfoliated so that the impressions of the outer plates may be seen on the internal mold of the brachial valve. They are long and almost parallel, reaching slightly more than one-third of the distance toward the anterior margin. Because of the exfoliation it is not possible to be certain whether or not low plications were present. The holotype measures 34.0 mm in length, 34.8 mm in width, and 25.4 mm in thickness and has a pedicle valve about twice as convex as the brachial valve. Other figured specimens include USNM 156856-156863.

Order Strophomenida

SUBORDER STROPHOMENOIDEA

DISCUSSION: The suborder is represented by *Leptaenisca* in the *Quadrithyris* Zone and *Spinoplasia* Zone and by *Leptaena* at all fossiliferous levels in the Siegenian and Emsian of Nevada, but *Leptaena* evidently disappeared in Nevada and possibly from all of western North America in the early Middle Devonian. The orthotetaceans are represented in all of the Siegenian and Emsian Zones of Nevada by a stock of lenticular, subplanate schuchertelloids with the most rudimentary dental lamellae, or with none at all. Their impunctate shell requires the eventual proposal of a new genus. *Aesopomum* represents a second impunctate stock which mimics the pseudopunctate Streptorhynchinae of the Carboniferous and Permian.

The stropheodontids are important members of the Siegenian and Emsian brachiopod fauna of Nevada and include a diverse number of forms belonging to the principal families or subfamilies of the group. *Strophonella* is represented by both costate and parvicostellate forms. The costate species (of the *punctulifera* type) must eventually receive a separate generic name. The so-called Devonian *"Brachyprion"* (that is, *"B." mirabilis*) is another form which needs a new generic name. Both of these will be proposed in a paper under preparation by Harper, Boucot, and the writer dealing with the stropheodontids. *Mesodouvillina* is well represented in the *Quadrithyris* Zone by forms resembling the Appalachian species *M. varistriata* and *M. varistriata arata,* but the flat lenticular mesodouvillinid species assigned to *Mclearnites* apparently were invaders derived from an Old World source. *Cymostrophia,* present in the *pinyonensis* Zone, may have had a similar origin, but uncertainty arises over the possible genetic connections of the homeomorphs in the Appalachian Province (*Cymostrophia patersoni* of the Onondaga). Early and simple *Megastrophia* and *Stropheodonta* both occur, and as generically assigned herein (excluding *S. magnacosta*) they differ only in external ornament and slightly in shape. *Megastrophia transitans* and *Stropheodonta filicosta* are found in the same collections and are identical internally. *Megastrophia iddingsi* is a distinctive species that needs close comparison with typical *Megastrophia* from the Appalachian Province. Leptostrophiids are represented in all of the Siegenian and Emsian Zones of Nevada by the genus *Leptostrophia*. The provincial Appalachian genus *Protoleptostrophia* is unknown in Nevada. *Pholidostrophia* is represented by a rather typical species in the *kobehana* Zone, but the other pholidostrophiid *Phragmostrophia* occurs rarely in the *kobehana* Zone and commonly in the *pinyonensis* Zone and is a shell which has no counterpart in the Appalachian Province. *Phragmostrophia* has a geographic distribution from Nevada to the Canadian northwest to Novaya Zemlya of the northern U.S.S.R., a distribution similar to that of the dalmanellid genus *Cortezorthis*. The most closely comparable form is the Australian genus *Nadiastrophia* Talent.

Superfamily STROPHOMENACEA King, 1846
Family STROPHOMENIDAE King, 1846
Subfamily LEPTAENOIDEINAE Williams, 1953
Genus *Leptaenisca* Beecher, 1890
TYPE SPECIES: *Leptaena concava* Hall, 1857, p. 47.
Leptaenisca sp.
(Pl. 17, figs. 15-21)

EXTERIOR: The shells are transversely subquadrate to shield-shaped in outline and concavo-convex in lateral profile. The ventral interarea is linear to trapezoidal, in the latter case approximating a triangle with its apex removed. The delthyrium is triangular and is closed by a plate of shell material of uncertain construction. The cardinal angles may be acute and slightly auriculate, and the maximum shell width is at the hinge line. However, on some specimens the cardinal angles are obtuse and on these shells maximum width is variable, either posterior or anterior to midlength. Anteriorly, the pedicle valve is geniculate and may bear a raised marginal ridge at the line of geniculation.

INTERIOR OF PEDICLE VALVE: There is a more or less well-developed broad and long longitudinal ridge medially. Two strong dental lamellae converge slightly medially to join the lateral edges of the longitudinal ridge. There is a more or less well-developed median septum between the dental lamellae.

INTERIOR OF BRACHIAL VALVE: The cardinal process of the only brachial valve available appears to be trilobate; however, the shell is poorly preserved. Laterally, there are marginal partitions close to the hinge line. The brachial ridges are raised above the floor of the valve and curve laterally from a posteromedial position and then recurve spirally toward midvalve where they are separated by a shallow median groove.

OCCURRENCE: *Leptaenisca* is present in Gedinnian beds of the upper Roberts Mountains Formation in the Roberts Mountains, in the *Quadrithyris* Zone at Coal Canyon, and in the lower McMonnigal Limestone. It is also present in the *Spinoplasia* Zone at Coal Canyon and at Antelope Peak.

FIGURED SPECIMENS: USNM 156875-156879.

Subfamily LEPTAENINAE Hall and Clarke, 1895
Genus *Leptaena* Dalman, 1828
TYPE SPECIES: *Leptaena rugosa* Dalman, 1828, Pl. 1, fig. 1.
Leptaena sp. A
(Pl. 16, figs. 1-4)

EXTERIOR: The shells are transversely shield-shaped in outline with the anterior margin nearly straight and subparallel to the hinge line. The hinge line is long and straight and is the place of maximum width. Cardinal angles of the larger specimens are moderately auriculate. The brachial valve is subplanar and the pedicle valve is very gently convex. It is geniculate dorsally along its anterior and lateral commissures. The brachial valve is also geniculate dorsally and fits within the ventral trail. The ventral interarea is short, low, triangular, flat, and apsacline. It is cleft medially by a broad triangular delthyrium, which opens into an apical foramen. The dorsal interarea is long, flat, and anacline. There is a medially grooved, transverse, crescentic chilidium showing growth lines.

The ornament consists of relatively strong, simple, U-shaped, concentric rugae crossed by numerous radial costellae. Commonly there are about 10 rugae in a distance of a centimeter, measured anterior to the eccentric center about which they are oriented. The valves abruptly become strongly geniculate anteriorly, commonly at approximately 10 to 12 mm. The line of geniculation is approximately parallel to the hinge line and interrupts the subcircular pattern of the rugae. On the largest specimen available, a pedicle valve, the trail and the length of the valve are each equal to 13 mm. The radial costellae are relatively regular and straight. They increase slightly in number anteriorly, but it was not determined whether this was due to bifurcation or to implantation. The costellae continue uninterruptedly on the trail of the valves.

INTERIOR OF PEDICLE VALVE: The teeth are relatively widely divergent anterolaterally and are supported basally by short dental lamellae defining small, conical umbonal cavities. The adductor muscle scars are not strongly impressed, but the diductor impressions are roughly circular or may be slightly transverse. They are defined by arcuate muscle-bounding ridges which are strongest anterolaterally. Generally there is a very slight discontinuity between their posterior ends and the anterior edges of the dental lamellae. In large and thick-shelled specimens the interarea becomes lamellose with growth lines. The hinge teeth become broad, stubby, and widely divergent, and are strongly crenulated. The adductor muscle scars are situated on slightly raised elongate areas in the anterior two-thirds of the muscle impression. Dental lamellae become completely obsolescent, and the diductor impressions are strongly elevated anterolaterally on a platform surrounded by strong muscle-bounding ridges that project flangelike dorsally. Small, thin-shelled specimens are crenulated by the impress of the concentric rugae, but in large specimens secondary shell material forms a smooth interior surface.

INTERIOR OF BRACHIAL VALVE: The cardinal process consists of a pair of lobes conjunct posteriorly, and built up of a series of continuous layers of shell material that are deeply cleft medially, preserving the appearance anteriorly of two lobes. The sockets consist of a pair of widely divergent, narrow, shallow grooves in the shell, bounded on their medial edges by low, straight, subangular socket plates. Anterior to the cardinal process there is a broad low myophragm which tapers sharply in a short distance anteriorly dividing only the posterior adductors. The posterior adductors are roughly elliptical in outline and are defined laterally by broad thickenings of shell material. The anterior adductors consist of a small elliptical pair of medial depressions, bounded by radially ridged and raised areas laterally.

Medially, a myophragm bisects the impression and in some specimens rises to the height of a true median septum (breviseptum) in the anterior portion of the shell. Anteriorly there is a raised marginal ridge along the line of geniculation. The anterior and lateral portions of the shells are pustulose, and the interior is more or less strongly crenulated by the impress of the concentric rugae.

OCCURRENCE: *Leptaena* sp. A is relatively common in the *Quadrithyris* Zone and is present both in the lower McMonnigal Limestone in the Toquima Range and in the Windmill Limestone at Coal Canyon.

COMPARISON: *Leptaena* sp. A is distinguished from *L.* cf. *acuticuspidata* and *L.* sp. C by the greater length at which the latter species become geniculate.

FIGURED SPECIMENS: USNM 156868, 156869.

Leptaena cf. *acuticuspidata* Amsden, 1958
(Pl. 16, figs. 8-13)
Leptaena acuticuspidata Amsden, 1958a, p. 83, Pl. 3, figs. 1-9.

EXTERIOR: Small shells are semicircular in outline. They become increasingly auriculate and more rectangular or shield-shaped in large shells. In lateral profile the valves are plano-convex, although the pedicle valve is only gently arched. The hinge line is long and straight and is the place of maximum width. The ventral interarea is long and low, nearly linear, and is flat and apsacline. It is cleft medially by a broad, triangular, obtuse delthyrium which opens into a semicircular foramen at its apex. There is a relatively well-developed, subtriangular, convex chilidium which covers approximately one-third to one-half of the posterior faces of the cardinal process lobes on medium- and large-sized specimens. The chilidium is deeply grooved medially. The dorsal interarea is long, linear, and anacline. The pedicle valve is strongly geniculate dorsally along its anterior and lateral commissure. Generally, the trail is nearly at 90 degrees to the plane of the commissure anteriorly, but slopes away at an angle of approximately 45 degrees laterally at about midlength. The brachial valve is also geniculate dorsally.

The ornament consists of strong, rounded, concentric rugae separated by deep U-shaped interspaces and crossed by low, narrow, closely set, radial costellae which continue anteriorly onto the trail of larger individuals. There are five to seven rugae on specimens posterior to the line of geniculation which commonly occurs at about 16 to 20 mm anterior to the hinge line measured along the midline of the pedicle valve. The trail is of variable length.

INTERIOR OF PEDICLE VALVE: The hinge teeth are low, widely divergent, and crenulated. They rest directly on the interior surface of the shell and, if supported by dental lamellae, as appears likely on the smallest and most thin-shelled specimen, they are almost in every case made obsolescent by the addition of secondary shell material. The diductor muscle scars are subrhombic in outline and in the smallest shells are bounded anterolaterally by low muscle-bounding ridges. The muscle-bounding ridges become increasingly more strongly developed and elevated on larger and thicker shelled specimens, forming flangelike, dorsally projecting plates anterolaterally on the largest ones. The thickest shelled individuals bear a relatively well-developed myophragm dividing the muscle impressions. Only the smaller and more thin-shelled specimens are corrugated by the impress of the concentric rugae.

INTERIOR OF BRACHIAL VALVE: The cardinal process consists of two lobes or plates conjunct posteriorly, narrowly divergent, and expanding in width anteriorly. The diductor attachment surfaces are flat and face posteroventrally. Laterally, the sockets are shallow, broad, and relatively strongly divergent. They are bounded posteriorly by the inner edges of the interarea and medially by low inner socket ridges. The sockets blend with the interior of the shell at their broad anterolateral ends. There is commonly a broad myophragm originating at the anterior edge of the base of the cardinal process lobes. It tapers in a short distance anteriorly to fade at about the midlength of the posterior adductor impressions. The posterior adductor scars are bounded posteriorly by low, well-defined, rounded muscle-bounding ridges. Muscle-bounding ridges are less well defined laterally and are only slightly elevated. The posterior adductors are moderately elongate-oval in outline. The anterior adductors consist of a pair of small, centrally located, elongate suboval pits in a medial thickening of shell material and are divided medially by a narrow myophragm. The posterolateral portions of the shell interior are pustulose, and the anterior portion of some specimens has strongly impressed radial vascular tracks. At the anterior margins of the valve there is a well-developed peripheral ridge along the line of geniculation.

OCCURRENCE: In central Nevada, *Leptaena* cf. *acuticuspidata* is found in the Rabbit Hill Limestone at Rabbit Hill where it is a relatively common species and in the *Trematospira* Zone in the Roberts Mountains and in the Sulphur Spring Range. The species is also present in the upper McMonnigal Limestone in the Toquima Range and in the Rabbit Hill at Coal Canyon.

DISCUSSION: Specimens from the *Costispirifer* Subzone at Willow Creek have very ponderous, nearly circular ventral muscle scars reminiscent of those of *Leptaena ventricosa* of the Oriskany Formation of New York.

FIGURED SPECIMENS: USNM 156871-156874.

Leptaena sp. C
(Pl. 16, figs. 5-7)

EXTERIOR: The shells attain relatively large size. Small specimens are subsemicircular, and large ones are shield-shaped with a gently curved to nearly straight anterior commissure that is parallel to the hinge line. Small shells are decidedly auriculate in some specimens and tend to have cardinal angles that are approximately right angles or even slightly obtuse when fully grown. In other specimens the small shells may be semicircular without auriculation, but the auriculation becomes marked in larger growth stages so that there is no consistent pattern of auriculation, or the lack of it. The hinge line is long and straight, and on most specimens is the place of maximum width; however, in several large specimens the cardinal angles are slightly obtuse, and the maximum width is between the hinge line and midlength. The valves are plano-convex in lateral profile, but there is some degree of variability in the degree of convexity of the pedicle valve from nearly flat to moderately strongly convex in the posterior two-thirds of the shells. The ventral interarea is long, linear, flat, and apsacline. The dorsal interarea is about half the height of that of the pedicle valve, but it is also long, linear, flat, and anacline.

The surface bears moderately strongly developed concentric rugae separated by relatively broad, U-shaped interspaces. The rugae and interspaces are crossed by numerous, closely set, low, rounded radial costellae which continue anteriorly onto the trail of the valve. There are 8 to 12 concentric rugae developed posterior to the line of geniculation, with 8 to 10 more common. The place of geniculation is somewhat variable between 19 and 25 mm anterior to the hinge line, measured along the midline of the pedicle valve. The trail may be deflected as much as 90 degrees from the plane of the commissure, but the angle decreases away from the midline.

INTERIOR OF PEDICLE VALVE: The muscle field is subcircular and surrounded by a low muscle-bounding ridge.

INTERIOR OF BRACHIAL VALVE: The sockets are widely divergent slots, impressed into the shell posteriorly, without muscle-bounding ridges. The cardinal process lobes are posteriorly conjunct, diverging broadly anteriorly, and separated by a triangular pit. A well-developed, medially cleft chilidium is present, covering about half the posteriorly facing myophore faces of the cardinal process lobes. The diductor muscle field consists of a pair of elongate, subtriangular areas, moderately impressed into the shell interior and separated medially by a pair of ridges and anteromedially by an elongate, elliptical central pit. The latter is bisected medially by a short breviseptum which extends just anterior to midlength. Most of the visceral disk is not affected by impression of the external rugae.

OCCURRENCE: *Leptaena* sp. C is fairly abundant in the *Acrospirifer kobehana* Zone on the north side of the Roberts Mountains and in the Sulphur Spring Range. It is also present in the *Eurekaspirifer pinyonensis* Zone in the Simpson Park Range. *Leptaena* occurs in numerous collections from the *Eurekaspirifer pinyonensis* Zone at other localities, but specimens generally are few in number, poorly preserved, and not clearly assignable to the species.

COMPARISON: *Leptaena* sp. C is distinguished from *Leptaena* cf. *acuticuspidata* of the underlying *Spinoplasia* and *Trematospira* Zones by more numerous concentric rugae and by the greater length of the shells. In addition, the rugae are slightly more angular and are not so strongly raised as in the earlier species.

FIGURED SPECIMEN: USNM 156870.

Superfamily DAVIDSONIACEA King, 1850
Family SCHUCHERTELLIDAE Williams, 1953
Subfamily SCHUCHERTELLINAE Williams, 1953
Genus *Schuchertella* Girty, 1904

TYPE SPECIES: *Streptorhynchus lens* White, 1862, p. 28.

DISCUSSION: Cooper (1962, p. 155) pointed out that some schuchertellids should be rejected from the family because they do not have pseudopunctae and in fact that all pre-Famennian species studied by him lack pseudopunctae (oral commun., 1964). Pseudopunctae were not observed in any of the Nevada Lower Devonian schuchertelloids, and thus they differ from the pseudopunctate *S. lens* (White, 1862, p. 28), the type species of *Schuchertella*. Contrary to the statement of Williams (1965) that *Schuchertella* is impunctate, Cooper reaffirmed the fact that *S. lens* is pseudopunctate (written commun., 1966).

It appears that during the Silurian to Middle Devonian the Davidsoniaceans were rapidly evolving. The Ordovician to Early Silurian genus *Fardenia* probably gave rise to the widespread *Coolinia* (=*Chilidiopsis fide* Boucot) which is the common Davidsoniacean of the upper Llandovery, Wenlock, and lower Ludlow. It overlaps in the Ludlow with the first appearance of *Morinorhynchus* (*M. attenuata* Amsden, 1951, Pl. 17). The type species is from the Kopanina Stage of Ludlow age of Bohemia. In central Nevada, *Morinorhynchus* is the common Davidsoniacean in the Ludlow and Pridoli, but no occurrences have been found in the Gedinnian.

With the advent of Gedinnian time, the group of genera representing the Davidsoniacea appears to have undergone a pronounced change. Some Silurian members disappeared and were replaced by the streptorhynchoid genus *Aesopomum*, by "*Schuchertella*" (as that term is used in the present paper), and by *Areostrophia*, and *Iridistrophia*. The latter three probably were derived from *Morinorhynchus*.

Iridistrophia is essentially a large, resupinate *Morinorhynchus*. It still has prominent dental lamellae. "*Schuchertella*" has obsolescent dental lamellae or completely lacks them in some cases. It has relatively prominent, closely spaced radial costae. *Areostrophia* probably is an early offshoot from the main *Morinorhynchus*-"*Schuchertella*" lineage. It completely lacks dental lamellae, commonly has a tumid and twisted pedicle valve and a prominently lobed cardinal process.

Aesopomum appears to be a completely separate sideline attaining full streptorhynchoid morphology, although lacking pseudopunctae.

"*Schuchertella*" sp. A
(Pl. 18, figs. 1-7)

EXTERIOR: Shells are transverse oval in outline and planate to slightly biconvex in lateral profile. Cardinal angles are slightly acute on small shells to obtuse and slightly rounded on larger ones. Maximum shell width is near midlength. The valves may be flat, or on some the anterior is geniculate. The interarea is flat, low, and apsacline and is crossed by strongly developed growth lines. The delthyrium is covered by a complete pseudodeltidium.

The ornament consists of radial costellae crossed by numerous, subdued growth lines. More strongly developed growth lines are found at widely spaced irregular intervals, commonly toward the anterior. Costellae increase in number anteriorly by intercalation along the sides of interspaces rather than in the center. The costellae are high and rounded (wiry). Parvicostellae are faintly developed on some specimens. There are 12, 13, or 14 costellae in a space of 5 mm, 10 mm anterior to the beak.

INTERIOR OF PEDICLE VALVE: The tracks of the teeth are seen as ridges along the inner edge of the delthyrium, but dental lamellae are not developed. The costellae are impressed on the anterior of the shell peripherally.

INTERIOR OF BRACHIAL VALVE: Socket plates are stout, widely set apart, and widely divergent. Medially there is a connecting plate which curves posteriorly and is medially cleft on its posterior face, forming two separate lobes. The myophores are built up on strongly curved, U-shaped plates which are concave medially. No chilidium was observed. The adductor scars are broad, slightly impressed, and appear to be faintly flabellate anteriorly. They are divided, in some shells, by a poorly developed, posteriorly situated myophragm. The interior is crenulated by the impress of the costellae.

OCCURRENCE: "*Schuchertella*" sp. A is present in the *Quadrithyris* Zone of the Windmill Limestone at Coal Canyon and in the lower Mc-Monnigal Limestone of the Toquima Range.

COMPARISON: "*Schuchertella*" sp. A has finer costellae than "*S.*" sp. B or "*S.*" sp. C. In addition, it is smaller than either of the latter species and has a thinner shell than "*Schuchertella*" sp. B. "*Schuchertella*" sp. C and "*Schuchertella*" *nevadaensis* are both larger forms that bear dental lamellae.

FIGURED SPECIMENS: USNM 156886-156889.

"*Schuchertella*" sp. B
(Pl. 18, figs. 15-20)

EXTERIOR: The shells are planate to biconvex and are semicircular to transversely suboval in outline. Small shells have decidedly acute cardinal angles while large ones are slightly obtuse. Brachial valves tend to become incurved anteriorly and nearly become geniculate on large specimens. The ventral interarea is high, flat, and apsacline to nearly cataclinic on some specimens. The dorsal interarea is short, linear, and anacline. There is a convex pseudodeltidium which is rounded to slightly angular, and which is concave dorsally on its dorsal edge.

The ornament consists of strong radial costellae that number 9 to 11 in a space of 5 mm, 10 mm anterior to the beak. The costellae increase in number anteriorly by intercalation.

INTERIOR OF PEDICLE VALVE: The teeth are strong and leave well-defined tracks on the edges of the delthyrium. Dental lamellae are absent or are present on some specimens as low, widely divergent ridges at the apex of the valve. On medium- and large-sized shells, the impress of the costellae is only developed along the periphery of the shells.

INTERIOR OF BRACHIAL VALVE: The sockets are widely divergent and are bounded by socket plates of oval cross section along their inner edges. Medially they are joined by a transverse plate which curves posteriorly. On larger specimens the cardinal process is developed into two separate lobes which are covered, in part, apically by a narrow chilidium. In small shells, no complete chilidium was observed; instead it is composed of two discrete plates. On the anterior face of the median plate, there is a small median ridge. Large specimens are very thick-shelled and develop strong ridges which bound the adductor muscle field posterolaterally. Anteriorly, the adductor field is not well defined, but is faintly flabellate. A myophragm is only uncommonly developed in this species.

OCCURRENCE: This species is present in the *Spinoplasia* Zone at Rabbit Hill, at Coal Canyon, and throughout the *Trematospira* Zone in the Sulphur Spring Range. A very thick-shelled, coarse-ribbed form, possibly a distinct subspecies, is present in the *Costispirifer* Subzone at Willow Creek.

COMPARISON: *"Schuchertella"* sp. B has a thicker shell than either sp. A or sp. C. It has coarser costellae than does *"Schuchertella" nevadaensis*.

FIGURED SPECIMENS: USNM 156890-156892.

"Schuchertella" sp. C
(Pl. 18, figs. 8-14)

EXTERIOR: Shells of small specimens are subcircular and transversely semicircular in large ones. They are plano-convex to biconvex in lateral profile. The umbo is slightly inflated. Small shells have obtuse cardinal angles in most specimens; larger shells have cardinal angles of about 90 degrees. The ventral interarea is low, triangular, flat, and apsacline to nearly cataclne. The dorsal interarea is hypercline and linear. It forms a small ridge along the inner edge of the hinge line. The beak is twisted slightly at the umbo on the specimens studied.

The ornament consists of well-defined, rounded radial costellae which increase in number anteriorly by intercalation. There are approximately 8 costellae in a space of 5 mm, 10 mm anterior to the beak on the specimens examined.

INTERIOR OF PEDICLE VALVE: The pseudodeltidium is large, broad, and convex outward. On the inner side of the pseudodeltidium of large shells, several layers of plates are built up, making the structure thick and lamellar. The teeth are thin and triangular in cross section and are supported near the apex of the valve by short, broad, widely divergent dental lamellae. The diductor muscle field is broad, subcircular, and flabellate anterolaterally. It is divided posteriorly by a thin myophragm.

INTERIOR OF BRACHIAL VALVE: The socket plates are broadly divergent and their inner edges are strongly curved posteriorly. Medially, between them, there is a solid plate of shell material which is directed strongly posteriorly as two separate lobes of trapezoidal cross section. Medially, there is a small ridge on the medial plate anterior to the point of bifurcation. A strong, curved, bandlike chilidium is present covering the outermost edges of the cardinal process lobes. On the interior surface of the chilidium, there is a medial swelling which is connected to the dorsal side of the cardinal plate at the midline. The adductor scars are broadly flabellate and are divided by a more or less well-defined myophragm.

OCCURRENCE: *"Schuchertella"* sp. C is present in the *Acrospirifer kobehana* Zone in the Sulphur Spring Range and in the Roberts Mountains.

COMPARISON: This species has coarser costellae than any of the others described here, except a form in the *Costispirifer* Subzone at Willow Creek, and a thinner shell than any other described in this paper, except *"Schuchertella"* sp. A.

FIGURED SPECIMENS: USNM 156893-156895.

"Schuchertella" nevadaensis Merriam, 1940
(Pl. 19. figs. 1-13)

Schuchertella nevadaensis Merriam, 1940, p. 80, Pl. 6, fig. 5.

EXTERIOR: The shells are broadly transverse-oval in small specimens and nearly subcircular in large ones. They are biconvex in lateral profile with the pedicle valve having greatest convexity in its posterior half and the brachial valve reaching greatest convexity slightly posterior to midlength. The hinge line is long and straight and equal to about two-thirds the maximum width of the shell. The ventral interarea is flat, triangular, and apsacline. The delthyrium is closed by a complete, convex pseudodeltidium. Maximum width of the valves is near midlength. The anterior commissure is rectimarginate.

The ornament consists of fine radial costellae which increase in number anteriorly by intercalation and which are crossed at irregular intervals by numerous, slightly elevated growth lines. On the few specimens suitably preserved, there are approximately 15 costellae in a space of 5 mm, 10 mm anterior to the beak. The holotype has 10 costellae in a distance of 5 mm along the anterior commissure. There is a poorly developed parvicostellation due to the fact that intercalated costellae are considerably smaller than adjacent costellae for noticeable distances.

INTERIOR OF PEDICLE VALVE: Short, broadly spaced, and very widely divergent dental lamellae define the posterolateral margins of the muscle field. The diductor scar is broad, transversely oval, and flabellate. The adductor scars are cordate, posteromedially situated, and enclosed anteriorly by the diductor impressions. There is a narrow myophragm separating the adductor scars. The costellae are variably impressed on the shell interior. Some specimens have very thick shells and are crenulated only peripherally by the impress of the costellae. In large shells the grooves corresponding to the costellae externally are very narrow and separate rectangular ridges which may have narrow medial depressions.

INTERIOR OF BRACHIAL VALVE: The socket plates are strong and widely divergent, defining broad sockets laterally between them and the hinge line. Medially there is a plate connecting the posterior edges of the socket plates and projecting backwards as two blunt lobes. Medially on the anterior face of the medial plate there is a short strong ridge. On large specimens, sockets are built up on thickened shell material anterolaterally. The adductor muscle impressions are well defined laterally by muscle-bounding ridges that originate at the socket plates. The adductor impressions are separated posteromedially by a well-defined myophragm. Anteriorly, they blend with the interior of the shell.

OCCURRENCE: "Schuchertella" nevadaensis is a rare species occurring in the *Eurekaspirifer pinyonensis* Zone in the Roberts Mountains and at Lone Mountain (Merriam, 1940, p. 54).

SHELL STRUCTURE: The shell is impunctate. Pseudopunctae are not developed.

FIGURED SPECIMENS: The holotype, USNM 96366, is from locality 105 (Merriam, 1940) at Lone Mountain. It has the following dimensions: length 35.6 mm, width 40.9 mm, and thickness 13.8 mm. Other figured specimens include USNM 156896-156901.

Genus *Aesopomum* Havlíček, 1965

TYPE SPECIES: *Strophomena aesopea* Barrande, 1879.

DIAGNOSIS: The genus is characterized by its irregular biconvex shape, its ornament of fine irregular striae, moderately high triangular interarea, and complete pseudodeltidium in the pedicle valve; and by its posteriorly directed bilobed cardinal process in the brachial valve.

DISCUSSION: *Aesopomum* has the appearance of a Permo-Carboniferous orthotetacean; however, at present the stratigraphic gap appears to be unbridged by any intermediate forms. Moreover, *Aesopomum* differs from Carboniferous genera (such as *Streptorhynchus*) which are pseudopunctate. The Bohemian Lower Devonian species *A. aesopeum* (Barrande, 1879, Pl. 92, case 4, Pl. 133, case 2) compares strongly with *A. varistriatus*. The author has not yet had the opportunity to study *aesopeum* at firsthand.

Aesopomum varistriatus Johnson n. sp.
(Pl. 17, figs. 1-14)

DIAGNOSIS: A small, elongate, deeply subconical species.

EXTERIOR: The shells are strongly biconvex and of irregular shape, neither decidedly transverse nor elongate. The pedicle valve conforms roughly to the shape of a low cone truncated from apex to base. The small relative width of the interarea is at variance with a pyramidal conformation. The beak of the only available pedicle valve is twisted in a manner not uncommon in orthotetaceans. The ventral interarea is apsacline and only slightly curved. Growth lines are developed parallel to the hinge line. The delthyrium is closed by a narrow pseudodeltidium which is convex outwards. The ventral interarea is about half the maximum width of the valves. There is no interarea on the brachial valve. Maximum width is well anterior to the hinge line.

The ornament consists of fine radial costellae which follow a somewhat irregular course. The costellae show the peculiar tendency to diverge and converge as groups, probably due to shell damage, requiring intercalation of numerous costellae at a narrow radius or conversely, the elimination of numerous costellae by rejoining. Divergence of groups of costellae commonly occurs in longitudinal depressions in the shell with new costellae being added at the base of the depression. Divergence also occurs around a hump or convex area, followed by convergence anterior to the hump with elimination of the innermost costellae of the two convergent groups. Rugose concentric corrugations may occur at irregular intervals along the shell.

INTERIOR OF PEDICLE VALVE: The tracks of the hinge teeth form ridges on the inner edges of the delthyrium. Dental lamellae are absent. Muscle impressions are not evident. The costellae are not impressed on the inner surface of the valve.

INTERIOR OF BRACHIAL VALVE: The cardinalia consist of a pair of closely set socket ridges that curve strongly posteroventrally. At the apex of the shell the socket ridges are united with a cardinal plate (or plates) of oval-transverse section, cleft by longitudinal grooves medially on both dorsal and ventral faces. At the distal end the grooves may effect complete bilobation. The *Quadrithyris* Zone specimens are medially grooved on the anterior face of the cardinal process and thus differ somewhat from the older Roberts Mountains form. There may be a short myophragm in the posterior part of the valve. The adductor impressions, where discernible, appear to be small subcircular pits that are slightly flabellate anteriorly.

SHELL STRUCTURE: Although the specimens are silicified, preservation of some of them is adequate to show that the shells are impunctate and did not develop pseudopunctae.

OCCURRENCE: Specimens of *Aesopomum* occur in the upper Roberts Mountains Formation (of Gedinnian age) in the Roberts Mountains, in the Windmill Limestone (*Quadrithyris* Zone) at Coal Canyon, and in the Wenban? Limestone in the Cortez Range and in unnamed beds at Lynn Window. The specimens from the Lynn Window occur with *Skenidioides*. Beyond Nevada, *Aesopomum* is known in North America from beds of Early Devonian age in the Yukon. This is the "triplesiid gen. indet." reported by Johnson and Boucot in Jackson and Lenz (1963, p. 752). The determination of the Yukon specimens was originally based on a single, free, silicified brachial valve. Lenz (1966, 1967b) reported *Aesopomum* at several Devonian horizons at Royal Creek, Yukon Territory.

Specimens present in the Boucot collection (loc. 56SC1) from the Upper Silurian or Lower Devonian Öved-Ramsåsa Group in small canal at Bjärspölagård parish of O. Kärstorp, Scandia, Sweden, may belong to the genus. A single small, exfoliated, pedicle valve exterior was seen along with two brachial valves, both of which are finely ribbed and have a cardinalia like *A. varistriatus*.

COMPARISON: *Aesopomum aesopeum*, the only other named species, is larger, less strongly biconvex, and more circular in outline.

FIGURED SPECIMENS: USNM 156880-156885. The pedicle valve, USNM 156885, illustrated in figures 11-14 on Plate 17 is the holotype. The paratype, USNM 156884, a brachial valve, was etched from the same rock sample from Gedinnian beds of the upper Roberts Mountains Formation.

Superfamily STROPHEODONTACEA Caster, 1939
[*nom. transl.* Sokolskaya, 1960, *ex* Stropheodontidae Caster, 1939]
Family STROPHONELLIDAE Caster, 1939
Genus *Strophonella* Hall, 1879

TYPE SPECIES: *Strophomena semifasciata* Hall, 1863b, p. 210.

Strophonella cf. *bohemica* (Barrande, 1848)
(Pl. 20, figs. 7-9)

Leptaena bohemica Barrande, 1848, p. 243, Pl. 23, fig. 1.
Strophomena bohemica Barrande, 1879, Pl. 39, fig. 1.
Strophonella bohemica Havlíček, 1967, p. 183, Pl. 39, figs. 7, 8, 12, 13.

DISCUSSION: The Bohemian specimens illustrated by Barrande (1879) and Havlíček (1967) and those present in Boucot's collection from the upper Koneprus show almost an identical external ornament to that on the form from the *Quadrithyris* Zone. These almost completely lack fine radial costellae between the prominent primary costellae.

EXTERIOR: The shells are large with respect to other brachiopods in the *Quadrithyris* Zone, but only of moderate size for the genus. They are flatly resupinate in lateral profile with no specimens showing any great degree of convexity. The hinge line is long and straight. The ventral umbo is very low and is mildly subcarinate. Pedicle valves at hand appear to be convex to a length of approximately 10 to 12 mm, then flatten out perceptibly.

The ornament consists of numerous, very fine, low costellae arranged in a parvicostellate pattern anteriorly. Over most of the valves the very fine costellae are lacking.

INTERIOR OF PEDICLE VALVE: The ventral process is low, broad, and triangular. It is most prominent posteromedially and bears a shallow longitudinal groove along the anterior portion of its length. It is continuous anteriorly with a broad low myophragm which tapers and becomes narrow and fades in a short distance anteriorly. The adductor scars are moderately elongate-oval and are slightly impressed posteriorly. Laterally, the diductor impressions flare broadly and are moderately impressed. They are flabellate in their anterior portions and are bounded anterolaterally by scalloped margins.

INTERIOR OF BRACHIAL VALVE: The cardinal process lobes are each pyriform in cross section with their broad ends facing anterolaterally. Their outer edges diverge anterolaterally. There is a minute, longitudinal median ridge between the lobes. The socket plates are situated posterolateral to the cardinal process lobes and diverge anterolaterally, nearly parallel to the outer edges of the lobes. The anterior edge of the cardinal process lobes is a broad, very gently elevated platform that abruptly narrows anteriorly, forming a low, rounded myophragm. The myophragm divides a pair of lightly impressed adductor scars that blend with the interior of the shell anterolaterally. The shell is lightly crenulated anteriorly by the impress of the parvicostellae.

SHELL STRUCTURE: The shell substance is pseudopunctate.

OCCURRENCE: This parvicostellate species of *Strophonella* is present, but not common, in the *Quadrithyris* Zone at Coal Canyon and in the lower McMonnigal Limestone in the Toquima Range.

COMPARISON: *Strophonella* cf. *bohemica* is the only parvicostellate species known from the Great Basin Lower Devonian and as such, is easily distinguished from the costate *Strophonella* cf. *punctulifera* characteristic of the younger zones.

FIGURED SPECIMENS: USNM 156902-156904.

Strophonella cf. *punctulifera* (Conrad, 1838)
(Pl. 20, figs. 1-6; Pl. 21, figs. 13, 14)

Strophomena punctulifera Conrad, 1838, p. 117.
Strophodonta punctulifera Hall, 1859, p. 188, Pl. 21, fig. 4; Pl. 23, figs. 4, 5, 7e.
Strophonella punctulifera Hall and Clarke, 1892, p. 291, Pl. 12, figs. 10-12.
Strophonella punctulifera Schuchert and others, 1913, p. 323, Pl. 59, figs. 8-10.
Strophonella punctulifera Schuchert and others, 1913, p. 323, Pl. 59, figs. 8-10.
Strophonella punctulifera Cooper, 1944, p. 339, Pl. 130, figs. 19, 20.

EXTERIOR: The shells are relatively large and subsemicircular to transversely shield-shaped in outline. The valves are resupinate in lateral profile. The ventral umbo is very low and only slightly arched. The valves are nearly biplanate in their posterior portions until attaining a length of an inch to approximately an inch and a half in length, where the curvature reverses. None of the specimens examined is strongly incurved anteriorly to the point of geniculation, but the peripheral portions of the shell commonly slope off at approximately 45 degrees to the plane of the commissure, as established in the posterior half of the valves. The ventral interarea is long, flat, and apsacline. The dorsal interarea is anacline. The cardinal margins are rounded and are approximatley right angles. Some specimens bear a broad, low sulcus in the anterior convex portions of the brachial valve.

The ornament consists of subangular radial costae which become relatively strong anteriorly, particularly on portions of the shell anterior to the region of reversal of curvature. The interspaces are relatively shallow and rounded in cross section and are of approximately the same width as adjoining costae. The costae increase in number anteriorly by bifurcation and implantation. Generally on the brachial valves, new costae arise by implantation along the flank of one of the two adjoining costae. On the pedicle valve, however, some cases of more or less equal splitting occur in which one costa gives rise to two of more or less equal height and width.

INTERIOR OF PEDICLE VALVE: The ventral process is a relatively small, moundlike protuberance in the apex of the valve. The adductor muscle scars are very lightly impressed and appear to be bounded laterally by somewhat anterolateral diverging impressions. Anteriorly, the adductor scars appear to be very faintly flabellate. The diductor muscle impressions are broad and transverse suboval. They are poorly impressed posterolaterally, but laterally and anteriorly their edges became raised on thin, scalloped plates which are re-entrant and join the floor of the valve medially. The interior of the valve is crenulated by the external costae.

INTERIOR OF BRACHIAL VALVE: The cardinal process lobes are disjunct and relatively widely separated basally. They originate along the posterior edge of a T-shaped platform. The cardinal process lobes project posteroventrally, but their attachment surfaces face posteriorly. Each lobe is trapezoidal in cross section. There commonly is a low transverse ridge of shell material which connects the two lobes anteriorly at their bases. Socket plates are present at the posterolateral edges of the transverse platform of shell material serving as base for the cardinal process lobes. They are essentially posterior to the bases of the cardinal process lobes and are short, thin, and widely set apart. They diverge from the hinge line at moderate angles. The medial portion of the posterior platform abruptly narrows anteriorly, forming a myophragm which fades anteriorly into the interior surface of the shell. The adductor muscle impressions are relatively well marked posteromedially. They are roughly oval in plan and diverge slightly anterolaterally. No sharp division of anterior and posterior adductors has been noted, and anteriorly the adductor impressions blend imperceptibly with the interior of the valve. The posterolateral portions of the valve are pustulose and the interior surface is only faintly crenulated by the impress of the external costae.

SHELL STRUCTURE: The shell substance is pseudopunctate. Well-preserved shell material, particularly in the posterior portions of some shells, has a nacreous luster. The structure of platelets and the absence of fibrous calcite is easily distinguished by inspection of exfoliated shells under the binocular microscope, where the platelets have the appearance of mica flakes. This is seen clearly in places where the shell is exfoliated, and it is clear that the shell material consists of lamellar flakes of calcite that lie parallel to a tangential plane.

It is interesting to note here the occurrence of the nonfibrous texture which gives rise to a nacreous luster and which has previously been recorded only among the pholidostrophiids. Williams (1953b, p. 5) regarded this replacement of the fibrous layer by one of "calcitic platelets" as a unique texture, but investigations of stropheodontid shell structure to determine the extent of development of pseudonacreous texture are still to be inaugurated.

OCCURRENCE: This costate species of *Strophonella* is present in the *Acrospirifer kobehana* Zone in the Roberts Mountains and in the *A. kobehana* and *Trematospira* Zones in the Sulphur Spring Range. It is present in the *E. pinyonensis* Zone at Lone Mountain and at Cortez. A single specimen is known from the *Spinoplasia* Zone at Rabbit Hill.

FIGURED SPECIMENS: USNM 156905-156910.

Family STROPHEODONTIDAE Caster, 1939
Genus *Brachyprion* Shaler, 1865
TYPE SPECIES: *Strophomena leda* Billings, 1860, p. 55, figs. 2, 3.
"*Brachyprion*" *mirabilis* Johnson, n. sp.
(Pl. 22, figs. 1-12)

cf. *Shaleria* sp. Boucot, *in* Boucot and others, 1960, p. 12, Pl. 2, figs. 23-26, Pl. 3, figs. 1-8.

DISCUSSION: Harper (*in* Boucot and others, 1966, p. 41) has recently attempted to clarify the relations of this genus and finds it to be a probable senior subjective synonym of *Protomegastrophia* Caster, 1939. However, neither the specimens described here nor those illustrated as *Brachyprion* in the brachiopod Treatise (Williams, 1965, Fig. 255, 1) are congeneric with *Protomegastrophia profunda* as is obvious from a comparison of the cardinalia of the forms in question. The species described below needs a new generic name, but the material at hand is not sufficient on which to base such a definition. No diagnosis is given here because no other named species truly congeneric with "*B.*" *mirabilis* is known to the writer.

EXTERIOR: The shells are plano-convex to gently concavo-convex in lateral profile. The pedicle valve is low and only moderately arched. Most shells are incomplete at their margins due to poor preservation, but early growth stages were semicircular and moderately auriculate at the cardinal margins. The hinge line is long and straight and is the place of maximum width. The ventral interarea is long, flat, and relatively high. Its inclination is apsacline. The dorsal edges of the interarea are strongly denticulate along most of their length. The dorsal interarea is relatively strong and is moderately anacline. Denticulations are present internal to the ventral edge of the interarea medially, as well as on the interarea proper.

The ornament consists of numerous fine costellae which develop a parvicostellate pattern anterior to the umbo. The radial ornament is superposed upon a concentric development of interrupted rugae. On some specimens, particularly brachial valves, the primary costellae are situated on slightly raised crests and separated by broad U-shaped furrows. In these furrows the concentric rugae become decidedly discontinuous and develop into arcuate radial groups which are convex posteriorly.

INTERIOR OF PEDICLE VALVE: The ventral process is relatively ponderous and subpyramidal. It extends posteriorly as a rectangular ridge and attaches to the inner side of the plate or plates which largely close the triangular delthyrium. On its anterior edge, the ventral process has a deep, long, medial groove which is bounded laterally by a pair of rounded ridges that merge anteriorly, forming a simple myophragm dividing the muscle field. The diductor muscle scars are relatively well impressed laterally and lack bounding ridges. Their lateral edges are strongly divergent and the diductor field merges with the interior of the valve anteriorly. The adductor scars are situated posteriorly within the diductor impressions. They are strongly impressed at their posterior edges where they are bounded by short, stout muscle-bounding ridges. They are elongate, but broaden perceptibly anteriorly where they merge with the interior of the shell. The surface of the interior is not crenulated by the impress of the external costellae.

INTERIOR OF BRACHIAL VALVE: The cardinal process lobes are disjunct and are directed ventrally and moderately posteriorly. There is some degree of variability in the distance the bases of the cardinal process lobes are set apart, indicating only a small divergence from the conjunct condition. The cardinal process lobes expand laterally on their distal ends to form bilobate posteroventral myophores. The chilidium consists of a relatively ponderous elongate ridge with a medial groove. The socket plates are long, low, V-shaped ridges that diverge widely and are set close to the hinge line. The adductor impressions are divided into two pairs. The posterior scars are broadly suboval, widely divergent anteriorly, and are bounded posteriorly by broad, low, muscle-bounding ridges. Medially, they are separated by a broad, low myophragm which tapers in a relatively short distance anteriorly and disappears. The anterior adductors are closely set medially within the posterior adductors and are elongate. Both pairs of adductor muscle impressions blend with the interior of the shell anteriorly.

OCCURRENCE: *"Brachyprion" mirabilis* is present in the *Quadrithyris* Zone at Coal Canyon.

DISCUSSION: The occurrence of a species of *"Brachyprion"* in the Windmill Limestone, along with typical *Mesodouvillina* of the *subinterstrialis* type, suggests a reassessment of specimens illustrated as *subinterstrialis* by Kozlowski (1929). There is, in figs. 7 and 7a of Plate 4 of his monograph, a specimen with decidedly more broadly spaced primary costellae, interrupted rugose concentric ornament, and auriculate cardinal margins. Viewed in light of the Nevada occurrences, this specimen, whose ornament was discussed by Kozlowski (1929, p. 96, 98), may be a *"Brachyprion"* as the term is used in this paper, but it probably does not belong to the subspecies *seretensis* which is the type species *Mesodouvillina*.

FIGURED SPECIMENS: USNM 156920 A and B, 156921-156926; the holotype is USNM 156920 A and B.

Genus *Mesodouvillina* Williams, 1950
TYPE SPECIES: *Stropheodonta (Brachyprion) subinterstrialis* var. *seretensis* Kozlowski, 1929, p. 96, 97, Figs. 28, 29; Pl. 4, figs 1-7.

Mesodouvillina cf. *varistriata* (Conrad, 1842)
(Pl. 23, figs. 1-15)

Strophomena varistriata Conrad, 1842, p. 255, Pl. 14, fig. 6.
Strophodonta varistriata Hall, 1859, p. 180, Pl. 8, figs. 2-16; Pl. 16, figs. 1-8.
Stropheodonta (Brachyprion) varistriata Hall and Clarke, 1892, Pl. 13, figs. 6-16.

EXTERIOR: The shells are generally subsemicircular in outline, but vary from moderately trigonal to transversely shield-shaped. They are gently to moderately concavo-convex in lateral profile, but relatively large shells may be geniculate anteriorly. The degree of convexity of the pedicle valve is variable; some small shells are only gently convex, although others of comparable size are relatively strongly inflated. The hinge line is long and straight and is the place of maximum width. The ventral interarea is orthocline or very gently apsacline. Approximately three-fifths to two-thirds of the central portion of the interarea is denticulate along the anterior third of its height. The delthyrium appears to be closed apically by a plate or plates level with the interarea; however, on the best preserved specimen there is a small, broad, V-shaped notch at the anterior edge of the palintrope.

The exterior is strongly parvicostellate. The primary costellae may be broadly spaced with numerous finer costellae grouped on the flat intervening sectors, or this simple pattern may be modified by variation in the strength of the secondary costellae.

INTERIOR OF PEDICLE VALVE: The ventral process is small and low, consisting of an inverted V-shaped ridge with its apex touching the inner side of the plate or plates closing the apex of the delthyrium. The muscle field is flanked posterolaterally by strong, rounded muscle-bounding ridges which project anterolaterally and curve very slightly medially. In some specimens there is a low, rounded ridge adjacent to the inner side of the palintrope and parallel to it, leaving a transverse groove along the posterior margin of the shell. The adductor impressions are relatively large and cordate, occupying the posteromedial portion of the muscle field. The diductor impressions are broadly oval to cordate, diverging anterolaterally, and enclosing the adductor scar medially. The anterior portion of the diductor impressions is bilobed due to strong incurving medially. Most shells are faintly grooved by the impress of the primary costellae, particularly around the anterolateral margins.

INTERIOR OF BRACHIAL VALVE: The cardinal process lobes are disjunct and project posteroventrally. The lobes are broad laterally, each developing a double posterior myophore. Muscle-bounding ridges are relatively well developed posterior to the adductor muscle field. The socket plates are situated relatively close medially on the posterior flanks of the muscle-bounding ridges. They diverge strongly from the hinge line and are short and stout. Anterior to the base of the cardinal process lobes there is a well-defined myophragm which tapers anteriorly. The posterior adductor scars are moderately impressed posteriorly, and are bounded posteriorly, and in some specimens laterally, by muscle-bounding ridges. The anterior adductors adjoin elongate, narrow, slightly divergent, rough-surfaced brace plates.

OCCURRENCE: *Mesodouvillina* cf. *varistriata* is present in the *Quadrithyris* Zone at Coal Canyon and questionably in the lower McMonnigal Limestone of the Toquima Range.

FIGURED SPECIMENS: USNM 156927-156936.

Mesodouvillina cf. *varistriata arata* (Hall, 1859)
(Pl. 23. fig. 16)

Strophodonta varistriata var. *arata* Hall, 1859, p. 183, Pl. 16, fig. 16; Pl. 18, figs. 1a-i.

Stropheodonta varistriata var. *arata* Hall and Clarke, 1892, Pl. 13, figs. 17, 18.

DISCUSSION: One specimen from the *Quadrithyris* Zone is strongly concavo-convex and bears relatively strong radial costellae of variable size. Finer secondary costellae have not been observed, but the specimen is coarsely silicified.

FIGURED SPECIMEN: USNM 156937.

Genus *Mclearnites* Caster, 1945

TYPE SPECIES: *Brachyprion mertoni* Mclearn, 1924, p. 61, Pl. 4, figs. 16-18; Pl. 28, fig. 12.

Mclearnites invasor Johnson, n. sp.
(Pl. 24, figs. 1-8)

DIAGNOSIS: Thin-shelled, subplanate, with parvicostellate ornament.

EXTERIOR: The outline is semicircular or shield-shaped, commonly with the cardinal angles acute and auriculate. In lateral profile the valves are plano-convex to flatly concavo-convex. The pedicle valve is not strongly inflated and the body cavity is very small. The hinge line is long and straight and is the place of maximum width. The ventral interarea is flat and apsacline. The dorsal interarea is almost linear and is hypercline.

The ornament is finely parvicostellate, commonly with three or four finer costellae between the raised primaries. Anteriorly they may increase in number to six or seven. The hinge line is denticulate about three-fifths of its length.

INTERIOR OF PEDICLE VALVE: The ventral process consists of two short ridges joining posteromedially and extending posteriorly as a short stem adjoining the shell material filling the delthyrium. The adductor scars are posteriorly situated and are moderately impressed posterolaterally within the anterior edges of the ventral process. The diductor muscle field is broad and enclosed laterally by muscle-bounding ridges which diverge strongly near their posterior end, but recurve slightly to diverge moderately from the midline throughout most of their length. The anterior edge of the muscle field blends imperceptibly with the interior of the shell. Medially, there may be an indistinct myophragm. On small shells the muscle field is only poorly discernible, and on slightly larger ones it becomes impressed and in addition is bounded by indistinct muscle-bounding ridges. The muscle-bounding ridges become sharp and well defined only on large specimens. The interior is not impressed by the external costellae.

INTERIOR OF BRACHIAL VALVE: The cardinal process lobes are of subcircular cross section at their bases and are disjunct, projecting posteriorly and slightly ventrally. Their posterior faces are wide, swollen laterally, and each is longitudinally cleft, forming bilobate myophores. The lobes arise from a slight elevation of shell at the apex of the valve. Socket plates are small, relatively well-elevated ridges that are widely set apart lateral to the bases of the cardinal process lobes. They diverge laterally only slightly from the hinge line and instead are more nearly subparallel to it. The adductor muscle scars are only impressed posteriorly and laterally where they are bounded by muscle-bounding ridges. There is a low, rounded myophragm, and the adductors are not clearly delineated into two pairs. The interior of the valve is pustulose, but is not impressed by the external costellae.

SHELL STRUCTURE: This species is pseudopunctate and relatively thin-shelled.

OCCURRENCE: This species of *Mclearnites* is at present known only from the *Acrospirifer kobehana* Zone in the Sulphur Spring Range, but specimens are fairly numerous.

COMPARISON: *Mclearnites invasor* is distinguished from *M.* sp. B by its much thinner shell.

FIGURED SPECIMENS: USNM 156938-156944. The holotype is USNM 156938.

Mclearnites sp. B
(Pl. 24, figs. 9-13)

EXTERIOR: The shells are relatively large and semicircular to shield-shaped. The valves are commonly transverse, but some have the width approximately equal to the length. They are plano-convex in lateral profile with the pedicle valve only moderately inflated. The hinge line is long and straight and commonly is the place of maximum width, or the cardinal angles may be slightly obtuse with maximum width between the hinge line and midlength. The ventral interarea is high, flat, and apsacline. The dorsal interarea is anacline.

The ornament consists of fine radial costellae developed in a strong parvicostellate pattern. There are commonly eight or ten fine costellae between the stronger primary costellae.

INTERIOR OF PEDICLE VALVE: The ventral process is strong and triangular. It is relatively short, measured along the midline, and is broader than long. Anteriorly, the adductor muscle scars are deeply impressed along their anterior edges. They are elongate-oval and are divided by a myophragm which arises at the anterior base of the ventral process. The diductor muscle field is broadly cordate and fairly well impressed. It is flanked laterally by ponderous muscle-bounding ridges. Anteriorly, the diductor muscle field may be slightly raised above the interior surface of the valve and slopes off abruptly along a bounding shelf. The interior is pustulose posterolaterally and is not impressed by the external costellae.

INTERIOR OF BRACHIAL VALVE: The cardinal process lobes are disjunct and project posteroventrally. They diverge slightly posteriorly and are broader on their distal ends. Socket plates are short, narrow ridges, widely set lateral to the cardinal process lobes and diverging at a moderate angle from the hinge line. The posterior adductors are elongate-semicircular and bounded laterally by ponderous muscle-bounding ridges. The anterior adductor scars are much more strongly elongate and are situated on two long, low, rounded ridges which diverge very slightly from the midline. They are pustulose on their anterior portions and are separated medially by a trough which may bear a short low myophragm. The posterolateral portions of the shell are pustulose. The interior is not crenulated by the impress of the external costellae.

SHELL STRUCTURE: The shell material is very thick and is pseudopunctate.

OCCURRENCE: *Mclearnites* sp. B is common in the *Eurekaspirifer pinyonensis* Zone in the Cortez Mountains, but no specimens have been collected at any of the other numerous localities in the Simpson Park Mountains, the Roberts Mountains, Lone Mountain, or in the Sulphur Spring Range. The species is associated with *Muriferella masurskyi* and *Bifida* sp. along with more typical species of the *Eurekaspirifer pinyonensis* Zone, such as *Phragmostrophia merriami* and *Leptocoelia infrequens*.

COMPARISON: The species differs from *M. invasor* by its very thick shell and by its extraordinarily ponderous muscle-bounding ridges.

FIGURED SPECIMENS: USNM 156945-156948.

Genus *Cymostrophia* Caster, 1939

TYPE SPECIES: *Strophomena stephani* Barrande, 1848, p. 230, Pl. 20, fig. 7, (see Havlíček, 1967, p. 127).

Cymostrophia sp.
(Pl. 35, figs. 1, 2)

Stropheodonta patersoni Walcott, 1884, p. 119.

DISCUSSION: Only a few fragmentary specimens have been found—too few on which to base a satisfactory description. The ornament is parvicostellate with superposed interrupted rugae. The latter are not as prominently developed as in *C. stephani*. The shape of brachial valves is deeply concave and hemispherical. The single pedicle valve internal mold seems, however, to be geniculate with a trial longer than the length posterior to the line of geniculation (as in *C. stephani*). On the other hand, the specimen is slightly crushed, so geniculation may be due only to deformation.

The ventral interior has a large muscle field and muscle-bounding ridges like those of *Mesodouvillina* and *Mclearnites*. The cardinalia consist of a pair of cardinal process lobes and a pair of widely divergent, ridgelike socket plates.

OCCURRENCE: So far found only in *Eurekaspirifer pinyonensis* Zone of the northern Roberts Mountains.

FIGURED SPECIMENS: USNM 156781, 156782.

Genus *Megastrophia* Caster, 1939

TYPE SPECIES: *Strophomena (Strophodonta) concava* Hall, 1857, p. 140, fig. 1.

Megastrophia transitans Johnson, n. sp.
(Pl. 25, figs. 1-19)

DIAGNOSIS: Small, thin-shelled *Megastrophia* with brace plates.

EXTERIOR: The shells are small- to medium-sized, subcircular to transversely shield-shaped in outline, and strongly concavo-convex in lateral profile. Cardinal angles vary from acute to approximately 90 degrees and in some shells, may be obtuse. In others, the cardinal angles are moderately auriculate. The place of maximum width is at the hinge line or slightly anterior to it. The ventral interarea is strongly apsacline to orthocline. It is denticulate along nearly its entire length. The delthyrium is closed except for a thin crescent-shaped opening at the hinge line. The dorsal interarea is long, linear, and hypercline.

The ornament is closely parvicostellate with the primary costellae raised slightly on crests and separated by broad, shallow furrows. Commonly, there are only one or two intervening secondary costellae between the primaries on small shells. In larger ones, this ratio is kept approximately the same due to elevation of the secondaries to equal height with the primary costellae.

INTERIOR OF PEDICLE VALVE: The ventral process is broadly triangular and connects posteriorly with the inner side of the shell material filling the delthyrium. The diductor muscle scars originate lateral to the ventral process and diverge widely laterally, curving slightly inward. The muscle field is impressed posterolaterally on small shells and develops low muscle-bounding ridges posterolaterally on large shells. The diductor field becomes broad and cordate anteriorly and merges with the inner surface of the valve. The adductor scars are deeply impressed posteriorly and laterally. They compose an elongate pair that blends anteriorly with the enclosing diductor scars. They are not impressed anteriorly, but are enclosed medially by the inner edges of the diductor impressions. The interior is only faintly impressed by the external costellae.

INTERIOR OF BRACHIAL VALVE: The dorsal interior of this species is known only from small specimens and so the following observations will be modified when large-size specimens become available. The cardinal process lobes are circular to suboval in cross section. They project strongly posteriorly and recurve dorsally posterior to the hinge line. They diverge only slightly and are not strongly swollen or thickened on their lateral edges. The cardinal process lobes are situated on a thickening of shell which extends laterally as two strong, rounded muscle-bounding ridges almost parallel with the hinge line. Socket plates are small and very widely set apart on the posterolateral edges of the muscle-bounding ridges. They diverge at a moderate angle from the hinge line. The adductor scars are relatively strongly impressed medially and posteriorly and are roughly triangular in outline. Brace plates, situated medially, are elongate, stout, slightly divergent ridges which become more or less pustulose and slightly broader at their anterior ends. Medially, there is a low, flat, elongate trough between the anterior adductor ridges. The interior of the valves is pustulose and slightly corrugated by the impress of the external costellae.

OCCURRENCE: This species of *Megastrophia* first appears in the upper beds of the Rabbit Hill Limestone at Rabbit Hill and at Coal Canyon. It is most common in the *Trematospira* Zone in the Sulphur Spring Range, and a few specimens which are assigned have been collected from the *Acrospirifer kobehana* Zone in the Sulphur Spring Range.

COMPARISON: *Megastrophia transitans* is smaller and has a thinner shell than *Megastrophia iddingsi*. In addition, the ornament is decidedly stronger than that developed on *M. iddingsi*. Outside Nevada, no comparable named species is known to the writer.

FIGURED SPECIMENS: The holotype is USNM 156949. Other figured specimens include USNM 156950-156957.

Megastrophia iddingsi (Merriam, 1940)
(Pl. 26, figs. 1-16)

Strophodonta demissa Walcott, 1884, p. 118, Pl. 2, figs. 9, 9a, 9b; *not* Conrad, 1842.
Stropheodonta iddingsi Merriam, 1940, p. 79, Pl. 6, fig. 6.

EXTERIOR: The shells are subsemicircular to transversely suboval in outline and plano-convex to moderately concavo-convex in lateral profile. The ventral umbo is strongly convex. Posterolateral slopes are concave on medium- and large-sized specimens, producing a slight auriculation, but on most shells the cardinal angles are slightly obtuse. Maximum width is attained between the umbo and midlength. The anterior commissure is rectimarginate, except in some large shells which have the commissure slightly bowed toward the pedicle valve. Interareas are denticulate about three-fifths to three-quarters of their total length. The ventral interarea is orthocline; the dorsal interarea is hypercline. The pseudodeltidium is broad, flat, and level with the adjoining surface of the interarea; it bears a very narrow, medially folded area.

The ornament is finely parvicostellate.

INTERIOR OF PEDICLE VALVE: The ventral process forms a triangular protuberance in the apex of the beak. On its anterior side there is a wide, shallow, V-shaped depression. The adductor scars are impressed flatly into the base of the ventral process. The posterior ends of the muscle-bounding ridges begin as well-defined plates that connect with the hinge line. The muscle-bounding ridges continue anterolaterally, almost in a straight line, diverging at about a right angle and then converging to enclose the broadly cordate diductor field. In some shells the diductor scars are slightly striated longitudinally.

The adductor scars compose a quadrate pair that is enclosed anteriorly by the diductor scars. At the anteromedial extremity of the adductors, there is a small, oval, elevated ridge between the diductors. A pair of impressions of the vascula media extend anteriorly. The ornament is slightly impressed on the anterior of the shell.

INTERIOR OF BRACHIAL VALVE: The cardinal process lobes project posteroventrally and are separated by a small cleft that widens slightly posteriorly. The lateral edges of the lobes flare laterally. Socket plates are situated along the hinge line and are widely set apart lateral to the cardinal process lobes. They diverge anterolaterally at a high angle. The posterior adductor scars are situated lateral to the anterior pair and are deeply impressed and roughly triangular with arcuate anterolateral margins defined by muscle-bounding ridges. The anterior adductors are elevated on a platform and widen anteriorly, but taper medially along their anterior edge. The lateral edges of the anterior adductors are elevated above the average height of the muscle platform. At about the center of the anterior adductor scars, there is a breviseptum that extends to a position about three-quarters of the length of the shell.

SHELL STRUCTURE: The shell material is pseudopunctate, except over the muscle scars. The anterior adductors of the brachial valve, however, are situated on pseudopunctate shell material.

OCCURRENCE: *Eurekaspirifer pinyonensis* Zone, Sulphur Spring Range, Roberts Mountains, Lone Mountain, and northern Simpson Park Mountains.

FIGURED SPECIMENS: The holotype is USNM 96370. Other figured specimens are USNM 156958-156963.

Genus *Stropheodonta* Hall, 1852
TYPE SPECIES: *Strophomena demissa* Conrad, 1842, p. 258, Pl. 14, fig. 14.
Stropheodonta filicosta Johnson, n. sp.
(Pl. 22, figs. 13-18)

DIAGNOSIS: Finely costate *Stropheodonta* with brace plates.

EXTERIOR: The shells are suboval to semicircular in outline, with width approximately equal to the length. In lateral profile the valves are moderately concavo-convex. Some specimens are shield-shaped. The hinge line is long and straight and in most specimens is slightly less than maximum width which is attained at midlength or posterior to midlength. The cardinal angles are right angles or are obtuse. The ventral interarea is long, flat, and apsacline to nearly orthocline. It is denticulate along approximately three-fifths of its length. The dorsal interarea is short, anacline, and poorly developed.

The ornament consists of narrow, closely set, conspicuously raised, subrounded costellae that increase in number anteriorly by bifurcation and by implantation. The costellae commonly appear to arise along one side of an interspace from the flank of an adjoining costellae, but none have been seen to split and give rise to two new costellae of equal size. Individual costellae swell and attenuate alternately along their course. Some apparently are hollow and give rise to small openings at the anterior edges of swellings.

INTERIOR OF PEDICLE VALVE: The ventral process consists of two, short, anterolaterally diverging ridges which are extended posteriorly as a medially cleft ridge which joins the apex of the valve. The adductor muscle scars are moderately well impressed posterolaterally and are divided medially by a myophragm which may bear a median groove. The diductor muscle field originates lateral to the posterior portion of the ventral process and extends anterolaterally. It is fairly well impressed laterally without development of muscle-bounding ridges. The lateral edges of the diductor muscle field are nearly rectilinear. The impressions are faint, but slightly flabellate anteriorly where they blend with the interior of the valve. The surface is lightly crenulated by the impress of the external costellae.

INTERIOR OF BRACHIAL VALVE: The cardinal process lobes are disjunct and are elliptical in cross section at their bases. Toward their extremities they become elongate parallel to the midline and are longitudinally grooved on their distal ends. They are directed slightly ventrally and relatively strongly posteriorly. Small linear socket plates are situated widely lateral to the cardinal process lobes and diverge from the hinge line at a moderate angle. The adductor scars are faintly impressed and bounded posteriorly by muscle-bounding ridges which join shell material at the base of the cardinal process lobes. They are roughly triangular in section and blend anterolaterally with the interior of the shell. Brace plates, situated medially on two low rounded ridges, diverge slightly from the midline and are pustulose anteriorly. Between them, there is a shallow furrow; no myophragm is present. The dorsal interior of these shells closely parallels the structure of *Megastrophia transitans* found in the same collection.

OCCURRENCE: This species first appears in the Rabbit Hill Limestone where it is represented by relatively large individuals, but is extremely rare. The species is most common, but principally in relatively small shells, in the *Trematospira* Zone in the Sulphur Spring Range. A few specimens closely comparable with this species are present in a collection from G.S.C. loc. 67145 in the Stuart Bay Formation of Bathurst Island.

COMPARISON: The fine wirelike costellae at once distinguish this species from *Stropheodonta magnacosta* which is coarsely costate. *Stropheodonta* sp. of Williams (1953b, Pl. 7, fig. 10) is similar, but its costellae are coarser. *S. micropleura* Imbrie (1959, Pl. 56, figs. 11-17) is very close, although it is relatively broader. *S. consobrina* (Barrande, 1879, Pl. 42, figs. 6-14) is similar, but has posteriorly concave costellae near the hingeline. *S. verneuili* (Barrande, 1879, Pl. 42, figs. 21-26) has slightly coarser, non-nodose costellae.

FIGURED SPECIMENS: USNM 156916-156919. The holotype is USNM 156916.

Stropheodonta magnacosta Johnson, n. sp.
(Pl. 21, figs. 1-12)

DIAGNOSIS: Coarsely costate *Stropheodonta* without associated parvicostellae.

EXTERIOR: The shells are semicircular to shield-shaped and generally slightly wider than long. In lateral profile they are moderately to deeply concavo-convex. The ventral interarea is relatively long, flat, and orthocline. It is denticulate along nearly its entire width. The dorsal interarea is poorly defined and linear. It appears to be hypercline.

The ornament consists of relatively coarse, strong, angular costae which increase in number anteriorly by bifurcation and by implantation.

INTERIOR OF PEDICLE VALVE: The ventral process is a relatively ponderous, elevated, triangular structure which is deeply grooved on its anterior edge. Posteriorly, it joins the apex of the valve. The diductor muscle field on large specimens is strongly impressed and may develop, in addition, low muscle-bounding ridges. The lateral margins of the diductor muscle impressions commonly diverge from the midline at a relatively small angle and recurve slightly inward to form a diductor impression which does not flare broadly laterally. The diductor scars are slightly flabellate anteriorly. On large shells, a posterolateral platform is developed separating the visceral disc from the cardinal extremities. Both regions are pustulose, and the interior of the valve is only slightly crenulated by the impress of the external costellae.

INTERIOR OF BRACHIAL VALVE: The cardinal process lobes are disjunct, roughly circular in cross section near their bases, and they project posteriorly. Their posterior faces are medially grooved forming bilobate myophores. The socket plates are small and short. They are situated posterolaterally, widely set apart relative to the bases of the cardinal process lobes, and they diverge from the hinge line at a low angle. The posterior adductor scars are roughly triangular and strongly impressed. On a single large specimen, they are elaborated medially with curved plates of shell material forming a dendritic pattern. The anterior edges of the posterior adductors may be elevated on thickenings of shell material. The anterior adductors are elongate-pyriform and are divided by a myophragm. There are strong, concentric, elevated ridges originating posterolaterally and encircling the interior peripherally. The interior is pustulose and is not crenulated by the external costellae.

OCCURRENCE: *Stropheodonta magnacosta* is present in the Rabbit Hill Limestone in the Monitor Range, in the *Trematospira* Zone in the Sulphur Spring Range and in the *Acrospirifer kobehana* Zone in the Roberts Mountains.

COMPARISON: No comparable species is known to the writer.

FIGURED SPECIMENS: USNM 156911-156915. The holotype is USNM 156915.

Family LEPTOSTROPHIIDAE Caster, 1939
Genus *Leptostrophia* Hall and Clarke, 1892

TYPE SPECIES: *Strophomena (Strophodonta) magnifica* Hall, 1857, p. 54.

DISCUSSION: *Leptostrophia,* and probably *Leptostrophia* s.s., occurs in all Siegenian and Emsian zones in Nevada. Four or possibly five species are present, but their limits are not yet well understood because of insufficient material. Specimens from the *Quadrithyris* Zone called *Leptostrophia* sp. A are unknown regarding external ornament and ventral interior (Pl. 23, figs 17, 18). *Leptostrophia inequicostella,* from the *Spinoplasia* Zone, is characterized by a unique ornament. *Leptostrophia* cf. *beckii,* from the *Trematospira* Subzone, is the only form with concentric rugae, but it is more faintly developed than in Appalachian specimens. Rugae have not been seen in specimens from the *Costispirifer* Subzone, and these may belong to *Leptostrophia* sp. D of the *A. kobehana* Zone (Pl. 28, fig. 10). Most of the latter are small, so comparison with large specimens is difficult. One large *Leptostrophia?* sp. has been seen in the *E. pinyonensis* Zone of the Roberts Mountains (Pl. 29, fig. 9), and another has been seen in that zone in the Simpson Park Range, but the external ornament and dorsal interiors of both *E. pinyonensis* Zone specimens is unknown.

Leptostrophia sp. A
(Pl. 23, figs. 17, 18)

DISCUSSION: There are two brachial valves present in the collections from the *Quadrithyris* Zone at Coal Canyon which belong to *Leptostrophia.*

The ornament on these specimens is not preserved, but preservation is adequate to assure that no coarse ornament was developed. They are nearly flat, without any noticeable curve in lateral profile. The cardinal process lobes are conjunct, and their posterior ends are continuous with a convex, medially cleft chilidium. Each lobe is elongate-subelliptical and narrower posteriorly. The myophores of the two lobes face posteroventrally. The socket plates consist of two, prominent, short ridges that are set closely lateral to the cardinal process lobes and which diverge anterolaterally at a high angle to the hinge line. The adductor muscle scars are deeply impressed posterolaterally where they tend to also develop broad, low, muscle-bounding ridges. The posterolateral portions of the interior of the shell are pustulose.

FIGURED SPECIMENS: USNM 156975-156976.

Leptostrophia inequicostella n. sp.
(Pl. 29, figs. 1-8)

DIAGNOSIS: Pedicle valve with finely parvicostellate ornament; brachial valve with costellate ornament.

EXTERIOR: The shells are small and thin for the genus. They are sub-semicircular in outline and biplanate to faintly plano-convex in lateral profile. The hinge line is long and straight and is the place of maximum width. The cardinal angles are approximately right angles, or they may be slightly acute. The ventral interarea is long, flat, and apsacline. It is denticulate approximately three-quarters of its length. The dorsal interarea is long, nearly linear, and anacline.

The ornament consists of very fine, closely spaced, radial costellae on the brachial valve and parvicostellae on the pedicle valve.

INTERIOR OF PEDICLE VALVE: The ventral process has roughly the form of an inverted T. It is relatively high and thick posteriorly and is attached to the apex of the shell. Anteriorly, it widens abruptly by the addition of two lateral extensions which serve as posterior muscle-bounding ridges to the adductor muscle impressions. The ventral process is continuous anteriorly with a rounded myophragm that bisects the adductor muscle impressions. On one specimen there is a deep, medial cleft at the posterior end of the ventral process. The adductor muscle field is strongly impressed, originating posterolateral to the ventral process. Its sides are impressed and diverge at about 45 degrees from the hinge line and then curve inward very slightly to enclose a relatively narrow, elongate muscle field. Large specimens develop muscle-bounding ridges posterolaterally, as well as impressed lateral edges which define the muscle field. Anteriorly, the diductor impressions blend with the interior of the valve. The posterolateral portions of the valve are pustulose. The interior is not crenulated by the impress of the costellae.

INTERIOR OF BRACHIAL VALVE: The cardinal process lobes are short, stubby, elliptical in cross section, and slightly conjunct posteriorly. They project ventrally and slightly posteriorly. Relatively ponderous socket plates are set close laterally to the cardinal process lobes and diverge from the hinge line at a high angle. There is a small, semicircular, medially cleft chilidium at the posteromedial ends of the cardinal process lobes. The cardinalia are situated on a low transverse thickening which curves around enclosing the adductor muscle field posterolaterally. Anteriorly, shell thickening and adductor scars both blend imperceptibly with the interior of the valve. The posterolateral portions of the shells are pustulose, and the surfaces are not crenulated by the impress of the external costellae.

OCCURRENCE: *Leptostrophia inequicostella* is known from the Rabbit Hill Limestone on the east side of the Monitor Range and at Coal Canyon in the Simpson Park Range.

Leptostrophia sp. 2 of Amsden (1958b, p. 59, Pl. 2, figs. 20-24) from the Fittstown Member of the Bois d'Arc Formation of Oklahoma may belong here if it proves to have the internal structures of *Leptostrophia*.

COMPARISON: Several Appalachian forms, such as the nonrugose specimens of *L. oriskania* Clarke (1900, Pl. 7) and *L. arctimuscula* Schuchert (*in* Schuchert and others, 1913, Pl. 58, fig. 1) resemble *L. inequicostella*, but neither of the former, nor any other species of which the writer is aware, has the external ornament of *Leptostrophia inequicostella*.

FIGURED SPECIMENS: The holotype is USNM 156977. Other figured specimens are USNM 156978-156982.

Leptostrophia cf. *beckii* (Hall, 1857)
(Pl. 29, figs. 10, 11)

Strophomena (*Strophodonta*) *beckii* Hall, 1857, p. 52, figs. 1-4.
Stropheodonta beckii Hall, 1859, p. 191; 1861, Pl. 22, fig. 1.
Leptostrophia becki Hall and Clarke, 1892, Pl. 13, figs. 23, 24.
Leptostrophia beckii Schuchert, *in* Schuchert and others, 1913, p. 314, Pl. 57, figs. 12, 13.
Rhytistrophia beckii Cooper, 1944, p. 341, Pl. 131, figs. 26-28.
Leptostrophia sp. Williams, 1953b, Pl. 9, figs. 12, 13.

EXTERIOR: The specimens are relatively large and thick shelled and are subsemicircular to transversely shield-shaped. On small- and medium-sized specimens, the cardinal angles are decidedly acute and auriculate. In lateral profile the valves are plano-convex.

The ornament consists of fine radial costellae evenly spaced across the valves. Posteriorly, there is a faint development of concentric rugae. The hinge line is long and straight and is the place of maximum width. Approximately the middle two-thirds of the hinge line is denticulate. The ventral interarea is long and flat. Its inclination is strongly apsacline.

INTERIOR OF PEDICLE VALVE: The ventral process is triangular and relatively ponderous. It is continuous with a short, stout myophragm anteriorly in some specimens. The diductor muscle field is strongly bounded laterally along edges that are impressed and nearly rectilinear, and that bear fairly well-developed muscle-bounding ridges. Some slight flabellation of the diductor impressions is developed anteriorly. The posterolateral portions of the shell are pustulose. The illustrated specimen lacks impression of one side of the diductor muscle field, evidently reflecting a congenital deformity of the individual. The interior is not impressed by the external costellae.

INTERIOR OF BRACHIAL VALVE: A single specimen from the *Trematospira* Subzone in the Sulphur Spring Range shows the dorsal interior (*not* illustrated). It has a pair of stout, platelike cardinal process lobes and closely adjacent socket plates characteristic of *Leptostrophia* s.s.

OCCURRENCE: *Leptostrophia* cf. *beckii* is present in the *Trematospira* Zone in the Sulphur Spring Range.

FIGURED SPECIMEN: USNM 156984.

Leptostrophia sp. D
(Pl. 28, figs. 1-10)

EXTERIOR: The valves are plano-convex in lateral profile with pedicle valves only faintly convex. The outline is semicircular (transverse) in small shells; some develop auriculations at the cardinal extremities (Pl. 28, figs. 7, 8). Transverse outline is maintained in some specimens, but not in others where length may equal width in large individuals. The hinge line is long and straight and is the place of maximum width. The ventral interarea is long, low, and apsacline; the dorsal interarea is nearly linear and anacline. Both are denticulate along the whole of their preserved length.

External ornament consists of fine, rounded radial costellae which on the pedicle valve increase by intercalation. Very faint concentric rugae are developed posteriorly on some specimens.

INTERIOR OF PEDICLE VALVE: Small specimens lack a ventral process, but larger ones have a ridgelike rhomboidal process. The muscle scar is prominent and triangular, enclosed by muscle-bounding ridges and further accentuated by the impression of the diductor scars in their posterior half. Adductor scars are not preserved on the available specimens. Smaller shells have the costellae impressed on the interior, but larger ones are smooth internally.

INTERIOR OF BRACHIAL VALVE: The cardinalia consist of a pair of subparallel, platelike cardinal process lobes that project ventrally and slightly posteriorly. They are longitudinally grooved on their distal faces and are flanked closely by a pair of ridgelike, subparallel socket plates. Anteriorly the thickened area at the base of the cardinal process lobes extends as a low, rounded myophragm dividing the adductor scars. The latter are impressed and bounded posterolaterally by muscle-bounding ridges that make an angle with the midline of about 45 degrees. The posterior and anterior adductor pairs are not differentiated and blend imperceptably with the interior of the valve anteriorly. Large specimens are impressed by the costellae only on the anterior portions.

OCCURRENCE: Known with certainty only from the *A. kobehana* Zone in the Sulphur Spring Range.

FIGURED SPECIMENS: USNM 156968-156974.

Leptostrophia? sp. D?
(Pl. 29, fig. 9)

EXTERIOR: This form is known only from a single internal mold of a pedicle valve. It is a relatively large specimen, transversely suboval with a long, moderately straight hinge line. In lateral profile it is only slightly convex.

Bits of shell material on one flank show the impressions of radial costellae, but preservation is not adequate to determine whether this species was merely finely costellate or parvicostellate. The shell substance is pseudopunctate.

INTERIOR OF PEDICLE VALVE: The ventral process is a relatively small triangular protuberance at the posterior end of the adductor muscle impressions. The adductor impressions are long and suboval with their lateral edges almost parallel to one another. The diductor scars are relatively deeply impressed posteriorly and are defined laterally by nearly straight-sided impressions diverging anterolaterally. Within the diductor muscle field, there is a strong development of radial ridging which shows that the muscle field was relatively large and probably was equal in length to about half the length of the valve. Anterior to the adductor impressions there is a broad, low myophragm that narrows in a short distance and continues anteriorly as a narrow ridge bisecting the anterior portion of the diductor impressions.

OCCURRENCE: The figured specimen is from the *Eurekaspirifer pinyonensis* Zone on the north side of the Roberts Mountains. One other was collected from the same zone in the Simpson Park Range.

COMPARISON: This form, questionably assigned to *Leptostrophia* due to the lack of a brachial valve, is similar in size and shape to *Leptostrophia* sp. D from underlying beds of the *kobehana* Zone, but bears a much more strongly radially ridged diductor muscle field in the pedicle valve.

FIGURED SPECIMEN: USNM 156983.

Family PHOLIDOSTROPHIIDAE Stainbrook, 1943
[*nom. transl.* Sokolskaya, 1960, *ex* Pholidostrophiinae Stainbrook, 1943]
Genus *Pholidostrophia* Hall and Clarke, 1892
TYPE SPECIES: *Strophomena (Strophodonta) nacrea* Hall, 1857, p. 144.
Pholidostrophia (Pholidostrophia) sp.
(Pl. 28, figs. 11-17)

EXTERIOR: This is a small, thin-shelled form with maximum dimension generally not exceeding a centimeter. The outline is subquadrate to shield-shaped, and the lateral profile is nearly planar; the brachial valve is flat, and the pedicle valve is very slightly convex. The hinge line is long and straight and is the place of maximum width. The cardinal angles tend to be auriculate. The ventral interarea is flat and very low. It is steeply apsacline, almost catacline, and its dorsal edge is denticulate about two-thirds of the distance from the midline to the cardinal extremities. Preservation is too poor to determine the presence or absence of deltidial structures.

The external ornament consists of a few distantly spaced concentric growth lamellae. Radial ornament is unknown.

INTERIOR OF PEDICLE VALVE: The ventral muscle scar is very faint and is slightly impressed posterolaterally, without bounding ridges. Its anterior edges are not discernable and probably blend imperceptibly with the interior of the valve. There appears to be a very faint myophragm in the posterior half of the muscle impression. No ventral process is present in the small specimens studied.

INTERIOR OF BRACHIAL VALVE: The inner edge of the dorsal hinge line is faintly raised as a low, rounded ridge along the posterior part of the valve. Medially, upon it, there is a pair of cardinal process lobes that project posteroventrally, but apparently do not diverge laterally. Socket plates are absent. There is a very faint myophragm emanating from the base of the cardinal process lobes and a lateral pair of elongate, slightly arcuate, posterior adductor scars impressed into the shell interior. The anterior adductors are not discernible.

OCCURRENCE: This species is best known only from a few specimens from the *Acrospirifer kobehana* Zone in the Sulphur Spring Range. All of the illustrated specimens come from this horizon and locality. A few additional specimens have been seen in the *Trematospira* Subzone in the Sulphur Spring Range. The latter are slightly thicker shelled, but no brachial valves have been found on which to confirm the generic identification. In addition, a few poorly preserved pedicle valves were noted in the *Spinoplasia* Zone from Rabbit Hill, but again no brachial valves are known from this horizon.

FIGURED SPECIMENS: USNM 156964-156967.

Genus *Phragmostrophia* Harper, Johnson, and Boucot, 1967

TYPE SPECIES: *Phragmostrophia merriami* Harper, Johnson, and Boucot, 1967.

DESCRIPTION: Small, transversely extended shells, quadrate in outline, and plano-convex to concavo-convex in lateral profile. The ornament is unequally parvicostellate. Diductor impressions in the pedicle valve are suboval to subpyriform. In the brachial valve the adductor impressions are situated on an elevated transverse platform. There is a central pit dividing the anterior adductors anteromedially. Anterior to the adductor platform, there is a more or less well-developed breviseptum that extends forward from the central pit. Brachial ridges are strong and S-shaped. Peripherally, there is an elevated ridge that is geniculate along its exterior edge.

DISCUSSION: This interesting little genus ranges geographically from Nevada to the Canadian Arctic Islands and Novaya Zemlya in the upper Siegenian, Emsian, and Middle Devonian. The only other named species is *P. latior* (Meyer, 1913) from the Middle Devonian Bird Fiord Formation of Ellesmere Island. Related genera are present in other faunal provinces as evidenced by *Nadiastrophia* in Australia and by *Teichostrophia* and *Ancylostrophia* in Europe. Harper and others (1967) have illustrated specimens of *Phragmostrophia* from Nevada and from Arctic Canada and thought that the genus was restricted to western and arctic North America. However, specimens of pholidostrophiids from the Eifelian or upper Emsian Yükiang Formation of southern China (Wang, 1956) may prove to belong to *Phragmostrophia* and not to *Nadiastrophia* as suggested by Talent (1963, p. 62) and by Harper and others (1967) following Talent. Wang illustrated three specimens, one named *Nervostrophia? gigantia,* and the other two under the same *Shaleria (Telaeoshaleria) yukiangensis*. All the specimens probably belong to the same genus judging from variations in specimens from Nevada observed by the writer. Wang's specimen illustrated in figures B1-4 of Plate 3 has a radial ornament of somewhat irregular costellae, but the surface is not parvicostellate. This type of ornament has only been observed on *Phragmostrophia merriami* specimens from central Nevada. However, the extremely low interarea of this same specimen distinguishes it from the Nevada species even though the interior of the Yükiang Formation form is unknown. The specimens illustrated in figures C1-5 externally are extremely close to *Phragmostrophia merriami*. They appear to have a very faintly developed ornament more like the Nevada species than like the strikingly parvicostellate specimens of *Nadiastrophia superba*, and it appears to the writer that the pedicle valve interior illustrated in Wang's figure C5, Plate 3 could belong to either *Phragmostrophia* or *Nadiastrophia*. Should the species *Shaleria* (*Telaeoshaleria*) *yukiangensis* prove to be a *Phragmostrophia*, the writer knows of no way at present to distinguish it from *Phragmostrophia merriami*, but it is hoped that the presentation of new information regarding variation of form and the nature of the internal structures of the brachial valve, when available, will solve both questions.

The earliest known representatives of *Phragmostrophia* occur in the upper Siegenian part of the *Gypidula* 1-*Biconostrophia* Unit at Royal Creek, Yukon Territory (Lenz, 1966).

Phragmostrophia merriami Harper, Johnson, and Boucot, 1967
(Pl. 26, fig. 17; Pl. 27, figs. 1-20)

Strophodonta calvini Walcott, 1884, p. 122, Pl. 13, fig. 6; not Miller, 1883.
Phragmostrophia merriami Harper, Johnson, and Boucot, 1967, p. 430, Pl. 7, figs. 5-8; Pl. 8, figs. 4, 5.

EXTERIOR: The shells are plano-convex to slightly concavo-convex with the pedicle valve strongly convex in most shells. The outline is transversely quadrate with the anterior margin more or less parallel to the hinge line. The hinge line is the place of maximum width of some shells, but others expand toward the anterior with a maximum width anterior to midlength. The ventral interarea is well defined, relatively high, apsacline, and varies from nearly orthocline to about the 45-degree position. The dorsal interarea is anacline. The interareas are denticulate nearly their entire length. The delthyrium is triangular and is almost completely filled with a flat pseudodeltidium that is slightly arcuate along its dorsal edge. The anterior commissure is rectimarginate, but in a few shells there is a wide, shallow depression in the anterior geniculate wall of the pedicle valve.

The exterior ornament consists of very fine costellae arranged in a parvicostellate pattern which is better defined on the brachial valve than on the pedicle valve.

INTERIOR OF PEDICLE VALVE: The ventral process consists of a triangular, raised area in the apex of the valve. Medially, on its dorsal side, there is a small U-shaped groove. The adductor scars are impressed into the anterior side of the ventral process. The diductor scars are impressed laterally and defined by lateral bounding ridges. The diductor impressions form a broadly cordate pair with lateral edges convex outward. Medially, they are separated by a more or less well-developed rounded median ridge.

Near the anterior edge of the shell there is commonly a low ridge that fits along the inner edge of the dorsal marginal ridge.

INTERIOR OF BRACHIAL VALVE: The cardinal process lobes project posteroventrally and diverge distally. Socket plates are set widely apart lateral to the cardinal process lobes and diverge at slight angles from the hinge line. The posterior adductor scars are situated lateral to the anterior adductor scars and both pairs may be longitudinally fluted or grooved on large specimens. The adductor impressions are elevated on a prominent transverse platform that is essentially rectangular, except for the projecting anterior adductors. The platform is cleft anteromedially by a deep, elongate-oval central pit. Anteriorly, adjoining the central pit, there is a breviseptum that extends about half way to the anterior margin. In most specimens the breviseptum is extended anteriorly as a strong median ridge that joins the peripheral ridge. This anterior median ridge, however, is only vestigial in large specimens. Beyond the anterolateral edge of the muscle-bearing platform, there is a well-developed pair of S-shaped brachial ridges. The interior surface of the shell is strongly pustulose, especially adjacent to the muscle impressions. Peripherally, there is a high ridge which, in most cases, is highest where it joins the median ridge.

OCCURRENCE: *Phragmostrophia merriami* is a common species in the *Eurekaspirifer pinyonensis* Zone. It is present in the Sulphur Spring Range in the *E. pinyonensis* beds. It is also present in the Roberts Mountains, at Lone Mountain, at Table Mountain, in the northern Simpson Park Mountains, and in the Cortez Range. In addition, a few specimens have been seen in a single collection from the *Acrospirifer kobehana* Zone of the Sulphur Spring Range.

FIGURED SPECIMENS: The holotype is USNM 140409. Other figured specimens are USNM 140408, 140410, 140411, and 140413-140420.

UNFIGURED SPECIMEN: USNM 140412. This specimen, however, was figured by Harper and others (1967, Pl. 8, fig. 4).

SUBORDER CHONETOIDEA

Superfamily CHONETACEA Bronn, 1862
Family CHONETIDAE Bronn, 1862

DISCUSSION: Chonetids are not plentiful in the Siegenian and Emsian of Nevada, and only occur in any abundance in the *Eurekaspirifer pinyonensis* Zone. The genera occurring are "*Strophochonetes,*" *Chonetes, Parachonetes,* and *Anoplia.* Species of "*Strophochonetes*" that occur in the Rabbit Hill Limestone and in the *Eurekaspirifer pinyonensis* Zone of the McColley Canyon Formation lack an enlarged median capilla on the pedicle valve and probably eventually will be removed from the genus *Strophochonetes. Chonetes* sp. is poorly illustrated in the present paper due to the inadequacy of available collections, but the form does occur in abundance in certain beds of the *E. pinyonensis* Zone. Its shell almost always is found in an altered, dark red-brown condition, and the external costellae are so poorly preserved that they tend to be nearly imperceptible.

The writer regards these early chonetid genera as representatives of three subfamilies of the Chonetidae, with *Anoplia* being representative of

ORDER STROPHOMENIDA

a separate family as was proposed by Boucot and Harper (1968). The writer has regarded *Parachonetes* as a derivative of *Protochonetes*, probably through the coarse-ribbed genus *Eccentricosta*, and originally followed Muir-Wood in alloting *Protochonetes* to the subfamily Devonochonetinae. Now it appears that because of the cardinalia, in particular the bifid internal nature of the cardinal process of *Protochonetes*, it is best regarded as an early representative of the subfamily Chonetinae, and it probably is the genus from which *Chonetes* s.s. was derived. *Strophochonetes* is regarded here as a member of a separate subfamily Strophochonetinae Muir-Wood, from which *Protochonetes* was probably derived. Inclusion of *Protochonetes* in the Chonetinae suggests no close connection between *Parachonetes* and the subfamily Devonochonetinae and, because of obvious differences of morphology, *Parachonetes* is here included in a new subfamily Parachonetinae.

Subfamily STROPHOCHONETINAE Muir-Wood, 1962
Genus *Strophochonetes* Muir-Wood, 1962

TYPE SPECIES: *Chonetes cingulata* Lindström, 1861, p. 374, Pl. 13, fig. 19 (Pl. 30, figs. 1-5 of this paper).

"*Strophochonetes*" *filistriata* (Walcott, 1884)
(Pl. 31, figs. 1-17)

Chonetes filistriata Walcott, 1884, p. 127, Pl. 13, figs. 15, 15a.
Chonetes cf. *filistriata* Merriam, 1940, Pl. 7, fig. 10.
Chonetes filistriata Merriam, 1940, Pl. 11, fig. 14.
Devonochonetes filistriata Muir-Wood, 1962, p. 44.

DISCUSSION: "*Strophochonetes*" *filistriata* was incorrectly assigned to *Devonochonetes* by Muir-Wood (1962, p. 44) as evidenced by the presence of a few long spines normal to the hinge line and of a bilobate cardinal process. *Devonochonetes* has short, more numerous spines diverging at low angles from the hinge line and a spatulate cardinal process. Muir-Wood (1962) did not illustrate the interior structures of *Strophochonetes cingulatus*, and it is not possible from her description to determine the basis for description of interior structures of the genus *Strophochonetes*. In order to eliminate any doubt about the structure of the genus, typical specimens from Gotland, including Lindström's original of *S. cingulatus*, were examined and internal molds of two brachial valves were prepared and are illustrated in Plate 30. These bear a short pair of anderidia and a simple, internally bilobate cardinal process. A median septum was not found to be developed.

"*Strophochonetes*" *filistriata* described below, differs morphologically from the type species in two respects. It lacks an enlarged median capilla and bears a dorsal median septum, at least in some specimens. The variable presence of a dorsal median septum suggests that that structure is not a reliable generic feature. Whether or not the presence or absence of an enlarged median capilla on the pedicle valve is of generic value is still to be determined. The taxon *Quadrikentron*, proposed as a subgenus of *Strophochonetes* by Boucot and Gauri (1966) may be more closely related to *Chonetes* s.s., as it has similar cardinalia. Both of the latter differ from "*Strophochonetes*" *filistriata* in this respect.

EXTERIOR: The shells are small- or medium-sized and flatly concavoconvex. In outline they are transversely shield-shaped. Maximum width of most specimens is slightly behind midlength. The cardinal angles are in most cases obtuse, but only slightly more than 90 degrees. In some shells there are small auricular projections at the cardinal angles making them acute. The ventral interarea is flat or only slightly incurved. Its inclination is apsacline, commonly steep and nearly catacline in its median half. The interarea is modified at intervals by spine bases developed along the beak ridges. The delthyrium is broadly triangular, making an angle of about 90 degrees. It is rounded at the apex. No deltidial covering was observed. The dorsal interarea is anacline and nearly linear.

The ornament consists of fine radial costellae that increase in number anteriorly by both bifurcation and implantation on each valve. On one well-preserved pedicle valve, there are approximately 20 costellae in a space of 5 mm at a distance of 5 mm anterior to the beak.

INTERIOR OF PEDICLE VALVE: There is a short, thin, undivided median septum posteriorly. The diductor muscle field is wide and flabellate, but only faintly impressed. The interior is crenulated by the impress of the costellae.

INTERIOR OF BRACHIAL VALVE: The cardinalia consist of socket ridges that join a bilobed cardinal process medially. The cardinal process is cordate in transverse cross section. The socket ridges laterally are nearly parallel to the hinge line. Posteriorly, there are two short anderidia. A breviseptum is commonly present near midlength.

DISCUSSION: The type specimens of this species (USNM 13811) are labeled "Lower Dev., Combs Mt., Eureka district, Nevada." They are larger than most specimens from the Roberts Mountains, measuring about 18 mm in width at the hinge line compared to about 12 mm for the specimens from the Roberts Mountains. The types also bear a faint and shallow sulcus in the anterior of the pedicle valve.

FIGURED SPECIMENS: Merriam (1940, Pl. 11) designated one specimen, USNM 13811, on the type slab as holotype. It is illustrated here as figure 1 on Plate 31. Two paratypes, USNM 13811A and 13811C, are figured. Hypotypes include USNM 156990-157001.

"Strophochonetes" sp.
(Pl. 30, fig. 6)

DISCUSSION: This chonetid species is rare in the available collections, and no interiors are known. A few specimens have been seen in the type exposures of the Rabbit Hill Limestone, and they are fairly common, especially in the lower part of that formation, on the east flank of Coal Canyon. They are small, nearly flat, and shield-shaped, commonly with four long spines on the ventral hinge line. The spines project slightly laterally at their bases, then turn and extend normal to the hinge line for a distance, in some specimens, equalling the length of the shell. The external ornament is finely costellate, without an enlarged median capilla.

OCCURRENCE: *Spinoplasia* Zone at Rabbit Hill and Coal Canyon, and rarely in the *Trematospira* Subzone in the Sulphur Spring Range.

FIGURED SPECIMEN: USNM 156985.

Subfamily CHONETINAE Bronn, 1862
Genus *Chonetes* Fischer de Waldheim, 1830
TYPE SPECIES: *Terebratulites sarcinulatus* Schlotheim, 1820, p. 256, 1822, Pl. 29, fig. 3.

Chonetes sp.
(Pl. 30, figs. 7-10)

EXTERIOR: The outline is decidedly transverse, subsemicircular to nearly shield-shaped. In lateral profile the valves are plano-convex to slightly concavo-convex, with the pedicle valve only slightly inflated. The spines and other hinge line structures are yet to be studied.

The exterior bears numerous, very fine, rounded radial costellae, but alteration processes that commonly produce a distinctive red-brown shell apparently are responsible for obliterating the external ornament on most specimens. Those illustrated in Plate 30, figures 9, 10 appear to be quite unusual in their excellent preservation.

INTERIOR OF PEDICLE VALVE: There is a short, prominent median septum dividing a broad posteriorly impressed muscle scar whose posterolateral edges diverge at over 100 degrees and are the only parts clearly discernible (Pl. 30, fig. 7).

INTERIOR OF BRACHIAL VALVE: The cardinal process is internally bilobate and is situated on a ponderous pair of inner socket ridges that parallel the hinge line almost to the lateral extremities. These are not well shown on the specimen illustrated in Plate 30, figure 8, but the structure is very marked on a few recently acquired specimens. Adductor scars are impressed between relatively short, slightly divergent anderidia and are divided by a low median septum. Accessory septa have not been seen in the few Nevada specimens that are exposed internally, but careful examination of all the specimens of *Chonetes sarcinulatus* in the Boucot collection by E. D. Gill in the fall of 1965 revealed to him (oral commun., 1966) that these structures are only uncommonly developed even amongst well-preserved specimens.

OCCURRENCE: The species has been seen in the *Eurekaspirifer pinyonensis* Zone in the Roberts Mountains, at Lone Mountain, and at Table Mountain in the Mahogany Hills.

DISCUSSION: On external grounds, at least, this form resembles *Chonetes burgeniana* Zeiler (Kayser, 1889, Pl. 22, figs. 7, 8) and *Chonetes subextensa* Khalfin (1948, Pl. 11, figs. 8, 9).

FIGURED SPECIMENS: USNM 156986-156989.

Subfamily PARACHONETINAE n. subfam.
DIAGNOSIS: Costate Chonetidae with prominent dorsal alveolus.
Genus *Parachonetes* Johnson, 1966

TYPE SPECIES: *Chonetes macrostriata* Walcott, 1884, p. 126.

DIAGNOSIS: Chonetids with coarse costellae originating along the hinge line of some specimens. Brachial valve cardinalia characterized by a pair of posteriorly conjunct plates forming an exteriorly quadrilobate cardinal process on large specimens and a more or less deep alveolus.

DISCUSSION: Boucot and Harper (1968) state that since, "the brachial valve of adult *Parachonetes* has a median septum as in *Protochonetes*, it appears likely that it is not a direct descendant of *Eccentricosta*, but instead shares a common ancestry with *Eccentricosta* from *Protochonetes*." Implicit in this conclusion is the assumption that morphological features once developed are always retained in a lineage, that is, evolution is strictly orthogenetic. The present writer cannot subscribe to the latter hypothesis, and would point out instead that as small *Parachonetes* lacks a median septum and as large specimens possess one, it is quite possible that *Parachonetes* developed from an aseptate ancestor, such as *Eccentricosta*, rather than a septate one, such as *Protochonetes*. Furthermore, the assumption that *Eccentricosta* lacks a median septum in all specimens cannot be sustained. Specimens of *Eccentricosta* are commonly very thin shelled, without secondary material, but even on these specimens it has been possible to distinguish a dorsal median septum on some specimens of *Eccentricosta* loaned by Z. P. Bowen, one of which is illustrated herein, Plate 33, figures 11, 12. In short, from either point of view the presence or absence of a median septum in the ancestral stock of *Parachonetes* appears to be of little consequence. The writer's examination of *Eccentricosta* and *Protochonetes* convinces him that they are internally virtually identical and differ principally in the presence of coarser costellae in *Eccentricosta*. In this feature it is intermediate between *Protochonetes* and *Parachonetes*, and thus is still a likely candidate for ancestor of *Parachonetes*.

AGE: The writer originally supposed that *Parachonetes* persisted into the Middle Devonian (Johnson, 1966c), based principally on the Receptaculites Limestone occurrence of southeastern Australia. However, the assertion that these beds are Lower Devonian (Philip and Pedder, 1964) has now been adequately demonstrated by Philip's conodont work (1966, 1967; also Philip and Jackson, 1967). *Parachonetes* appears to be widely distributed in Emsian age beds of the Old World province.

Parachonetes macrostriatus (Walcott, 1884)
(Pl. 32, figs. 1-18; Pl. 33, figs. 1-5)

Chonetes macrostriata Walcott, 1884, p. 126, Pl. 2, fig. 13(?); Pl. 13, figs. 14, 14a, 14b, 14c.

Chonetes macrostriata Merriam, 1940, p. 55, Pl. 6, fig. 4.

Chonetes macrostriatus Cooper, 1944, p. 345, Pl. 134, fig. 16.

Longispina macrostriatus Muir-Wood, 1962, p. 47.

Parachonetes macrostriatus Johnson, 1966c, p. 367, Pl. 62, figs. 1-17; Pl. 63, figs. 1-3.

OCCURRENCE: *Parachonetes macrostriatus* first appears in the *Acrospirifer kobehana* Zone on the north side of the Roberts Mountains. It is present in the *Eurekaspirifer pinyonensis* Zone in the Sulphur Spring Range, in the Roberts Mountains, at Lone Mountain, at Table Mountain in the Mahogany Hills, and in the northern Simpson Park Range. It has also been noted in the Spotted Range of southern Nevada (Johnson, 1966c).

FIGURED SPECIMENS: USNM 13809, 13809A, 140435-140446. The holotype is USNM 13809.

DISCUSSION: The widely reported species *Parachonetes verneuili* (Barrande) is illustrated for comparison (Pl. 33, figs. 6-10). *Parachonetes macrostriatus* was described and compared with *P. verneuili* in another paper (Johnson, 1966c).

Family ANOPLIIDAE Muir-Wood, 1962
Subfamily ANOPLIINAE Muir-Wood, 1962
Genus *Anoplia* Hall and Clarke, 1892
TYPE SPECIES: *Leptaena nucleata* Hall, 1857, p. 47.
Anoplia sp. A
(Pl. 34, figs. 1-4)

EXTERIOR: The valves are of minute size and semicircular in outline with a long hinge line. Cardinal angles are slightly auriculate on some specimens. Pedicle valves are moderately convex in lateral profile. No brachial valves have been seen. The ventral umbo is very low and unobtrusive. The beak is minute. The ventral interarea is relatively long and orthocline to anacline. There is a narrow triangular delthyrium.

The surface of the valves is smooth, apparently without radial or concentric ornament.

INTERIOR OF PEDICLE VALVE: There is a short, stout median septum dividing the muscle field posteriorly. On a single specimen, which is among the larger ones available, the median septum splits anteriorly and projects a very short distance anterolaterally as two short ridges. The diductor muscle field is broad and is impressed posterolaterally. Anteriorly, it blends with the inner surface of the valve.

OCCURRENCE: *Anoplia* sp. A occurs in the *Spinoplasia* Zone at Coal Canyon and at Rabbit Hill, and in the *Trematospira* Zone in the Sulphur Spring Range.

COMPARISON: *Anoplia* sp. A is smaller than *A. nucleata* and most closely resembles the species *helderbergiae* Schuchert and *pygmaea* Amsden which Amsden (1963) placed in *Anopliopsis*. Amsden's assignment to *Anopliopsis* was made because of the presence of spines, but E. D. Gill (oral commun., 1965) discovered spines on Camden specimens of *Anoplia nucleata*. No spines nor spine bases were observed on the Nevada specimens and no brachial valves are available for study so the possibility of assignment to *Anopliopsis* cannot be considered.

FIGURED SPECIMENS: USNM 157002-157004.

Anoplia elongata Johnson, n. sp.
(Pl. 34, figs. 5-15)

DIAGNOSIS: *Anoplia* with a broad umbo and an elongate outline. Shells deeply concavo-convex.

EXTERIOR: The shells are minute, suboval to elongate, pentagonal in outline, and concavo-convex in lateral profile. The ventral umbo is broad and low, merging imperceptibly with the posterolateral flanks. Anteriorly the pedicle valve may be slightly carinate. Cardinal angles are commonly obtuse or are approximately right angles. Maximum width of the shell is attained slightly anterior to the hinge line. The ventral interarea equals about two-thirds to three-fourths of the maximum width. It is low, triangular, and anacline. The ventral beak is minute, and there is a small subtriangular open delthyrium. The brachial valve appears to lack an interarea.

The ornament consists of fine concentric growth lines. No radial ornament has been observed.

INTERIOR OF PEDICLE VALVE: The hinge teeth are relatively large and stubby and are attached directly to the inner side of the palintrope along the sides of the delthyrium. A median septum divides the muscle field. The diductor muscle impressions are broadly cordate and deeply impressed posterolaterally. Anteriorly, they blend with the interior of the shell.

INTERIOR OF BRACHIAL VALVE: There is a strong pair of accessory septa originating posteriorly and diverging slightly from one another. These septa become higher anteriorly and reach a maximum height at about three-quarters of the length of the shell. They continue and become lower toward the anterior edge of the shell. There is no median septum. Flanking the accessory septa at their place of greatest height, there appears to be two pairs of more or less poorly developed radial ridges.

OCCURRENCE: *Anoplia elongata* is present in the *Acrospirifer kobehana* Zone in the Roberts Mountains and in the Sulphur Spring Range. It is also abundant in the lower *Eurekaspirifer pinyonensis* Zone at Lone Mountain.

COMPARISON: *Anoplia elongata* has a much larger length-width ratio than *A. nucleata* and is considerably smaller than that species. The length-width ratio also allows *Anoplia elongata* to be distinguished from *Anoplia* sp. A as well as the *Anopliopsis*-like species *helderbergiae* and *pygmaea*.

FIGURED SPECIMENS: The specimen USNM 157005, illustrated in figures 11, 12 on Plate 34, is the holotype. It is 3.5 mm long and 3.3 mm wide. Other figured specimens are USNM 157006-157010.

SUBORDER PRODUCTOIDEA

Superfamily PRODUCTACEA Gray, 1840
Family PRODUCTELLIDAE Schuchert and LeVene, 1929
Subfamily PRODUCTELLINAE Schuchert and LeVene, 1929
Genus *Spinulicosta* Nalivkin, 1937

TYPE SPECIES: *Productus spinulicostae* Hall, 1857, p. 173.

DISCUSSION: A few specimens described here as *Spinulicosta* appear to be as old or older than any previously illustrated from North America. For this reason it seems relevant to review the earliest reports of productids on a world-wide basis.

Erben (1962, p. 69) recently recorded the lower range of *Productella* in western Europe as beginning in the upper Emsian. This report is based on the occurrence of *Productella subaculeata* from the Heisdorf beds and is according to Solle (1942, p. 19; Erben, written commun., August 5, 1966). Further, according to W. Struve (written commun., Nov. 10, 1966) the earliest (west European) productellid certainly identified is that recorded by Solle (1942, p. 19) and that, "this record has been checked recently by R. Werner (Senckenberg), affirmed, and very exactly located. By this it is evident, that *Productella* occurs in beds of Heisdorf (upper Emsian) age." This occurrence, although checked by competent observers, has recently become complicated by the conodont evidence relating to the age of the Heisdorf beds which indicates that at least in part they are Eifelian rather than high upper Emsian (Klapper and Ziegler, 1967). Another German occurrence, that of *Productella antiqua* Schmidt (1939) is not at all satisfactory, because the illustrated specimens seem to be doubtful representatives of *Productella* or even of productids in the broad sense. These need to be restudied.

As far as the writer knows, there are no reports of productids in the Lower Devonian of Bohemia. In Russia, the earliest occurrences appear to be in the Salairka beds (*Paraspirifer gurjevskensis* Zone) at the western margin of the Kuznetzk Basin (Rzhonsnitskaya, 1960a, p. 131, 1962). The Salairka beds are concluded to be of late Emsian age. In the Urals, productids have been described and illustrated by Khodalevich (1951, Pl. 4, figs. 1-3) from "Eifelian" beds, but some of these beds at least occupy the Salairka interval and are probably of late Emsian age.

Productids are present in the Floresta beds of Colombia. The illustrated specimen (Caster, 1939, Pl. 7, figs. 19, 20) probably belongs to *Spinulicosta* as the spines emanate from radial ridges The Floresta fauna almost certainly correlates with the Camden-Schoharie of the eastern United States because of the abundance of *Eodevonaria* (Caster, 1939, Pl. 7), and because of the presence of *Prionothyris* (=*Meganteris australis* Caster, 1939, Pl. 13, figs. 11-13), and, therefore, is of Emsian age. In the United States, productids have been reported in possible Camden-Schoharie equivalents by Kindle (1938) and by Boucot and Johnson (1968).

In their monograph on the Productoidea, Muir-Wood and Cooper (1960, p. 10) list *Spinulicosta* as the oldest known productid, and cite the earliest occurrence as the "upper part of the Lower Devonian, Columbus limestone, (Onondaga formation) of Ohio." However, on page 153 of the same volume, the occurrence of *Spinulicosta* is listed as Middle and Upper Devonian. This latter citation apparently was the correct one at the time because the Columbus Limestone is of Onondaga age, younger than the highest Lower Devonian, of Camden-Schoharie age, and which has been independently dated as Eifelian on the basis of its goniatites (House, 1962, p. 253).

The first stropholosiids, although they are not of direct concern here, are cited by Muir-Wood and Cooper (1960, p. 10) as occurring in the Lower Devonian of Gaspé, represented by *Devonalosia*. However, the range of that genus (Muir-Wood and Cooper, 1960, p. 84) is given as Middle Devonian. The latter citation possibly is the erroneous one as it fails to take into account the Gaspé occurrence mentioned earlier, but the latter is not proved to be Lower Devonian.

In the brachiopod Treatise, the range of the Productellidae is given as upper Lower Devonian to Upper Mississippian, and the range of *Spinulicosta*, as Lower Devonian to Upper Devonian. The lower range appears to be correct in the light of the review given above, although Muir-Wood gave no notation of new evidence for a range of *Spinulicosta* below beds equivalent to the Columbus Limestone, that is, Eifelian (Muir-Wood, 1965, p. 464, 465).

Spinulicosta? sp.
(Pl. 35, figs. 3-13)

DISCUSSION: Only four specimens are known to the writer from Nevada and three of these appear, on external grounds at least, to be assignable to *Spinulicosta*. The fourth specimen (Pl. 35, figs. 3, 4) appears slightly less likely to belong to *Spinulicosta*, but all are listed with a query because the internal structures are unknown.

EXTERIOR: Only pedicle valves are available. They are very strongly convex and strongly curved, through about 180 degrees, along the midline. Slight differentiation of the umbo from the flanks occurs on the two smallest specimens because of the development of gentle auriculations. The ventral beak is blunt, but relatively prominent.

The ornament consists of very fine, closely spaced, concentric growth lines seen on one well-preserved specimen (Pl. 35, fig. 8) and of relatively numerous, raised spine bases that are situated, for the most part, on radial ridges or corrugations that are most strongly developed anteriorly (Pl. 35, fig. 6).

The largest specimen, which seems more like *Productella* and less like *Spinulicosta* by the absence of radial ridges, is broader and is a relatively shallower shell than the others described above. In addition, it possesses relatively prominent concentric growth lines at a few intervals and possibly represents another genus such as *Productella*.

OCCURRENCE: The four specimens are from the *Eurekaspirifer pinyonensis* Zone at Lone Mountain and at McColley Canyon. The McColley Canyon locality was the source of a single specimen found in association with abundant *Parachonetes macrostriatus* illustrated elsewhere in the present paper (Pl. 32). Two of the specimens from Lone Mountain are from unknown positions within the zone, but one specimen (Pl. 35, figs. 5-7) was collected in 1966 from approximately the middle of the *Eurekaspirifer pinyonensis* Zone, in the interval 485-535 feet above the base of the McColley Canyon Formation at Lone Mountain, by Mr. Eric Gronberg. It occurs with *Schizophoria nevadensis, Gypidula loweryi, Megastrophia iddingsi, Chonetes* sp. *Trigonirhynchia occidens, Atrypa nevadana*, and *Brachyspirifer pinyonoides*.

FIGURED SPECIMENS: USNM 157011-157014.

Order Rhynchonellida

SUBORDER RHYNCHONELLOIDEA

DISCUSSION: Rhynchonellids are one of the most stratigraphically useful groups of brachiopods in the Lower Devonian of Nevada. In every case the rhynchonellid species prove to be restricted to a single zone or pair of sequential zones in which they have considerable value as index fossils. In the Gedinnian faunas of the upper part of the Roberts Mountains Formation, being investigated by the writer, A. J. Boucot, and M. A. Murphy, the characteristic rhynchonellids are species of *Ancillotoechia* and *Sphaerirhynchia*, neither of which is known at any other level. In the *Quadrithyris* Zone, *Machaeraria* sp. is the common rhynchonellid, and it occurs in no other zone in central Nevada. In the next younger *Spinoplasia* Zone and the basal portion of the succeeding *Trematospira* Zone (that is, *Trematospira* Subzone), *Pleiopleurina anticlastica* is abundant. A single specimen of *Pegmarhynchia* was found in the *Spinoplasia* Zone. *Ancillotoechia aptata* is restricted to the *Trematospira* Subzone. No rhynchonellid is common in the *kobehana* Zone, but *Trigonirhynchia occidens* is widespread in its occurrence in the *pinyonensis* Zone. Other rhynchonellids that occur in the *pinyonensis* Zone are also restricted to that zone, but are uncommon.

None of the Lower Devonian rhynchonellid species from Nevada seems to be related to those of other horizons, indicating that they are invaders from other regions.

Superfamily CAMAROTOECHIACEA Schuchert and LeVene, 1929
[*nom. transl.* Havlíček, 1960 *ex* Camarotoechiidae Schuchert and LeVene, 1929]

Family RHYNCHOTREMATIDAE Schuchert, 1913
Subfamily ORTHORHYNCHULINAE Cooper, 1956
Genus *Machaeraria* Cooper, 1955

TYPE SPECIES: *Rhynchonella formosa* Hall, 1857, p. 76, figs. 1-5.

DISCUSSION: This genus and closely related forms presently assigned to several other genera are widespread geographically in beds of Early Devonian age. The type species is from the Becraft (Siegenian) of New York. Bowen (1967) has described a species from the Keyser Limestone.

The writer examined the types of "*Rhynchonella*" *carolina* Hall (1867b,

Pl. 54) which proves to be a *Machaeraria*. Its probable Eifelian age makes it the youngest known species of *Machaeraria*.

Stenochisma althi Kozlowski (1929) probably belongs to *Machaeraria* because it has similar cardinalia to *M. formosa*, including crural plates (Kozlowski, 1929, p. 147, Fig. 44C). The species is from the lower Gedinnian (Tajna beds) of Podolia. *Latonotoechia, Sicorhyncha,* and *Zlichorhynchus* from the Lower Devonian of Bohemia (Havlíček, 1961, 1963) appear to be closely related to *Machaeraria*, but *Latonotoechia* has a large amount of shell material obscuring the details of structure in the brachial valve (Havlíček, 1961, p. 25). *Sicorhyncha* may lack crural plates, and *Zlichorhynchus* has long ones (Havlíček, 1963, Pl. 1, figs. 3, 5).

Johnson and Boucot (*in* Jackson and Lenz, 1963) have reported *Latonotoechia?* from Yukon Territory, but this form has been recently described by Lenz (1967a) as a new genus *Thliborhynchia*.

Machaeraria sp.
(Pl. 36, figs. 1-14)

EXTERIOR: The shells are commonly transverse-oval in outline, although one small shell is elongate (*Sicorhyncha*-like, *see* Pl. 36, figs. 12, 13). In lateral profile the valves are unequally biconvex with the brachial valve more convex. The ventral beak is blunt and rounded. The delthyrium encloses an angle of about 60 degrees, and the foramen is within the delthyrium in small shells, but apical in larger ones. The pedicle valve has a flat-bottomed sulcus, and the brachial valve bears a corresponding fold.

The ornament consists of radial costae of rounded to subangular cross section. Anteriorly, on large specimens, the costae may be sharply angular. There are commonly 7 to 9 costae on each flank of the valves and 4 or 5 in the sulcus or on the fold. One or two parietal costae or pairs of costae are present on some shells, and in one large pedicle valve (Pl. 36, figs. 1, 2) there is a single pair of parietal costae that becomes obsolescent anteriorly. No costae have been seen to bifurcate.

INTERIOR OF PEDICLE VALVE: The interior is poorly known due to lack of well-preserved material. There are two short subparallel dental lamellae set close against the posterolateral walls of small- and medium-sized shells, and the umbonal cavities are partially filled with secondary shell material. Dental lamellae become obsolescent in large shells. The muscle scar is subcircular and slightly impressed; there are no muscle-bounding ridges.

INTERIOR OF BRACHIAL VALVE: The sockets are bounded posterolaterally by the edge of the shell. Socket plates form the bases and inner edges of the sockets. The anteromedial edges of the socket plates are extended ventrally beyond the average level of the edges of the sockets. The crura originate from thickenings on the inner sides of the socket plates and project ventrally as ribbon-like structures. The crura appear to be connected to the floor of the valve by crural plates, but the space between the crural plates and the inner walls of the socket plates is largely filled by shell material.

Medially there is a bladelike cardinal process which may or may not be continuous with a low myophragm anteriorly.

DISCUSSION: This species will probably prove to be new when more carefully studied, but that will only be possible when more material becomes available. In the meantime it is worth noting that there is a definite affinity with the type species as evidenced by the long parietal slopes of both forms. The Nevada form might be a geographic subspecies of *M. formosa* since the two are the same age.

OCCURRENCE: *Machaeraria* sp. is known in the Great Basin only from the *Quadrithyris* Zone at Coal Canyon and in the upper Vaughn Gulch Limestone, Independence quadrangle, southeastern California.

FIGURED SPECIMENS: USNM 157017-157024.

Family HEBETOECHIIDAE Havlíček, 1960
Genus *Pleiopleurina* Schmidt, 1964

TYPE SPECIES: *Atrypa pleiopleura* Conrad, 1841, p. 55.

DIAGNOSIS: Internally like *Plethorhyncha*. Externally it bears bifurcating costae and a ventral sulcus in contrast to *Plethorhyncha* which has simple regular costae and a subcuboidal shape.

Schmidt (1964, p. 506) proposed *Pleiopleurina* without a discussion of its relation to other genera or how it may be distinguished from other related genera. Her diagnosis is as follows:

Large, but not especially thick. Pedicle valve gently convex, brachial valve strongly convex. Sulcus and tongue broad. Fold strongly accentuated anteriorly. Ribs numerous, relatively fine, rounded, beginning at the beak. Shell margins serrated. Ventral muscle field situated far posteriorly, divided longitudinally by a low ridge. Dorsal septum short. Hinge plates joined by a stout cardinal process which consists of two funnel-shaped shafts. Crura thick. (J.G.J./tr).

Internally the structure of *Pleiopleurina* is the same as in *Plethorhyncha*. The two genera thus differ essentially in the form of the shell and in the type of costae.

Pleiopleurina anticlastica Johnson n. sp.
(Pl. 36, figs, 15-18; Pl. 37, figs. 1-16)

EXTERIOR: The shells are elongate to transverse-subpentagonal in outline. In large shells the brachial valve is about twice as convex as the pedicle valve. The ventral beak is small and incurved and touches the dorsal umbo. There is an apical foramen, but no interarea. Maximum width is commonly near midlength. Anteriorly, there is a strong sulcus on the pedicle valve and a corresponding fold on the brachial valve. The sulcus and fold may be relatively wide or only about half the maximum width and may be rectangular or smoothly rounded. The commissure is crenulate.

The ornament consists of radial costae which bifurcate anteriorly. On several pedicle valves there are about 16 costae in a space of 10 mm, at a distance 10 mm anterior to the beak. Costae become larger anteriorly even though there is bifurcation.

INTERIOR OF PEDICLE VALVE: The hinge teeth are small and knoblike and rest directly on the inner wall of the valve, unsupported by dental lamellae. The pedicle cavity is pentagonal and is separately defined posterior to the diductor muscle impressions. The diductor muscle field is elongate-oval, nonflabellate, and strongly impressed on large specimens. Medially, it is divided by a more or less well-developed myophragm. In some specimens the adductor muscle scars are present as a pair of arcuate ridges at the center of the diductor muscle field. The umbonal cavities may be pustulose.

INTERIOR OF BRACHIAL VALVE: The cardinalia consist of socket plates which diverge widely laterally and define the sockets along the posterolateral edges of the interior of the shell. On the inner edges of the socket plates there is a pair of outer hinge plates that unite medially with a septalium which is supported by a stout median septum. On medium- and large-sized shells, the septalium is filled with shell material which forms the base of a ventrally directed, bilobed cardinal process. In transverse (horizontal) section, the cardinal process has the shape of a pair of longitudinally truncated cylinders joined along one edge. The concave side of this pair of semicylindrical plates faces posteriorly. The adductor muscle field is composed of separate pairs of posterior and anterior impressions. The posterior pair lies lateral to the anterior pair, and both sets are elongate.

OCCURRENCE: *Pleiopleurina anticlastica* is common in the Rabbit Hill Limestone at Rabbit Hill (*Spinoplasia* Zone) and in the *Trematospira* Subzone at the base of the McColley Canyon Formation at McColley Canyon, Sulphur Spring Range.

COMPARISON: The writer examined specimens of *Pleiopleurina pleiopleura* on loan from the U.S. National Museum. The specimens are from the Glenerie Formation, New York highway 9W, 1 mile north of Glenerie, New York, and one mile south of Cockburn. The New York species is much less sulcate than *P. anticlastica* even though the specimens examined are appreciably larger than the largest specimens of *anticlastica*. The ribbing pattern on both species is closely similar.

FIGURED SPECIMENS: USNM 157025-157034. The holotype is USNM 157032.

Family TRIGONIRHYNCHIIDAE Schmidt, 1965
Genus *Ancillotoechia* Havlíček, 1959
TYPE SPECIES: *Rhynchonella ancillans* Barrande, 1879, Pl. 36.
Ancillotoechia aptata Johnson, n. sp
(Pl. 38, figs. 1-21; Text-Fig. 5)

DIAGNOSIS: Cuneate to subpentagonal *Ancillotoechia* with three or four angular costae on the dorsal fold and two or three angular costae in the ventral sulcus. Flanks of medium- and large-sized shells have five to seven angular costae.

EXTERIOR: Small shells (4 to 5 mm maximum dimension) are cuneate to broadly subpyriform in outline and approximately equally biconvex in lateral profile. Pedicle valves attain greatest convexity at the umbo

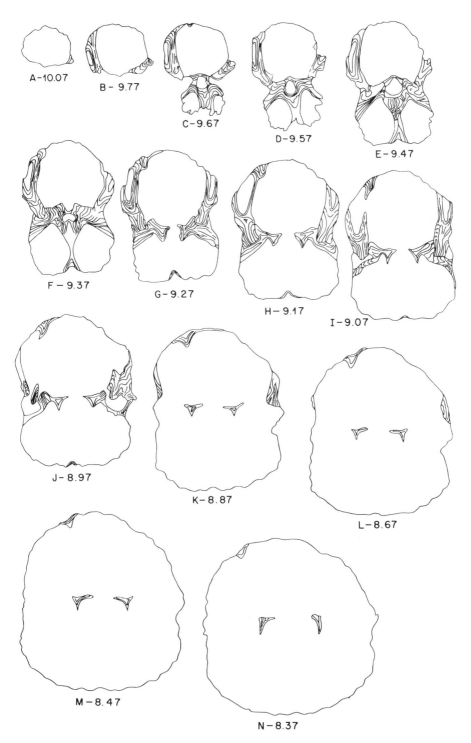

Figure 5. Serial sections × 10 of *Ancillotoechia aptata* Johnson, n. sp., USNM 157039A. Numbers indicate distance from anterior commissure, expressed in millimeters.

while brachial valves are most convex at about midlength. Larger shells lose the cuneate outline upon development of the dorsal fold and ventral sulcus and commonly are transversely pentagonal and unequally biconvex in lateral profile with the brachial valve much more strongly convex than the pedicle valve. The ventral beak is small and pointed and projects only a slight distance beyond the posterior of the brachial valve. The beak is only slightly incurved. The fold and sulcus on most specimens are well defined with the fold rising relatively abruptly from the flanks, but on some specimens parietal costae intervene irregularly. The base of the sulcus may be flat or gently curved.

The ornament consists of prominent angular radial costae, three or four on the fold of the brachial valve and two or three in the sulcus of the pedicle valve. Costae on the flanks of each valve vary from five to seven except on small shells. No bifurcation of costae has been observed, and prominent concentric growth lines are not present.

INTERIOR OF PEDICLE VALVE: Thin, short, but well-defined dental lamellae are present. They are slightly convex laterally and are set closely to the inner walls of the shell. The muscle scar is not impressed. The interior is crenulated by the impress of the costae.

INTERIOR OF BRACHIAL VALVE: The sockets are widely divergent with their outer edges bounded by the posterolateral edges of the shell. Medially, the inner edges of the socket plates connect with a small septalium which is supported by a low median septum. The height of the median septum is equal only to the depth of the septalium which is, therefore, elevated only slightly above the floor of the valve. The septalium is partially covered by crural flanges which meet along the midline posteriorly. The interior of the valve is crenulated by the impress of the costellae.

OCCURRENCE: The species is common in the *Trematospira* Zone in the Sulphur Spring Range. It is also present in the *Trematospira* Zone at the base of the McColley Canyon Formation at Coal Canyon, northern Simpson Park Range.

COMPARISON: *Ancillotoechia aptata* is a homeomorph of several *Ancillotoechia* species. Silurian species lack a prominent sulcus and in this respect only small specimens of *A. aptata* resemble the type species. *A. aptata* externally at least, appears to very closely resemble *A. cumberlandica* (Rowe) from the Ridgely Sandstone of Cumberland, Maryland (Schuchert and others, 1913, Pl. 62, figs. 23, 24). Rowe's species also occurs in Virginia in beds of Port Jervis Limestone (earliest Deerpark) age (Swartz, 1929, Pl. 8, figs. 23-25). Swartz called these beds Becraft, but Woodward (1943) showed that they certainly are younger. The same fauna is found high in the Licking Creek Limestone of Pennsylvania, which was shown by Swartz (*in* Willard and others, 1939, Table 14) to be a Port Jervis equivalent.

Possibly *A. aptata* is no more than a geographic variant of *A. cumberlandica*. They differ essentially in the slightly more acute beak angle of *A. cumberlandica*.

FIGURED SPECIMENS: USNM 157035-157041. The holotype is USNM 157035, illustrated in figures 1-5 of Plate 38. It comes from the *Trematospira* Zone in the Sulphur Spring Range.

Genus *Trigonirhynchia* Cooper, 1942
TYPE SPECIES: *Uncinulina fallaciosa* Bayle, 1878, Pl. 13, figs. 13-16.

Trigonirhynchia occidens (Walcott, 1884)
(Pl. 38, figs. 22-26; Pl. 39, figs. 6-28)

Rhynchonella? occidens Walcott, 1884, p. 152, Pl. 15, figs. 3, 3a, 3b.

EXTERIOR: The shells are subcuboidal to elongate-subpentagonal or subpyriform in outline and unequally biconvex in lateral profile. The brachial valve is more strongly convex. The ventral beak is not prominent and projects only slightly beyond the dorsal umbo which rests beneath it. The delthyrium encloses an angle of about 60 degrees anterior to an apical foramen. There is a very small plane area composed of two narrow bands adjacent to the edges of the delthyrium. Anteriorly, there is a fold of rectangular to rounded cross section on the brachial valve and a sulcus and strongly deflected tongue on the pedicle valve. The fold, however, is not strongly elevated.

The ornament is of radial costae of rounded cross section separated by U-shaped interspaces of approximately the same width as the costae. The costae are drawn out to points which interlock along the commissure with the spaces between costae on the opposite valve. The radial costae number 21 to 23 on pedicle valves with 5 of this total being in the sulcus. On the brachial valve there are 22 to 24 costae of which 6 are on the fold. The number of costae o nthe fold and in the sulcus is a consistent feature of numerous medium- and large-sized shells from several localities. In specimens in which the sulcus is uncommonly deep, the first lateral costae may be parietal, but in most shells the fold and sulcus are well defined.

INTERIOR OF PEDICLE VALVE: There are two well-developed, thin dental lamellae present. They diverge slightly anteriorly, but are approximately parallel in transverse section. The muscle scar is elongate-oval and only slightly impressed.

INTERIOR OF BRACHIAL VALVE: The sockets are bounded laterally by the shell walls and on their inner edges by socket plates which curve up above the level of the plane of commissure. The crura are T-shaped in cross section and attach to the inner edges of the socket plates. The dorsal edge, or base of the "T," is continuous with the crural bases which unite medially to form a septalium supported by a long, high median septum. The crural flanges, or medial upper limbs of the "T" join medially over the anterior part of the septalium to form a cover (conjunct inner hinge plates), but a small foramen is left open posteriorly.

OCCURRENCE: The species is uncommon, but is widely distributed geographically. In addition to the type locality at Combs Peak, it is present in the *Eurekaspirifer pinyonensis* Zone in the Sulphur Spring Range, Roberts Mountains, Lone Mountain, northern Simpson Park Mountains, and in the Cortez Mountains. A single locality near Cortez has yielded abundant silicified specimens.

FIGURED SPECIMENS: USNM 157042-157052. In addition, the holotype, USNM 13855, is refigured in Pl. 39, figs. 6-10.

Genus *Astutorhyncha* Havlíček, 1961

TYPE SPECIES: *Rhynchonella proserpina* Barrande, 1847.

Astutorhyncha cf. *prosperpina* (Barrande, 1847)
(Pl. 40, figs. 1-4)

Terebratula proserpina Barrande, 1847, p. 420, Pl. 19, fig. 4.
Astutorhyncha proserpina Havlíček, 1961, p. 107, Pl. 12, figs. 3-7; Text-Figs. 40, 41; (additional synonymy *in* Havlíček, 1961, p. 107).

DISCUSSION: Only a single specimen is known to the writer. The description below is, therefore, a description of this single individual and not of a species.

EXTERIOR: The shell is subquadrate in outline and unequally biconvex in lateral profile. The brachial valve is two to three times as deep as the pedicle valve and is more evenly convex across the flanks. The pedicle valve curves evenly from posterior to anterior along the midline, but the ventral flanks flatten out anterolaterally so that there is a prominent deflection of the shell at the juncture of the ventral sulcus and the flanks anterolaterally. The brachial valve bears a corresponding prominent fold. The hinge line is relatively long and gently curved. The beaks of both valves appear to be short and stubby, although difficult to characterize with certainty because the specimen is an internal mold.

The external ornament consists of a few strong, rounded radial costae separated by broad, U-shaped interspaces. No costae bifurcate. There are three costae in the ventral sulcus and presumably there were four on the dorsal fold. A pair of parietal costae is present, but becomes obsolescent anteriorly, on the fold, a little past midlength. The ventral flank, which is preserved, bears five radial costae. The preserved dorsal flank bears six radial costae.

INTERIOR OF PEDICLE VALVE: There is a pair of thin, platelike dental lamellae, subparallel in cross section, which diverge at an angle a little less than 90 degrees. The dental lamellae are limited to the apex of the valve. The muscle field is relatively small, poorly impressed and subcircular in outline. The diductor field is traversed by the internal expression of the costae and is neither impressed nor defined by muscle-bounding ridges. The adductor impression is small, elongate-cordate, and situated centrally within the muscle field.

INTERIOR OF BRACHIAL VALVE: There is a short, thin, bladelike median septum supporting a small V-shaped septalium. The cardinalia are not satisfactorily exposed on which to base a more complete description.

OCCURRENCE: This species is represented by a single occurrence in the *Eurekaspirifer pinyonensis* Zone in the northern Roberts Mountains.

FIGURED SPECIMEN: USNM 157054.

Family PUGNACIDAE Rzhonsnitskaya, 1956
Genus *Corvinopugnax* Havlíček, 1961

TYPE SPECIES: *Terebratula corvina* Barrande, 1847, p. 426, Pl. 20, fig. 5.

DISCUSSION: The type species is strongly costate only on its anterior. Havlíček (1961, p. 36) noted that *Corvinopugnax* differs from *Hypothyridina* by interlocking of the alternating costae and interspaces at the anterior commissure.

Corvinopugnax sp.
(Pl. 39, figs. 1-5)

EXTERIOR: The following description is based on a single well-preserved internal mold, the only one available to the writer. The outline is transversely subpentagonal, the valves are strongly biconvex with the brachial valve more convex than the pedicle valve. The maximum width is at about midlength. The hinge line is moderately long and curved. No interarea is developed. The pedicle valve is relatively flat in its posterior portion and bears a median sulcus which projects dorsally as a geniculated tongue. The brachial valve bears a more or less rectangular fold which accommodates the tongue of the ventral sulcus in its anterior portion.

The ornament consists of subangular radial costae. There are six costae on the fold and five in the sulcus. No parietal costae are developed. There are nine or ten costae on one flank of the brachial valve and ten costae on one flank of the pedicle valve.

INTERIOR OF PEDICLE VALVE: There is a pair of short, slightly divergent dental lamellae which bound the posterior portions of the muscle impressions. The adductor muscle field is not well impressed, but the diductor field is moderately impressed and bounded laterally by nearly straight, slightly divergent, lateral edges.

INTERIOR OF BRACHIAL VALVE: The cardinalia were not observed. A median septum is not present.

OCCURRENCE: The only specimen known to the writer is believed to come from the *Eurekaspirifer pinyonensis* Zone in the Sulphur Spring Range.

FIGURED SPECIMEN: USNM 157053.

Family LEIORHYNCHIDAE Stainbrook, 1945

DISCUSSION: Schmidt and McLaren (1965, p. H580) have relegated Leiorhynchidae to the position of junior synonym within a family Camarotoechiidae. However, there are reasons to doubt that *Leiorhynchus* and *Camarotoechia* are related at the family level. As illustrated by Sartenaer (1961b, Pl. 1), there is a marked external homeomorphy, especially in regard to the pattern of rib development. However, the configuration of the dental lamellae is very different in that they diverge toward the base of the valve instead of converging as in *Leiorhynchus*. The latter configuration is characteristic of various species of *Leiorhynchus* that are contemporaneous with *Camarotoechia congregata*. The muscle impressions of *Camarotoechia congregata* (Sartenaer, 1961b, Pl. 1) and those of *Leiorhynchus quadracostatus* (Sartenaer, 1961a, Pl. 112) show little similarity. Lastly, the critical dorsal internal structures (cardinalia and crura) are unknown in *Camarotoechia congregata,* and on this basis alone it seems premature to suggest any close phylogenetic connection between *Camarotoechia* and *Leiorhynchus.*

In an attempt to clarify the concept of a family centered around *Leiorhynchus* as is employed here, the writer suggests that in addition to the obvious external characteristics of *Leiorhynchus*, considerable weight should be placed on the disposition of the dental lamellae as illustrated by

Sartenaer (1961a, Text-Fig. 1; McLaren, 1962, Text-Fig. 22, p. 78). In addition, and most important in the writer's view, is the construction of the crura which are formed in *Leiorhynchus* as rods of circular cross section to which are attached ventral flanges of considerable prominence. This is well illustrated in *Leiorhynchus quadracostatus* by Sartenaer (1961a, Text-Fig. la, sections 1.3 to 1.75) and by the writer (Johnson, 1966a, Text-Fig. 3, p. 167). Far too little is known about the consistent development of various types of crura in Paleozoic rhynchonellids, but the writer thinks that they will eventually prove to be significant. The present study at least convinces the writer that the crura of the Trigonirhynchiidae represent a fundamental evolutionary divergence between its members and members of the Leiorhynchidae, with crura as described above.

Genus *Leiorhynchus* Hall, 1860

TYPE SPECIES: *Orthis quadracostata* Vanuxem, 1842, p. 168, Text-Fig. 42, no. 2.

"*Leiorhynchus*" sp.
(Pl. 40, figs. 5-15)

DISCUSSION: This is a rare fossil which comprises two forms in the *E. pinyonensis* Zone of central Nevada. A large, subcircular, gently sulcate form is found in the Cortez Range, and the more pentagonal, deeply sulcate form occurs at Lone Mountain. It seems likely that congeneric rhynchonellids will eventually be separated off from *Leiorhynchus* s.s. on the basis of external form and ribbing, but close relation to *Leiorhynchus* s.s. seems undeniable to the writer. The illustrated specimen from Lone Mountain (Pl. 40, fig. 5) shows the typical leiorhynchid conformation of the dental lamellae, converging toward the base of the valve. The silicified specimens from Cortez show the same characteristics of the dental lamellae (Pl. 40, fig. 10) and have a leiorhynchid cardinalia with crura of circular cross section (Pl. 40, fig. 15), but none of the specimens is so well preserved that it can be observed whether or not ventral flanges were attached to the crura.

EXTERIOR (SPECIMENS FROM LONE MOUNTAIN): The outline of the shell is rounded-subpentagonal. In lateral profile the valves are unequally biconvex, with a deep brachial valve. In the umbonal regions the valves are of about equal convexity, but the pedicle valve flattens out, and because of the prominent sulcation, becomes somewhat concave anteriorly. The brachial valve has its greatest convexity approximately two-thirds of the distance toward the anterior margin. Maximum width is anterior to midlength. The pedicle valve bears a broad, deep, relatively flat-bottomed sulcus that affects the anterior half of large specimens. The sulcus is extended dorsally as a prominent tongue which is accommodated by a prominent, subrectangular fold developed anteriorly on the brachial valve. The valves are anteriorly costate, both in the fold and sulcus and on the anterolateral flanks. The costae are low, but relatively broad, numbering four on the fold and three on the sulcus. Three or four low costae are developed on each flank in the anterolateral regions adjoining the fold and sulcus.

EXTERIOR: (OF SPECIMENS FROM CORTEZ): The shells are subcircular to broadly suboval, slightly wider than long. In lateral profile the valves are unequally biconvex with the brachial valve about three times as deep as the pedicle valve. The dorsal beak is strongly incurved beneath a very short, inconspicuous, incurved ventral beak. The latter protrudes posteriorly only very slightly further than the dorsal umbo. There is a broad, shallow sulcus developed in the anterior third of the pedicle valve, and it appears to be faintly costate, but costae were not detected on the anterolateral regions. However, the silicified specimens from Cortez are few in number and poorly preserved.

INTERIOR OF PEDICLE VALVE: Hinge teeth are not preserved. The delthyrium is broad and open, encompassing more than 90 degrees. The delthyrial cavity is short, broad, and rhomboidal, bisected from the apex by a narrow median ridge. Dental lamellae are very short, close to the valve walls, and convergent toward the base of the valve, but they do not meet. They are anteriorly continuous with low ridges that converge slightly toward the midline, then expand broadly at the posterior end of the muscle impressions. The diductor impression is elongate-elliptical and gently impressed, without prominent bounding ridges. No flabellation is developed anteriorly. A pair of elongate incisions and bounding ridges form the ventral adductor scars in the central part of the muscle field.

INTERIOR OF BRACHIAL VALVE: The sockets are broadly divergent cylindroidal grooves adjoining the posterolateral margin and the outer hinge plates, which are triangular horizontal plates connecting low socket ridges and the crural bases. Crura are situated close together medially, separated by a narrow triangular fissure, and projecting anteriorly as rods of circular cross section. A true septalium is not formed because the median septum is not continuous with either the outer hinge plates or the bases of the crura in the apex. The median septum is thick posteriorly, thinning anteriorly toward about midlength where it continues only as a low ridge.

OCCURRENCE: *"Leiorhynchus"* sp. occurs in the *Eurekaspirifer pinyonensis* Zone on the west flank of the Cortez Mountains and at Lone Mountain. Two of the three specimens known from Lone Mountain were collected by Mr. Eric Gronberg at a horizon 685 to 690 feet above the base of the McColley Canyon Formation at that locality.

The form from Lone Mountain bears a resemblance to *Felinotoechia? audax* from the Vinařice Kalk (Siegenian) of Czechoslovakia (Havlíček, 1961, Pl. 15, figs. 2, 3). Similar specimens from the *Sieberella* Unit at Royal Creek, Yukon Territory (Lenz, 1966) have also been shown to the writer by Dr. A. C. Lenz. Others have been seen by the writer from G.S.C. locality 25835, high in the Stuart Bay Formation of Bathurst Island.

FIGURED SPECIMENS: USNM 157055-157058.

Order Spiriferida

SUBORDER ATRYPOIDEA

DISCUSSION: This suborder has a diverse and important representation in the Lower Devonian of central Nevada due especially to the abundance of representatives of the superfamily Dayiacea. Atrypacean genera are generally less important numerically and less diverse in the range of morphology exemplified by the taxa present. *Atrypa* of the *reticularis* type is the common atrypid in the Gedinnian and in the lower (*Quadrithyris* Zone) part of the Siegenian. *Atrypina* is present in the same interval. *Spirigerina* is present near the base of the Gedinnian in the Roberts Mountains, but is not common throughout the Gedinnian interval of that formation. On the other hand *Spirigerina supramarginalis* is the most common atrypid of the *Quadrithyris* Zone at Coal Canyon, although it has not been found in that zone in the McMonnigal Limestone of the Toquima Range. A few poorly preserved specimens of *Spinatrypa* have been found at the latter locality. *Spirigerina* and *Toquimaella kayi*, present in the *Quadrithyris* Zone, represent the Old World influence within the superfamily Atrypacea in central Nevada.

Atrypa of the *reticularis* type virtually disappears above the *Quadrithyris* Zone, possibly due to unsuitable ecologic conditions in the succeeding beds of the Rabbit Hill Formation. Only a single specimen of *Atrypa* was found among the several thousands of specimens recovered from silicified residues of the Rabbit Hill Limestone in its type area. However, specimens appear to be more common in the *Spinoplasia* Zone of the Toquima Range where that zone is developed in the higher beds of the McMonnigal Limestone. No *Atrypa* has yet been seen in the *Trematospira* Zone, but large, robust, and frilly specimens of *Atrypa nevadana* occur for the first time in some profusion in the *Acrospirifer kobehana* Zone, and the species is represented in great abundance in the *Eurekaspirifer pinyonensis* Zone. A large species somewhat resembling *A. nevadana* occurs in the *Elythyna* beds above the *pinyonensis* Zone, especially in the southern Roberts Mountains.

The rather ordinary representation of the superfamily Atrypacea is in contrast with the rich series of faunas in which the superfamily Dayiacea is abundantly represented by the Leptocoeliidae and the Anoplothecidae. *Leptocoelia* appears in great abundance in the *Spinoplasia* Zone, represented by *L. murphyi* and coexists with the frilly and faintly plicate genus *Leptocoelina*. The latter is restricted to the *Spinoplasia* Zone and apparently to the Rabbit Hill lithofacies, but *Leptocoelia murphyi* continues as a prominent element of the *Trematospira* Zone of the lower part of the Mc-Colley Canyon Formation, both in the Sulphur Spring Range and in the Roberts Mountains. *Leptocoelia murphyi* is more common in the lower or *Trematospira* Subzone, but is still abundantly represented in one collection with *Costispirifer* at Willow Creek in the northern Roberts Mountains. *Leptocoelia* is much less abundant in the *kobehana* Zone where it is represented at several localities by *Leptocoelia* aff. *murphyi*. In the *Eurekaspirifer pinyonensis* Zone, *Leptocoelia* is commonly abundant (*L. infrequens*), but only at certain horizons. The genus apparently became extinct in Nevada during the time of the *E. pinyonensis* Zone. None has been found either in the *Elythyna* beds or in the Middle Devonian.

Coelospira first becomes abundant in the same beds with *Leptocoelia* in the *Spinoplasia* Zone, where it is represented by individuals of the type species, *Coelospira concava*. *Coelospira* is also abundant in the lower part of the *Trematospira* Zone, represented by *C. pseudocamilla*, and is less common in the succeeding assemblage with *Costispirifer*. *Coelospira* sp. occurs in a few collections from the *Acrospirifer kobehana* Zone and uncommonly in the available collections from the *pinyonensis* Zone. These occurrences have not yet been well studied. The diminution in numbers of *Coelospira* specimens of the *pinyonensis* Zone is in contrast to the entry of relatively abundant *Bifida* at some horizons. At Lone Mountain, *Bifida* is abundant above the typical development of the *pinyonensis* Zone, although its occurrence is judged to be still within the limits of that zone as presently defined. Neither *Coelospira* nor *Bifida* has been found in the Middle Devonian *Leptathyris circula* Zone or higher beds, although a probable new species of *Coelospira* occurs as high as the *Elythyna* beds at Willow Creek in the northern Roberts Mountains.

Leptocoelia, *Leptocoelina*, and *Coelospira* of the *concava* type, which is the only type of *Coelospira* represented in the Lower Devonian of central Nevada, indicate strong ties with the Appalachian faunal province of which central Nevada was a part during *Spinoplasia* Zone and *Trematospira* Zone time. The entry of the Old World genus *Bifida* during *pinyonensis* Zone time indicates communication with faunas of Old World aspect that were present north of Nevada.

Superfamily ATRYPACEA Gill, 1871
Family ATRYPIDAE Gill, 1871
Subfamily ATRYPINAE Gill, 1871
Genus *Atrypa* Dalman, 1828
TYPE SPECIES: *Anomia reticularis* Linnaeus 1758, p. 702.
Atrypa sp.
(Pl. 41, figs. 1, 2)

DISCUSSION: *Atrypa* of the *reticularis* type is relatively common in the *Quadrithyris* Zone, both at Coal Canyon and in the Toquima Range. The *Quadrithyris* Zone of the Windmill Limestone has yielded a few specimens of relatively large size from the upper breccia, but they are uncommon.

Some small, finely ribbed forms are common in the lower breccia and also in the lower part of the McMonnigal Limestone in the Toquima Range, and a single small specimen of *Atrypa* sp. is known from the Rabbit Hill Limestone (*Spinoplasia* Zone).

FIGURED SPECIMEN: USNM 157059.

Atrypa sp. A
(Pl. 41, figs. 3, 4)

Atrypa sp. A Johnson, 1962a, p. 166.

EXTERIOR: Small specimens are elongate and suboval in outline, but larger specimens are somewhat variable in their outline and length-width ratio. Most are longer than wide and have a modified subpentagonal outline which varies according to the place of maximum width. In the largest specimens, the hinge line is relatively long and nearly straight; maximum width may be slightly posterior to midlength. On other large specimens where the hinge line is more strongly curved, however, the maximum width may be anterior to midlength. On some specimens the lateral margins are subparallel along their edges between the distance of one-fifth to three-fifths of the distance toward the anterior commissure. On others, the large beak angle and relatively straight posterolateral edges make the posterior margin angular, but the anterior portion of the outline may be smoothly elliptical. The valves are unequally biconvex with the brachial valve about twice the convexity of the pedicle valve at midlength. The pedicle valve attains greatest convexity at the umbo, and the brachial valve attains it near midlength or slightly posterior to midlength. The anterolateral margins of the pedicle valves tend to be relatively flat and without any strong undulation of a median sulcus. The brachial valves are relatively smoothly curved in longitudinal and in transverse sections. A very low, triangular, almost linear interarea is developed on the larger specimens. It is cleft medially by a triangular delthyrium which opens into an apical foramen. The beak is only slightly incurved and is nearly straight.

The ornament consists of well-defined, even, radial costellae which increase in number anteriorly by bifurcation and by intercalation. There commonly are about 11 to 13 costellae in a space of 5 mm, 10 mm anterior to the beak. The costellae are crossed by relatively few, widely spaced, poorly defined, growth lamellae (tubular-lamellar of Copper, 1967), which develop frills only at the anterior of medium- to large-sized shells. The interior structures of this species are at present not known.

OCCURRENCE: *Atrypa* sp. A is common only in the *A. kobehana* Zone on the north side of the Roberts Mountains with *Gypidula praeloweryi* and *Strophonella* cf. *punctulifera*.

COMPARISON: *Atrypa* sp. A has considerably finer costellae than *Atrypa nevadana* from the overlying *Eurekaspirifer pinyonensis* beds. It also differs from the latter species in lacking frills on the posterior portion of the shells. *Atrypa* cf. *nieczlawiensis* from the upper Roberts Mountains Formation has fine costellae of about the same size as *Atrypa* sp. A, but it develops strong frilly growth lamellae over most of the shell.

FIGURED SPECIMEN: USNM 157060.

Atrypa nevadana Merriam, 1940
(Pl. 41, figs. 5-17)

Atrypa reticularis Meek, 1877, p. 38, Pl. 1, figs 7, 7a; *not* Linnaeus, 1758.
Atrypa nevadana Merriam, 1940, p. 83, Pl. 7, figs. 18, 19.
Atrypa nevadana Cooper, 1944, p. 319, Pl. 121, figs. 6, 7.

EXTERIOR: This species attains medium to large size for the genus and is somewhat variable in its outline. Most specimens have length about equal to width, or the length may be slightly the greater of the two. Anteriorly, the outline is suboval or in some specimens nearly semicircular, however, the beak angle is large, and on some specimens the hinge line is relatively long and nearly straight. On these forms the outline becomes somewhat subtrigonal, although the cardinal angles are in all cases strongly rounded. The maximum width of almost all individuals is posterior to midlength, and on large specimens the relatively strong development of a ventral sulcus produces a continuous narrowing anteriorly. The valves are strongly unequally biconvex with the brachial valve much more strongly convex than the pedicle valve. The brachial valve attains a relatively strong curvature near midlength. In transverse section, it may be somewhat carinate anteriorly to accommodate the extended tongue of a broad shallow sulcus in the anterior of the pedicle valve. The ventral beak is small, pointed, and strongly incurved, nearly touching the dorsal umbo.

The ornament consists of relatively strong radial costellae interrupted at regular intervals by well-developed growth lines that produce anterior frills. There are commonly 7 or 8 radial costellae in a space of 5 mm, 10 mm anterior to the beak.

INTERIOR OF PEDICLE VALVE: The hinge teeth are elliptical in cross section and are directed almost at right angles to the midline, especially on large shells where the hinge line is relatively straight. The hinge teeth of large specimens each bear a deep anterolateral groove which articulates with the corrugated center ridge within the socket in the brachial valve. Small specimens have short, thin, widely divergent dental lamellae, but they are obsolescent on large specimens. The adductor muscle scars are moderately impressed posteriorly and are elongate with their lateral edges nearly parallel to one another. They lie generally within the posterior half of the diductor muscle field which is suboval and relatively broad and flabellate. On the largest specimens the muscle field is relatively strongly impressed laterally and may be longer than wide. The lateral portions of the shell interior are pustulose. Only the smallest specimens are crenulated by the impress of the costellae over most of the shell; however, most specimens are corrugated peripherally.

INTERIOR OF BRACHIAL VALVE: Posteriorly, the sockets are bounded by the inner edges of the brachial valve. Medially, they are defined by relatively strong socket plates which curve ventrally and bear a pair of elongate triangular crural lobes. The inner edges of the crural lobes, however, diverge anterolaterally, defining discrete socket plates. The sockets are longitudinally bisected by a median ridge which expands anterolaterally within the socket. Posteromedially, the ridge begins along the inner edge of the valve and curves inward to the medial part of the socket. The ridges are strongly crenulated as are the medial parts of the sockets adjacent to them. There is no crenulation posterolateral to the outer edges of the ridge. Large specimens become thickened posteriorly and are completely filled with shell material to the anterior edge of the elevated triangular notothyrial cavity. Small- and medium-sized specimens bear a short myophragm beginning at the anterior edge of the notothyrial cavity. The site of diductor muscle attachment is a small striate triangle in the apex of the notothyrium. The adductor muscle scars are broadly flabellate and deeply ridged radially. Small- and medium-sized shells are slightly corrugated in their anterior halves and strongly corrugated peripherally by the impress of the costellae.

OCCURRENCE: *Atrypa nevadana* occurs in the *Eurekaspirifer pinyonensis* Zone and in the *Acrospirifer kobehana* Zone. It is common in the Sulphur Spring Range, Roberts Mountains, Lone Mountain, northern Simpson Park Range, and in the Cortez Range.

COMPARISON: *Atrypa nevadana* is distinguished from *Atrypa* sp. A of the *Acrospirifer kobehana* Zone, by its much coarser costellae, and by the presence of frilly growth lamellae on the posterior portions of the valves.

FIGURED SPECIMENS: USNM 93671 is the holotype. Other figured specimens are USNM 157061-157063.

Subfamily CARINATININAE Rzhonsnitskaya, 1960
Genus *Spirigerina* d'Orbigny, 1849

TYPE SPECIES: *Terebratula marginalis* Dalman, 1828, p. 143, Pl. 5, figs. 6a-e.

Spirigerina supramarginalis (Khalfin, 1948)
(Pl. 42, figs. 1-10)

Atrypa supramarginalis Khalfin, 1948, p. 159, Pl. 2, fig. 10; p. 176, Pl. 4, figs. 4-7.

Spirigerina supramarginalis Kulkov, 1963, p. 70, Pl. 5, figs. 11-13.

EXTERIOR: Specimens are somewhat irregular in outline, varying from subpyriform, with the length only slightly greater than the width, to broadly subpentagonal. The outline of the brachial valves may be slightly transversely suboval. In lateral profile the valves are unequally biconvex with the brachial valve about 50 to 100 percent more strongly convex than the pedicle valve. The flanks of the pedicle valve, particularly on larger specimens, tend to flatten out laterally. Brachial valves, however, are strongly convex in their posterior portion and curve anteriorly at a slightly decreasing angle. They are relatively strongly curved in transverse section as well as from front to back. The anterior part of medium- and large-sized pedicle valves develops a shallow flaring sulcus which is moderately deflected at the anterior commissure. Brachial valves bear a broad, low, rounded fold anteriorly to accommodate the projecting tongue of the ventral sulcus.

The ventral beak is short and nearly straight. The beak ridges are relatively well defined and bound a low crescentic interarea. The anterior concave arc of the crescent fits closely around the umbo of the brachial valve. Medially, there is a triangular delthyrium which is closed in its basal two-thirds by conjunct deltidial plates defining a hypothyrid or possibly submesothyrid circular foramen.

The external ornament consists of strongly defined, rounded to subangular, radial costae separated by deep U-shaped interspaces. The costae increase in number anteriorly by bifurcation and by intercalation. Newly formed costae tend to thicken in a short distance anteriorly to a comparable width with the adjoining costae. The formation of new costae is irregular, and the total number on any one valve may vary in comparison with the number on other specimens of the same size. There are commonly 16 to 18 costae on specimens with a length of approximately 15 mm. No concentric ornament was observed on the posterior portions of the specimens; however, a few more or less well-developed concentric growth lines appear anteriorly on some of the larger size shells.

INTERIOR OF PEDICLE VALVE: The hinge teeth are broad and elliptical in cross section and are directed almost at right angles to the midline. There is a pair of well-developed grooves which accommodate the posterior edge of the brachial valve along the posterior edges of the hinge teeth. The hinge teeth are supported by short, thin dental lamellae in small shells; however, the umbonal regions are relatively strongly thickened by secondary shell material making the dental lamellae obsolete on most of the available specimens. The ventral muscle field is trigonal and fairly strongly impressed. Its sides are straight and diverge slightly anterolaterally. Anteriorly, the muscle field blends with the interior of the valve and is nonflabellate. The interior is lightly crenulated by the impress of the costae.

INTERIOR OF BRACHIAL VALVE: The sockets are short and deep and diverge strongly anterolaterally. They are bounded posteriorly by the inner edge of the valve and on their inner edges by a pair of socket plates which diverge anterolaterally as well as basally. The socket plates are moderately thickened medially by the addition of crural lobes. Preservation of the available dorsal interiors is too poor to determine the configuration of the adductor muscle impressions. The interior is moderately to strongly crenulated by the impress of the costae.

OCCURRENCE: In Nevada, *Spirigerina supramarginalis* is present in the *Quadrithyris* Zone. So far its only occurrence is at Coal Canyon in the Windmill Limestone. Lenz (1966) reported this species to be the guide to his *Spirigerina* Unit at Royal Creek.

DISCUSSION: This relatively coarse-ribbed species of *Spirigerina* may be the successor of *S. marginaliformis* of the Gedinnian. It seems to be a little more coarsely costate than Stauffer's species *Spirigerina bicostata* from the Vaughn Gulch Limestone of southeastern California. *Spirigerina bicostata* and *S. marginaliformis* may be synonymous, but although the types of *S. bicostata* were examined by the writer the available specimens are insufficient on which to base an adequate idea of the range of variation of the species.

Coarse-ribbed species of *Spirigerina* occur in the Silurian, including the Ludlow of Nevada, but as a rule these older species tend to have reflexed anterolateral ventral margins, in contrast to the evenly curved to flattened ones of *S. supramarginalis*.

FIGURED SPECIMENS: USNM 157064-157068.

Subfamily ATRYPININAE McEwan, 1939
Genus *Atrypina* Hall and Clarke, 1893
TYPE SPECIES: *Leptocoelia imbricata* Hall, 1857, p. 108.
Atrypina simpsoni Johnson n. sp.
(Pl. 43, figs. 5-16)

DIAGNOSIS: Small pentagonal *Atrypina* with a nearly straight hinge line and lacking a median rib on the fold of the pedicle valve.

EXTERIOR: The shells are transversely pentagonal in outline and planoconvex in lateral profile. The curvature of the pedicle valve is relatively small front to back, and the beak is not strongly incurved, but is straight. It projects only a short distance posteriorly beyond the hinge line. An apical foramen is present. The cardinal angles are rounded and are close to 90 degrees. The maximum width is commonly in the posterior one-third or one-fourth of the length.

Ornament consists of radial plications of rounded cross section crossed by lamellose concentric growth lines. The development of growth lines is variable. They may be sparse and irregular in the interval of their occurrence, or more commonly they are numerous and occur at regular intervals. The basic pattern of radial plication is two plications on the medial region of the pedicle valve which originate either close to the beak or anterior to it by splitting of the medial fold. These plications are elevated above the average level of those on the flanks to form an indistinct fold on the pedicle valve. On some shells, two smaller plications originate on the lateral flanks of the medial plications and anterior to the umbo. There are two simple plications on each flank of the pedicle valve. The brachial valve bears a medial plication in the sulcus bounded by plications that may or may not bifurcate. The bounding plications tend to become obsolescent anteriorly in specimens on which they do not bifurcate. Laterally, there is an additional single pair of plications.

INTERIOR OF PEDICLE VALVE: The teeth are small and pointed and are unsupported by dental lamellae. Muscle scars were not seen.

INTERIOR OF BRACHIAL VALVE: The socket plates are closely set medially, leaving only a very small notothyrial cavity. The crural bases form part of the inner edges of the socket plates. There is a strong myophragm posteriorly. The brachidium is not preserved on the specimens examined.

OCCURRENCE: *Atrypina simpsoni* is known with certainty only from the *Quadrithyris* Zone at Coal Canyon. Specimens from the Gedinnian of the Roberts Mountains Formation are similar (Pl. 43, figs. 1-4), but not enough specimens are available to enable a conclusive comparison.

COMPARISON: The new species most closely resemble the type, *A. imbricata*, but that species is about twice as large and has more widely spaced growth lamellae (Cooper, 1944, Pl. 121, figs. 30, 32). *Atrypina hami* Amsden (1958a, Pl. 7) is a closely related species, but it also is much larger than *A. simpsoni*.

The Silurian species *A. disparilis* is a much more elongate species with a considerably shorter hinge line. *Atrypina barrandei* Davidson is another Silurian species with a ribbing pattern like *A. simpsoni* and the other species discussed above, but it has a much narrower hinge line than does *A. disparilis*.

The species *Atrypina clintoni* Hall and Clarke (1895), if properly assigned generically, differs from *A. simpsoni* because it has a median plication on the pedicle valve (Hall and Clarke, 1895, Pl. 53, fig. 19).

The Uralian Siluro-Devonian species *Atrypa sublepida* is the source of the brachiopod Treatise citation of *Atrypina* in the Urals. However, O. I. Nikiforova (written commun., 1967) showed the writer photos of the interior of this species which indicate relation to *Gracianella* (Johnson and Boucot, 1967).

FIGURED SPECIMENS: USNM 157071-157076. The holotype is USNM 157073. The specimen is 4 mm long, 5 mm wide, and 2 mm thick.

Atrypina cf. *simpsoni* Johnson, n. sp.
(Pl. 43, figs. 1-4)

DISCUSSION: A few specimens of *Atrypina* have been collected from the upper Roberts Mountains Formation in the Roberts Mountains. The ribbing pattern is similar to that of *A. simpsoni*, except that the plications are lower and a split of the ventral fold into two distinct plications does not occur. This form attains greater size than *simpsoni*, and in these larger specimens the shells are elongate. Small shells are transverse like A. *simpsoni*, but are not so pentagonal.

FIGURED SPECIMENS: USNM 157069, 157070.

Family PALAFERELLIDAE Spriestersbach, 1942
Subfamily KARPINSKIINAE Poulsen, 1943
Genus *Toquimaella* Johnson, 1967

TYPE SPECIES: *Toquimaella kayi* Johnson, 1967, p. 878, Pl. 111, figs. 1-5, 7-30.

Toquimaella kayi Johnson, 1967
(Pl. 42, figs. 11-23; Text-Fig. 6)

DISCUSSION: *Toquimaella* is the fine-ribbed precursor of the Lower and Middle Devonian atrypid genus *Vagrania*. They have similar, if not identical, internal structures and differ principally in form and ribbing.

The known occurrences of *Toquimaella kayi* in western North America indicate a limited stratigraphic distribution in association with the conodont *Icriodus pesavis* at a position just above *Monograptus hercynicus* and just below *Monograptus yukonensis*. It occurs in the latter position in the *Spirigerina* Unit at Royal Creek (Lenz, 1966, 1967b) and in the Stuart Bay Formation of Bathurst Island (Johnson, 1967).

OCCURRENCE: In Nevada *Toquimaella kayi* is known only in the *Quadrithyris* Zone and occurs in both localities of the lower McMonnigal Limestone north and south of Ikes Canyon in the Toquima Range and in *Quadrithyris* Zone collections at Coal Canyon.

FIGURED SPECIMEN: The holotype is USNM 155508. Other figured specimens are Columbia University, Department of Geology nos. 28790, 28791.

Superfamily DAYIACEA Waagen, 1883
Family ANOPLOTHECIDAE Schuchert, 1894
Subfamily ANOPLOTHECINAE Schuchert, 1894
Genus *Bifida* Davidson, 1882

TYPE SPECIES: *Terebratula lepida* D'Archiac and De Verneuil, 1842, p. 368, Pl. 35, figs. 2, 2a, 2b, 2c.

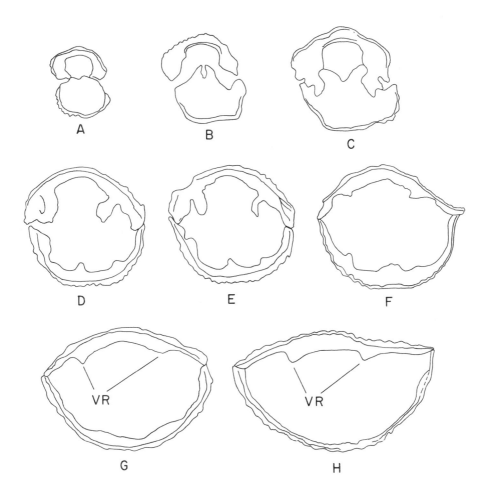

Figure 6. Serial sections × 6 of *Toquimaella kayi* Johnson, 1967, Columbia Univ., Department of Geology no. 28790. VR = vascular ridges.

Bifida sp.
(Pl. 43, figs. 17-27)

EXTERIOR: The shells are elongate suboval to short pyriform in outline and strongly unequally biconvex in lateral profile. The brachial valve is very gently convex and the pedicle valve is strongly convex. The place of maximum width is slightly anterior to midlength. The ventral beak is short and stubby and is not strongly incurved. The pedicle valve bears a low, narrow, indistinct fold and the brachial valve bears a relatively strong narrow sulcus.

The ornament consists of sparse radial plications crossed by regularly spaced lamellose growth lines. The pedicle valve bears two low plications on the fold that are stronger in the posterior half of the valve and one plication on each flank closely adjoining the fold and generally developed only in the posterior half of the valve. The brachial valve bears one narrow rib in the middle of the sulcus and two bordering plications.

INTERIOR OF PEDICLE VALVE: The teeth are widely set apart and are attached directly to the inner walls of the valve. They are short and elongate and are approximately elliptical in horizontal section. The medial edges of the hinge teeth bear deep longitudinal fossettes. Dental lamellae are absent. The muscle field consists of two deeply impressed and strongly elongate tracks separated by a stout myophragm. The diductor muscle tracks merge with the interior of the valve anteriorly. No brachial valves showing the internal structures are available for study.

OCCURRENCE: *Bifida* sp. is known in Nevada from the *Eurekaspirifer pinyonensis* Zone of the Cortez Mountains, Sulphur Spring Range, and Lone Mountain. The occurrence at Lone Mountain is at several horizons between 665 and 685 feet above the base of the McColley and above typical development of the *pinyonensis* Zone fauna.

FIGURED SPECIMENS: USNM 157077-157080.

Subfamily COELOSPIRINAE Hall and Clarke, 1895
Genus *Coelospira* Hall, 1863

TYPE SPECIES: *Leptocoelia concava* Hall, 1857, p. 107.

Coelospira concava (Hall, 1857)
(Pl. 44, figs. 1-11)

Synonymy is given by Boucot and Johnson, 1967a, p. 1235.

EXTERIOR: Pedicle valves are slightly elongate-subpyriform, and brachial valves are subcircular in outline. The valves are plano-convex in lateral profile. The ventral beak is very small and straight, or nearly straight. It bears an apical foramen. There is a very low, relatively narrow, flat ventral interarea consisting of a pair of smooth surfaces in the median third of the valve posterior to the adjacent edge of the brachial valve. The ventral interarea is orthocline to anacline. It is cleft medially by an open triangular delthyrium which encloses an angle of approximately 45 degrees. Maximum width is about at midlength or very slightly anterior to midlength on brachial valves and is more decidedly anterior to midlength on pedicle valves. The pedicle valve bears a low rounded fold that expands in width in a short distance anteriorly. The brachial valve bears a corresponding broad shallow sulcus.

The ornament consists of very low, rounded, radial costae which decrease slightly in their strength anteriorly. There commonly is a flat median groove along the midline of the fold on the pedicle valve flanked by a pair of costae that split on the umbo, giving rise to four. Lateral to these there are generally three or four costae on the flank which bifurcate anteriorly on relatively large specimens. The brachial valve bears a median costa in its sulcus. There are commonly three lateral costae which originate at the beak, plus an additional pair which are intercalated between the medial costae and the relatively widely flaring first lateral pair. On some shells there is a second intercalated pair lateral to the first lateral primary pair, originating at the umbo. A few relatively strong concentric growth lines appear at the anterior of larger specimens.

INTERIOR OF PEDICLE VALVE: The hinge teeth are very small and elliptical in cross section. They are attached directly to a thickening of shell material along the posterolateral edges of the valve. There is a pair of deep crural fossettes medially which posteriorly join the transverse groove that accommodates the posterior edge of the brachial valve. There is a flat rim of shell material posteriorly which narrows slightly anterolaterally and continues about to the midlength of the valve. The diductor scars are elongate cordate in outline and consist of a pair of long pyriform impressions divided by a myophragm. Some large specimens develop a short, low, median septum of crescentic outline at the anterior edge of the diductor muscle field. The interior of some specimens is faintly crenulated by the impress of the costae.

INTERIOR OF BRACHIAL VALVE: The sockets are elongate narrow grooves diverging anterolaterally. They are bounded posterolaterally by the inner edges of the valve and on their medial sides are bounded by strong socket ridges that splay anteriorly and extend anterolaterally as rimlike ridges around the margin of the valve. The posteromedial edges of the socket ridges are almost subparallel to the hinge line and are joined by a plate of shell material at the posterior extremity of the valve making a somewhat spathulate structure. Medially, there is a short broad myophragm which narrows posteriorly and extends into the notothyrial cavity between the posterior ends of the socket ridges. Adductor muscle impressions consist of small triangular pits relatively widely separated and impressed lateral to the broad myophragm. The interior of small specimens is lightly corrugated by the impress of the costae.

OCCURRENCE: *Coelospira concava* is relatively common in the *Spinoplasia* Zone at Rabbit Hill and at Coal Canyon.

FIGURED SPECIMENS: USNM 157081-157088.

Coelospira pseudocamilla Johnson, n. sp.
(Pl. 44, figs. 12-25)

DIAGNOSIS: Small plano-convex *Coelospira* with shallow pedicle valve and simple costae.

EXTERIOR: The shells are transversely suboval in outline and plano-convex in lateral profile. Maximum width is at about midlength. The ventral beak is very small, short, and pointed and is nearly straight. The posterolateral margins of the valves tend to be relatively straight, encompassing a large beak angle. The anterior three-fifths of the shell is approximately semicircular. There is a poorly defined ventral fold and a very broad, shallow dorsal sulcus.

The ornament consists of well defined, rounded, radial costae separated by narrow U-shaped interspaces. The pedicle valve bears a median groove which may bear a very narrow median costa on some specimens. On most specimens the median groove is bounded by two costae that originate at the beak and there are commonly three well defined lateral costae on each flank. In addition, on most specimens, there is a pair of secondary costae which arise on the umbo between the first and second costae adjacent to the midline. The brachial valve commonly bears a somewhat flat median costa and three simple lateral costae on each flank. A pair of secondary costae arise by intercalation between the median costa and the first lateral pair in the posterior part of the valve. The addition of new costae on the flanks of the valves by intercalation or by bifurcation is rare.

INTERIOR OF PEDICLE VALVE: The hinge teeth are small and narrow and diverge anterolaterally. They rest directly on the inner edges of the valve and are grooved medially by deep crural fossettes. The transverse groove for accommodation of the posterior rim of the brachial valve is relatively shallow. The diductor muscle scars consist of a pair of elongate pyriform impressions divided by a myophragm and not elevated on a platform. The shell substance is relatively thin and the interior is commonly slightly crenulated by the impress of the costellae.

INTERIOR OF BRACHIAL VALVE: The sockets are bounded posterolaterally by the unthickened inner edges of the valve. They are bounded on their inner edges by strong, plate like socket ridges which are joined at the apex of the valve by a transverse, concave, spathulate plate which is directed posteriorly. On some specimens there is a slight elevation of shell material between the socket ridges posteriorly; however, it does not connect with any definite myophragm. Some specimens bear a short divergent pair of ridges at the anterior edge of the notothyrial platform dividing the adductor muscle scars anteriorly; however, the adductor scars are very poorly defined. The interior of most specimens is slightly crenulated by the impress of the costae.

OCCURRENCE: *Coelospira pseudocamilla* is present in the *Trematospira* Zone in the Sulphur Spring Range. It is especially abundant in the *Trematospira* Subzone in beds also carrying abundant *Leptocoelia murphyi*.

COMPARISON: *Coelospira pseudocamilla* is thinner shelled and more transverse than *Coelospira concava*. In addition, the latter species has less well defined and more radial costae which commonly bifurcate on the flanks of the valves. *Coelospira pseudocamilla* is not comparable to Silurian species of the genus (*see* Boucot and Johnson, 1967a) including *C. virginia*. It differs from *C. dichotoma* in being much smaller and from *C. camilla* in being flatter and in having only rarely bifurcating costae.

FIGURED SPECIMENS: USNM 157089-157094. The holotype is USNM 157089.

ORDER SPIRIFERIDA

Family LEPTOCOELIIDAE Boucot and Gill, 1956

DISCUSSION: The representatives of the atrypoid family Leptocoeliidae are important and abundant elements of the Siegenian and Emsian brachiopod faunas of central Nevada where they are represented by two genera, *Leptocoelia* Hall and *Leptocoelina* n. gen. *Leptocoelina*, represented by the type species *L. squamosa*, is a very abundant member of the brachiopod fauna of the Rabbit Hill Limestone to which it is restricted; it is not known to range outside of the *Spinoplasia* Zone. *Leptocoelina squamosa* has not been seen out of the relatively rich argillaceous facies of the Rabbit Hill Limestone and is noticeably absent (although the sample studied was small) from the dark bioclastic limestones of the upper part of the McMonnigal Limestone in the Toquima Range and at Antelope Peak. Some preference for facies or limitation of animal community is probably indicated.

Even more abundant and widespread stratigraphically is the genus *Leptocoelia* represented by two species, *L. murphyi* in the Siegenian and *L. infrequens* which occurs in Emsian beds (*Eurekaspirifer pinyonensis* Zone). *Leptocoelia murphyi* represents an extreme in the degree of convexity attained by its brachial valve. It is large and strongly biconvex compared to the Appalachian species *Leptocoelia flabellites*, but small specimens of *L. murphyi* more closely resemble the latter in convexity and ornament, indicating derivation from the Appalachian stock. *Leptocoelia murphyi* occurs abundantly in the *Spinoplasia* Zone of the Rabbit Hill Limestone and in the succeeding *Trematospira* Zone, both in its lower *Trematospira* Subzone and in its upper subzone characterized by *Costispirifer*. A related form, listed in the present paper as *Leptocoelia* aff. *murphyi*, occurs in the *Acrospirifer kobehana* Zone, but it is less abundant than is *L. murphyi* in older beds.

In the *Eurekaspirifer pinyonensis* Zone *Leptocoelia* is represented abundantly at some levels by *L. infrequens*. It is a coarsely plicate species compared to other known members of the genus (with the exception of *L. biconvexa* from Kazakhstan) and differs from *flabellites* and *murphyi* in being lamellose externally. This is shown especially on the strongly biconvex subspecies *L. infrequens globosa*, common at localities in the Cortez Range.

Genus *Leptocoelia* Hall, 1857
TYPE SPECIES: *Atrypa flabellites* Conrad, 1841, p. 55.
Leptocoelia murphyi Johnson, n. sp.
(Pl. 45, figs. 13-19; Pl. 46, figs. 1-21)

DIAGNOSIS: Large, strongly biconvex *Leptocoelia* with few growth lamellae over most of the valve and with a relatively narrow ventral diductor field.

EXTERIOR: *Leptocoelia murphyi* is a large and strongly biconvex species. It is suboval to subtrigonal in outline and commonly has width about equal to length. However, forms with the length greater or with the width greater are both common. Some small specimens are unequally biconvex or nearly plano-convex, but large specimens are approximately equally biconvex. The posterior edges of the valves are gently curved and are evenly continuous with the lateral margins. The anterior margin is somewhat less curved than the posterolateral or anterolateral edges and may be nearly at right angles to the midline along a short distance anteromedially. The beak is relatively strong and pointed and is erect to slightly incurved on large specimens. It is pierced by a small apical foramen. No interarea or delthyrium is present.

Anteriorly, there is a relatively well-developed, broadly flaring, ventral sulcus and a corresponding dorsal fold. The sulcus bears a single plication and the fold bears a pair of plications separated on most specimens by a subangular median groove. Some specimens, particularly larger and more transverse forms, have a more or less strongly developed median groove on the plication in the middle of the sulcus on the pedicle valve and a complementary narrow medial rib in the groove down the middle of the fold on the brachial valve. Two specimens have been seen that bear an extra plication in the sulcus on one side of the median plication, but not on the other. In the brachial valve, one of the median pair of plications splits slightly past midlength. The flanks bear subangular radial plications and relatively deep V-shaped interspaces. There are three to five plications on each flank of the brachial valves. Most commonly there are four. Some specimens have an asymmetric distribution of plications, however, and one was seen with two plications on one flank and four on the opposite flank of the same valve. The plications are crossed by a few strongly developed concentric growth lamellae. The first imbrication commonly does not develop until the shell reaches a length of about 7 to 12 mm and generally succeeding imbrications occur at intervals of two or more millimeters except at the anterior of large specimens where they become closer together increasing the convexity of the shell without appreciably increasing its length.

INTERIOR OF PEDICLE VALVE: Small shells bear elliptical hinge teeth which diverge only slightly from the midline and are supported by thick, but short dental lamellae. The hinge teeth are bounded posteriorly by shallow transverse grooves for the accommodation of the posterior margin of the brachial valve. The muscle scars are only very faintly impressed. On medium- and large-sized specimens which have thicker shells, the hinge teeth become relatively ponderous and somewhat flattened on their distal ends, and the dental lamellae become obsolescent. There commonly is only a narrow slit between the inner edge of the pedicle valve and the posterolateral edge of the hinge teeth. The diductor muscle scar is generally strongly impressed and may be relatively straight sided or flabellate. Specimens with a flabellate scar are, however, in the minority, and the most common tendency is for deep impression of the lateral edges of the diductor muscle tracks. On some specimens the wall of shell material bounding the muscle impression is grooved longitudinally, in effect affording more surface area which would otherwise be gained by the utilization of a broadly flabellate scar. The anterior edge of the muscle impression is moderately to strongly elevated on a transverse subtriangular or subcrescentic platform. The adductor scars are situated centrally within the diductor scars and are strongly impressed, in general more deeply so posteriorly, and are separated medially by a short, thin myophragm. The interior of the valve is crenulated on small specimens by the impress of the plications, but only on the anterior portions of large specimens.

INTERIOR OF BRACHIAL VALVE: The sockets are short and diverge anterolaterally. They are bounded posterolaterally by the unthickened inner edges of the valve. Their inner edges are bounded by slightly curved socket plates that are relatively strongly extended toward the pedicle valve. On small specimens there is a thickening of shell material, between the bases of the socket plates, which fills part of the nothothyrial cavity between them. The posterior portion of the nothothyrial cavity thickening bears a small bilobate cardinal process and the anterior portion may bear a short median ridge. The adductor scars are not strongly impressed, but are divided by a narrow myophragm throughout their length. Large specimens are deeply thickened medially by shell material filling the nothothyrial cavity. The medial portion of the thickened shell material bears a short, strongly rounded shaft with a subtriangular, trilobed posterior face. The sockets on large specimens are crenulated suboval pits bounded by a thickening of shell material at their anterolateral ends as well as on their inner edges. The crural bases are attached to the medial sides of the socket plates and are suboval at their junction with the socket plates, but become thin, slightly arcuate ribbons in about 1 mm anteriorly. Maximum length of the crura is about 2 or 3 mm; on their extremities they become much broader by the presence of a pair of ventromedial projections. The elevated anterior edge of the nothothyrial platform is continuous with a short thick myophragm that divides the posterior adductors, but terminates abruptly at their anterior edge. The posterior adductors, on the best preserved specimen, consist of a pair of subcircular impressions with the middle portions at approximately the same elevation as the surrounding shell material. The anterior adductors are subtriangular to cordate and are moderately impressed posteriorly. The myophragm extends part way through the anterior adductors as a very slender ridge that has only a fourth or a fifth of its width between the posterior adductors. The interior of the shell is lightly crenulated by the impress of the plications.

OCCURRENCE: *Leptocoelia murphyi* is common in the *Spinoplasia* Zone at Rabbit Hill and in the northern Simpson Park Range. It is also abundant in the *Trematospira* Subzone in the Sulphur Spring Range and in the northern Roberts Mountains. It is present, but somewhat less common, in the *Costispirifer* Subzone in the Roberts Mountains and in the Sulphur Spring Range. A few poorly preserved specimens collected from the *Acrospirifer kobehana* Zone in the northern Roberts Mountains are also provisionally assigned to this species. Those shells differ slightly by having a narrower myophragm that does not terminate abruptly at the anterior of the posterior adductor scars.

COMPARISON: *Leptocoelia murphyi* differs from *Leptocoelia flabellites* in having a larger and more strongly biconvex shell. In addition, *Leptocoelia flabellites* has a pedicle valve consistently more convex than the brachial valve. *Leptocoelia murphyi* differs from *Leptocoelia nunezi* Amos and Boucot in having stronger plications and less regularly developed concentric growth lines. *Leptocoelia murphyi* differs from *Leptocoelia infrequens* (Walcott) in having more angular plications, a more strongly developed median plication in the sulcus, and a different outline.

Amsden (1962, Fig. 4) has illustrated specimens of *Leptocoelia* from the Cravatt Member of the Haragan (Helderbergian) of Oklahoma which invite comparison with *L. murphyi*. The Oklahoma specimens (O.G.S. 5054, 5055) were examined on loan. They are biconvex and have about the same number of costae or slightly more (four or five on each ventral flank, 3 or 4 on each dorsal flank), but are flatter and have a deeper sulcus. The ventral bounding costae flare outward. The median ventral costa is very narrow.

FIGURED SPECIMENS: USNM 157101-157109, 140382. The holotype is USNM 157105.

Leptocoelia infrequens (Walcott, 1884)
(Pl. 47, figs. 1-27)

DISCUSSION: It is the writer's opinion that the species *Leptocoelia biconvexa* Bublitschenko (*in* Rzhonsnitskaya, 1960b, Pl. 56, fig. 14) and best illustrated by Kaplun (1961, Pl. 12) is not separately distinguishable from *L. infrequens*. It probably may best be regarded as a subspecies of *Leptocoelia infrequens*. It is biconvex, relatively lamellose over most of the valves, is coarsely plicate and has the broad ventral sulcus and depth of anterior commissure deflection that is usually seen in *Leptocoelia infrequens*. It is more circular (to very slightly transverse-oval) than *L. infrequens infrequens*, which is relatively broad and tends to be pentagonal, and it has a slightly stronger median plication in the ventral sulcus than is typical of *L. infrequens infrequens*. Small specimens of *L. infrequens globosa* closely resemble *L. biconvexa*, but the latter is more lenticular than specimens of *L. infrequens globosa* in the same size range.

EXTERIOR: The shells are transversely suboval in outline and subequally biconvex in lateral profile. There is a relatively strong subangular sulcus on the pedicle valve which flares widely anteriorly and is bounded by strong plications. The sulcus bears a median plication which is narrower than the lateral plications. The brachial valve bear a fold with a shallow median groove. The plications are crossed by well-defined concentric growth lamellae which are strongly lamellose over most of the shell and occur at regular intervals.

INTERIOR OF PEDICLE VALVE: No interarea nor delthyrium is present. The hinge teeth are elliptical and diverge anterolaterally at a small angle from the midline. The hinge teeth are supported basally by thin, short dental lamellae which become obsolescent on medium- and large-sized shells by deposition of secondary shell material in the umbonal cavities. The hinge teeth are relatively widely set apart and are lightly grooved on their medial sides by crural fossettes. The diductor muscle scars are moderately impressed and relatively well defined along their bounding edges. They are cordate and nonflabellate in outline with their re-entrant sides concave anteriorly. The adductor muscle scars consist of a pair of suboval impressions medially that are bisected by a short, narrow, bladelike myophragm. The interior of the shell is lightly crenulated by the impress of the plications.

INTERIOR OF BRACHIAL VALVE: The sockets of small specimens are divergent and expand considerably in a short distance anterolaterally. Their posterior edges are bounded by the unthickened inner edges of the valve. Their distal ends open freely into the interior of the valve. Larger specimens attain a relatively thick posterolateral margin of the built-up growth layers of the shell into which the sockets are impressed. On these thick-shelled specimens, they form a pair of subcrescentic pits. The sockets on both small and large shells are crenulated. The socket plates defining the medial edges of the sockets are thin and greatly extended ventrally. The notothyrial cavity is largely filled between the socket plates by shell material forming an elevated notothyrial platform. The platform is gently cleft medially on its anterior face and is extended as a relatively strong bilobed process posteriorly. The notothyrial platform joins a short, broad, low myophragm at its anterior edge. The myophragm separates a circular, impressed, posterior adductor impression, but narrows perceptibly, bisecting the cordate anterior adductors as a much narrower and lower ridge. The interior is crenulated lightly by the impress of the plications.

OCCURRENCE: *Leptocoelia infrequens* is present in the *Eurekaspirifer pinyonensis* Zone at Lone Mountain, in the Roberts Mountains, the northern Simpson Park Range, and in the Cortez Range.

COMPARISON: *Leptocoelia infrequens* has an outline distinct from other lamellose species of *Leptocoelia* such as *L. nunezi* Amos and Boucot (1963). The latter has much less prominent costae. The subspecies *infrequens infrequens* has stronger plications and fewer of them than any species of *Leptocoelia* known to the writer. Both of the subspecies bear a markedly narrow median plication in the sulcus of the pedicle valve relative to the size of the plications on the flanks.

Leptocoelia infrequens infrequens (Walcott, 1884)
(Pl. 47, figs. 1-14)
Trematospira infrequens Walcott, 1884, p. 151, Pl. 4, figs. 3, 3a, 3b.
Leptocoelia infrequens Merriam, 1940, p. 54.
Leptocoelia infrequens Amos and Boucot, 1963, p. 449, Pl. 63, figs. 5-9.

EXTERIOR: The subspecies is broadly transversely oval and biconvex with the brachial valve slightly more convex than the pedicle valve. The beak angle is large and on large specimens the outline may be subpentagonal with the median portion of the anterior edge nearly at right angles to the midline. In most specimens, maximum width is attained posterior to midlength. The ornament consists of coarse, rounded radial plications on both valves, most commonly numbering two on each flank of brachial valves and three on each flank of pedicle valves. A few specimens show a faint third plication on each flank of the brachial valve. At the other extreme, some shells show only two plications on each flank of the pedicle valve.

OCCURRENCE: From the data presently available, it appears that the subspecies *L. infrequens infrequens* is restricted to the light-colored lithofacies of the McColley Canyon Formation and to the *Eurekaspirifer pinyonensis* Zone.

DISCUSSION: The holotype is Walcott's specimen (USNM 13843) from the *Eurekaspirifer pinyonensis* Zone at Lone Mountain. It has the following dimensions expressed in millimeters: length 14.2, width 18.1, thickness 8.1. The writer has studied specimens from the Roberts Mountains and from the northern Simpson Park Range. At the latter locality, a relatively large and well-preserved collection was made and this suite of specimens compares favorably with the holotype from Lone Mountain. A few specimens that were obtained in the Roberts Mountains also agree with Walcott's specimen.

FIGURED SPECIMENS: USNM 157110-157112. The holotype is USNM 13843.

Leptocoelia infrequens globosa Johnson, n. subsp.
(Pl. 47, figs. 15-27)

EXTERIOR: The subspecies is small to medium sized and transversely suboval in outline. In lateral profile it is strongly biconvex with the brachial valve of some specimens slightly more convex than the pedicle valve. The radial ornament consists of rounded to subangular plications on both valves, numbering three or four on each flank of pedicle valves and three or four on each flank of brachial valves. The plications bounding the sulcus of the pedicle valve are relatively strongly elevated and are considerably higher than adjacent lateral plications. The ventral sulcus is strongly convex anteriorly and the dorsal fold is relatively distinct.

OCCURRENCE: *Leptocoelia infrequens globosa* is very common in the *Eurekaspirifer pinyonensis* Zone assemblage in the Wenban Limestone of the Cortez Range and is the most abundant element in the Cortez Range fauna. It has not yet been possible to determine the exact stratigraphic level of the abundantly fossiliferous beds, and so it is still to be determined whether *L. infrequens globosa* is a geographic subspecies confined to the peculiar Cortez Range facies or whether the two subspecies of *L. infrequens* have stratigraphic value.

COMPARISON: *L. infrequens globosa* may be distinguished from *L. infrequens infrequens* by its greater relative convexity as well as by its more numerous and slightly smaller and more angular plications. *L. infrequens infrequens* attains larger size, but no specimens are as strongly convex as *L. infrequens globosa*. *Leptocoelia infrequens globosa* differs from *L. infrequens biconvexa* Bublitschenko principally in being more strongly biconvex in a comparison of full-size specimens.

FIGURED SPECIMENS: USNM 157113-157116. The holotype is USNM 157113.

Genus *Leptocoelina* Johnson, n. gen.

TYPE SPECIES: *L. squamosa* Johnson, n. sp.

DIAGNOSIS: Leptocoeliid brachiopods with relatively numerous, evenly spaced, concentric growth lamellae and sparse, low, radial plications. Ventral diductors small, narrow, and anteriorly platformed; cardinalia includes a simple, knoblike cardinal process.

DISCUSSION: The few and very low radial plications and the well-developed growth lamellae covering the whole of the valves makes recognition of *Leptocoelina* relatively easy on external grounds in the presence of well-preserved specimens; however, the rule is not without exception. *Leptocoelia infrequens* can have a strong development of concentric growth lamellae, and that feature is particularly evident on the subspecies *L. infrequens globosa* from the Cortez area. Thus it is possible to confuse *Leptocoelina squamosa* with *Leptocoelia infrequens globosa*, but with well-preserved specimens the differences are obvious. *Leptocoelia infrequens globosa* is a true *Leptocoelia* internally; externally it is much more deeply convex, its lateral plications are slightly stronger, and its sulcus does not widen anteriorly with the abrupt flare—nor is it deflected so strongly toward the brachial valve at the anterior commissure as is the case in *Leptocoelina squamosa*. The type of ventral diductor scar appears to be a good generic feature. Specimens of *Leptocoelina acutiplicata* from the Needmore Shale of the Appalachian region have a scar that is virtually identical. Inside the brachial valve, the simple cardinalia are characteristic. Many well-preserved specimens were examined, but none shows the relatively large moundlike site of diductor attachment that is developed in *Leptocoelia* and which commonly has three small lobes on its posterior face.

Two or possibly three species belong to *Leptocoelina*. In addition to *L. squamosa*, the genus includes the specimens illustrated in figures 4-15 of Plate 6 of Kindle (1912), which were called *Anoplotheca acutiplicata*. Kindle also re-illustrated some of the New York specimens of James Hall, but they may not belong to *Leptocoelina*. From a comparison of Kindle's figures only, the specimens from Pennsylvania, Maryland, Virginia, and West Virgina may not belong to *acutiplicata* and may need a new name. A second species that may belong to *Leptocoelina* is *Atrypa? biconvexa* Maynard (*in* Schuchert and others, 1913, p. 393, Pl. 68, figs. 1-3) from the Keyser Limestone of Keyser, West Virginia. However, the whereabouts of the types of *biconvexa* are unknown. The matter will need investigation in the future when topotypical specimens can be obtained.

The species *Leptocoelia nunezi* from Argentina, named by Amos and Boucot (1963, Pl. 62, figs. 17-26) mimics *Leptocoelina* in the presence of low radial plications and relatively numerous, evenly spaced, concentric growth lamellae. However, the dorsal interior of *Leptocoelia nunezi* has a wholly different cardinalia than that of *Leptocoelina squamosa*. The species called *Leptocoelia nunezi texana* by Amos and Boucot is not closely related to *Leptocoelia nunezi* from Argentina, as is obvious from a comparison of the published figures (that is, Amos and Boucot, 1963, Pl. 63, figs. 10-17). After studying a large suite of silicified specimens from the Stribling Formation of central Texas the writer is of the opinion that they are best assigned to *Australocoelia* because they possess the terminally expanded cardinal process characteristic of that genus.

Leptocoelina squamosa Johnson, n. sp.
(Pl. 45, figs. 1-12)

EXTERIOR: The shells are transversely suboval to subpentagonal in outline. The valves are subequally biconvex in lateral profile, and the brachial valve may be slightly more convex, especially posteriorly, than the pedicle valve. The posterior edges of the valves are straight or may be very slightly concave posteriorly. The cardinal angles are obtuse and strongly rounded. Lateral margins are curved relatively smoothly, but are deflected inward at the anterior portion of the valve due to the effect of the sulcus. There is a broadly flaring sulcus on the pedicle valve. It first becomes distinct at about 5 mm anterior to the beak. Anteriorly on larger shells, the sulcus is relatively strongly geniculate toward the brachial valve. There is a corresponding fold which becomes distinct at above 5 mm anterior to the beak on the brachial valve and then becomes relatively strongly raised anteriorly. Due to the elevation of the fold, the brachial valve is almost flat in its transverse section anterior to the beak.

The ornament consists of a few low, rounded plications separated by shallow U-shaped interspaces crossed by numerous imbricate growth lamellae which occur at regular intervals. A peculiar feature of the growth lamellae on some specimens is recurvature inward along the margin of many of the growth lamellae and continuity with the portions of the shell formed during the next growth stage. Other lamellae project shingle-like in a more common fashion out over the lamellae produced by later growth stages. There is a single low, rounded plication in the ventral sulcus and two relatively strong plications forming the dorsal fold and separated by a narrow U-shaped furrow. Each valve has one strong plication and one markedly weaker plication on each flank. The dorsal umbo has a very broad shallow sulcus due to the depression of the medial pair of plications relative to those on the flanks of the valve.

INTERIOR OF PEDICLE VALVE: The hinge teeth are blunt and triangular and are unsupported by dental lamellae. They are attached directly to the interior of the valve posterolaterally and are grooved on their medial sides by relatively deep crural fossettes. They are limited posteriorly by the transverse groove which accommodates the posterior edge of the brachial valve. The posterior margin of the brachial valve directly adjoins the base of the apical foramen of the pedicle valve, and its transverse groove in the pedicle valve is deeply incised, making interarea and delthyrium obsolete. The diductor muscle scars together form an elongate, narrow, triangular scar which is straight-sided or slightly curved and is strongly impressed laterally. The anterior edge of the muscle impression is elevated forming a subtriangular to subcrescentic platform with its convex side posterior. The adductor muscle scars are deeply impressed centrally within the diductor scar and are separated by a thin bladelike myophragm. The interior of the shell is faintly crenulated by the impress of the plications.

INTERIOR OF BRACHIAL VALVE: The sockets are relatively broad and diverge anterolaterally. The bases of the sockets are transversely corrugated. The socket plates define essentially the bases and the inner edges of the sockets and are commonly supported by deposits of secondary shell material in the umbonal cavities. The inner edges of the socket plates are strongly extended toward the pedicle valve and diverge from one another anteriorly at a relatively small angle. The crural bases are attached along the inner ventral edges of the socket plates. Crural plates are not developed. There is a stout, rounded myophragm in the posterior part of the valve which extends into the notothyrial cavity and is thickened forming a triangular platform that more or less fills the notothyrial cavity. Medially, it is elevated forming a more or less narrow triangular lobe. The myophragm narrows abruptly at the anterior edge of the posterior adductor impressions which consist of a pair of suboval pits on either side of the myophragm. The anterior adductors consist of a subtriangular or elongate pair that is undivided through most of its length by a myophragm although a dividing line or faint ridge is commonly seen. The interior is faintly crenulated by the impress of the plications.

OCCURRENCE: *Leptocoelina squamosa* is common in the *Spinoplasia* Zone and is restricted to it. The species is abundant in the Rabbit Hill Limestone at Rabbit Hill, in the northern Simpson Park Range, and at Dobbin Summit. In addition, a few specimens have been found in the Cortez Range.

COMPARISON: *Leptocoelina squamosa* has fewer ribs and a wider, more flaring sulcus and a more strongly deflected anterior commissure than the New York specimens illustrated as *Anoplotheca acutiplicata* by Kindle (1912). The other figured specimens on Kindle's Plate 12 (figs. 4-15) more closely resemble *L. squamosa*. They appear on the average to be slightly more circular, to have a less prominent ventral tongue, and appear to have three relatively well-defined plications on each flank of the valve compared to only one or two on *Leptocoelina squamosa*. If *Atrypa? biconvexa* should prove to belong to *Leptocoelina*, it can be distinguished from *L. squamosa* by the presence of three or four plications on each flank of the valves compared to the smaller number on *L. squamosa*.

FIGURED SPECIMENS: USNM 157095-157099. The holotype is USNM 157095.

SUBORDER ATHYRIDOIDEA

Superfamily ATHYRIDACEA M'Coy, 1844
Family MERISTELLIDAE Waagen, 1883
Subfamily MERISTELLINAE Waagen, 1883
Genus *Meristina* Hall, 1867

TYPE SPECIES: *Meristella maria* Hall, 1863b, p. 212.

DISCUSSION: The type of *Meristina* is a Silurian species and the genus commonly has been regarded as ancestral to the Lower Devonian genus *Meristella*. Accordingly, it has been recorded mostly from Silurian rocks. However, Boucot and others (1964), reasoning from studies of the musculature in the pedicle valve of the better known species of *Meristina* and *Meristella*, extended recognition of the former genus well into the Devonian where they consider it to be the probable true receptacle of *Meristella* species reported beyond the Appalachian Province. *Meristella* itself was restricted to species having close structural similarity to its type species, *M. laevis*, from the Appalachian Lower Devonian.

Meristina sp. A
(Pl. 48, figs. 1-6)

EXTERIOR: The outline is elongate subpyriform to pentagonal. The valves are subequally biconvex in lateral profile with the pedicle valve having greatest convexity posteriorly and the brachial valve being more convex in its anterior half. Anteriorly, the pedicle valve develops a very broad low sulcus which involves most of the width of the valve. The brachial valve lacks a distinct fold, but accommodates the deflected anterior commissure of the pedicle valve within its convex anterior flanks. The ventral beak is pointed, but not prominent, and it is only moderately incurved.

The ornament consists of strong, concentric growth lines which appear at sparse irregular intervals on most specimens. No radial ornament has been observed.

INTERIOR OF PEDICLE VALVE: Relatively short, thin dental lamellae are present in the apex of the valve and define the delthyrial cavity laterally. Anterior to the delthyrial cavity the muscle scars are poorly to strongly impressed and in this respect show considerable variability. In some shells, particularly the smaller ones, the muscle impressions are barely discernible. In the strongly impressed forms, the muscle field is narrow and elongate with strongly impressed lateral edges that diverge only slightly anteriorly. In addition, the tracks are exceedingly long, reaching to a considerable extent posterior to the midlength of the valves. The muscle field may be longitudinally grooved and thus divided into pairs of bilobate tracks with a narrow myophragm dividing them medially The umbonal cavities may bear a few chevron-like ridges which diverge *en echelon* anterolaterally.

INTERIOR OF BRACHIAL VALVE: The sockets are defined posterolaterally by the interior edges of the valve and basally by thin socket plates. Medially, there is a septalium supported by a median septum. In some specimens the median septum penetrates ventrally through the base of the septalium and lies in its medial trough. The median septum is high and continues anteriorly to a point slightly posterior to midlength.

OCCURRENCE: *Meristina* sp. A is present in the *Quadrithyris* Zone of the Windmill Limestone at Coal Canyon and the lower McMonnigal Limestone in the Toquima Range.

COMPARISON: This species is much less pyriform in outline than *Meristina* sp. B and has a more faintly defined fold and sulcus.

FIGURED SPECIMENS: USNM 157117-157121.

Meristina sp. B
(Pl. 48, figs. 7-21)

EXTERIOR: The shells are elongate suboval to pyriform in outline and subequally biconvex in lateral profile. Small specimens generally have the pedicle valve more convex, but the brachial valve becomes relatively more strongly convex in large specimens. The hinge line is narrow and curved. Maximum width is most commonly developed anterior to midlength, but on one specimen the reverse is true. The ventral umbo is relatively strongly convex, narrow, and strongly incurved. The beak is perforated by a circular foramen. A relatively strong fold is developed anteriorly on some brachial valves, but pedicle valves commonly lack a well-developed sulcus. Instead they are extended anteriorly along the midregion forming a tonguelike projection which is accommodated by the dorsal fold. The exterior is smooth, except for fine concentric growth lines.

INTERIOR OF PEDICLE VALVE: The hinge teeth are narrowly pyriform and are divergent anterolaterally. They are supported basally by relatively well-developed, thin dental lamellae. The dental lamellae diverge only slightly anteriorly and are convex outward. The umbonal cavities are of variable depth, but generally are only slightly infilled with shell material. In one specimen there is a raised, elongate-trapezoidal platform medially, but commonly the muscle scars are moderately impressed with straight lateral edges that diverge anterolaterally. The diductor impressions may be moderately radially grooved anteriorly, but otherwise blend anteriorly into the interior surface of the shell.

INTERIOR OF BRACHIAL VALVE: The sockets are long and narrow and diverge anterolaterally. A septalium is present and may extend relatively strongly anteriorly along the median septum. The median septum is high and thin and reaches approximately to midlength.

OCCURRENCE: *Meristina* sp. B is a relatively rare fossil, present in the *Acrospirifer kobehana* Zone and the *Eurekaspirifer pinyonensis* Zone. Specimens are known from the *Acrospirifer kobehana* Zone in the Sulphur Spring Range. *Eurekaspirifer pinyonensis* Zone occurrences are known from the Coal Canyon area, northern Simpson Park Range, north side of the Roberts Mountains, and from the Sulphur Spring Range.

COMPARISON: *Meristina* sp. B has a pronounced pyriform outline, not present in *Meristina* sp. A and is relatively more strongly convex than that species.

FIGURED SPECIMENS: USNM 157122-157125.

Genus *Meristella* Hall, 1859
TYPE SPECIES: *Atrypa laevis* Vanuxem, 1842, p. 120.
Meristella robertsensis Merriam, 1940
(Pl. 49, figs. 1-6)

?*Meristella (Whitfieldia) nasuta* Walcott, 1884, p. 148, Pl. 3, figs. 8, 8a, 8b; *not* Conrad, 1842.
Meristella robertsensis Merriam, 1940, p. 84, Pl. 6, fig. 1.
Meristella robertsensis Cooper, 1944, p. 333, Pl. 127, fig. 23.

DISCUSSION: The holotype of *M. robertsensis* comes from a locality in the southern Roberts Mountains which Merriam assigned to the *Acrospirifer kobehana* Zone (1940, p. 53). The holotype is refigured on Plate 49, but little adequate additional material was available to the writer which could be confidently assigned to the species and thus give some idea of its variation. The holotype has a markedly subrhomboidal outline and is not strongly convex. In addition, the ventral sulcus is very poorly developed.

Meristella is a common genus in the *Spinoplasia* Zone and in the *Trematospira* Zone, but those specimens are not readily separated from *Meristella robertsensis*. Indeed, Merriam (1940, Pl. 6) regarded specimens from the *Trematospira* fauna as comparable to *M. robertsensis*.

FIGURED SPECIMENS: USNM 96369 (the holotype) and USNM 157126.

Meristella cf. *robertsensis* Merriam, 1940
(Pl. 49, figs. 7-25; Pl. 50, figs. 1-16)
Meristella cf. *robertsensis* Merriam, 1940, Pl. 6, figs. 2, 13-16.

EXTERIOR: Small specimens are of almost equal width and length and are broadly subpyriform in outline. Larger specimens tend to have a more nearly rhomboidal outline, but with the posterolateral edges slightly straighter than the anterolateral edges which are gently rounded at the commissure. For the most part, large specimens show only the slightest inflection of a broad shallow sulcus anteriorly. Brachial valves are commonly slightly carinate, but do not develop a distinct median fold. The best preserved specimens have the beak very strongly incurved touching the dorsal umbo so as to effectively cover the foramen at the apex of the beak. A number of the most well-preserved specimens illustrate this condition and must have lacked a functional pedicle late in life. A less common form is developed by some specimens which may possibly be regarded as gerontic shells. These have a markedly subpentagonal outline and are greatly increased in relative thickness of the valves due to more equal addition of new shell material around the whole length of the commissure along strong, well-defined growth lines. These specimens have a somewhat less strongly incurved ventral beak and an open apical foramen. The dorsal fold becomes relatively well marked anteriorly and has roughly subparallel lateral edges. The pedicle valve develops a rather shallow, flat-bottomed sulcus and is somewhat drawn out anteromedially.

INTERIOR OF PEDICLE VALVE: Small specimens bear a pair of relatively short, thin dental lamellae. The muscle scars are not impressed; however, medium- to large-sized specimens become relatively strongly thickened with secondary shell material in the umbonal cavities and in the posterior regions in general. The diductor scars are strongly impressed and triangular. Some flare very strongly anterolaterally and are radially striate. Anteriorly, the muscle impression is relatively sharply defined, but is not strongly impressed. Some specimens that are relatively thick-shelled bear shell material filling in the posteromedial regions. In these specimens, the posterior end of the diductor impression is a narrowly acute angle. Thickening of shell material on either side of the midline in the pedicle cavity may fill up a third or more of its depth. The diductor impression can be strongly impressed on medium- and large-sized shells; however, it is relatively common to find even large shells where the muscle scars are not deeply impressed.

INTERIOR OF BRACHIAL VALVE: The sockets are relatively small and narrow, diverging anterolaterally along the inner edges of the shell. The inner edges of the socket plates arise almost directly from the posterolateral shell wall and are moderately thickened and elevated ventrally. Medially, there is a U-shaped septalium supported by a bladelike median septum. In some specimens there is an elevated median ridge or extension of the median septum in the base of the septalium. In other specimens the septalium is largely filled with secondary shell material. The median septum reaches to about midlength. The adductor muscle scars may be relatively strongly impressed on either side of the median septum posteriorly, but not anteriorly and laterally.

OCCURRENCE: *Meristella* cf. *robertsensis* is relatively common in the

ORDER SPIRIFERIDA 177

Spinoplasia Zone at Rabbit Hill and in the *Trematospira* Zone in the Roberts Mountains and the Sulphur Spring Range. One specimen, with a sulcus, is probably from the *pinyonensis* Zone.

FIGURED SPECIMENS: USNM 157127-157136, 157137A and B.

Family NUCLEOSPIRIDAE Davidson, 1881
Genus *Nucleospira* Hall, 1859

TYPE SPECIES: *Spirifer ventricosa* Hall, 1857, p. 57. *not* figs. 1, 2.

Nucleospira sp. A

EXTERIOR: The species is small and thin shelled. Most of the available specimens are fragmentary, making it difficult to fully describe their outline and convexity. One well-preserved brachial valve is transversely oval in outline. Both valves are convex. The beaks of both valves are very small and short. No fold nor sulcus is present on either valve.

INTERIOR OF PEDICLE VALVE: The hinge teeth are small and elliptical with their long axes nearly perpendicular to the midline. They are widely set apart and are unsupported by dental lamellae. The muscle scars are poorly impressed, but there is a long, thin, low, median septum originating posteriorly but not extending to the apex of the valve.

INTERIOR OF BRACHIAL VALVE: The sockets are relatively long and narrow and are directed almost straight laterally at right angles to the midline. They are incised in a thickening of shell material along the interior of the valve. There is a relatively ponderous semicircular cardinal plate which arises from the base of the valve between the sockets and is directed ventrally initially and then recurves strongly to project posteriorly. The anterior face of the cardinal plate is made bilobed by the presence of a pair of triangular grooves on either side of the midline. There is a stout myophragm originating basally, but it does not connect with the cardinal plate or its median ridge. The myophragm is long and tapers somewhat anteriorly. The adductor muscle scars are relatively strongly impressed forming an elongate pair closely adjacent to the myophragm.

OCCURRENCE: *Nucleospira* sp. A is present in the *Quadrithyris* Zone at Coal Canyon and in the Toquima Range.

COMPARISON: *Nucleospira* sp. A is a relatively thin-shelled species and differs in this respect from *N.* sp. B and *N. subsphaerica* of the overlying zones.

Nucleospira sp. B
(Pl. 51, figs. 1-16)

EXTERIOR: Small shells are nearly circular or may be slightly transversely suboval. Larger specimens tend to become slightly elongate and deeply biconvex. The ventral umbo is fairly prominent for the genus. The beak is strongly incurved touching the umbo of the brachial valve. The hinge line is short and curved. Maximum width is attained at about midlength. The anterior commissure is rectimarginate.

No radial ornament is developed; however, there are fine concentric growth lines which commonly became marked toward the anterior on large shells. The surface of the exterior of some specimens bears very numerous, small, hairlike spinules which are directed anteriorly at a high angle to the anterior curvature of the shell. Along the midline the fine spinules and groups of spinules are directed straight anteriorly. On the lateral flanks the spines are directed somewhat anterolaterally, but none is so strongly inclined laterally that it is perpendicular to a plane tangent to the commissure. The average diameter of the individual spinules appears to be very slightly less than a tenth of a millimeter.

INTERIOR OF PEDICLE VALVE: The hinge teeth are transversely elliptical and unsupported by dental lamellae. There is a small triangular convex plate in the apex of the delthyrium. Both the medial and lateral portions of the posterior part of the shell may be slightly thickened with secondary shell material, and in these specimens the adductor muscle scars form a small cordate pair medially. The diductor scars are very faintly impressed and moderately flabellate. They are divided medially by a long slender myophragm.

INTERIOR OF BRACHIAL VALVE: The sockets are narrow, deep, and widely divergent laterally. The socket plates, forming their inner edges, are united by a cardinal plate which projects ventrally and then posteriorly. It may be medially grooved on its ventral face. On the largest specimens there is a stout myophragm supporting the cardinal plate and connecting with the base of the valve where it extends anteriorly as a long, slender myophragm which reaches about to the midlength of the valves. The adductor scars are not strongly impressed except in the largest specimens and on those they constitute a narrow elongate pair closely adjacent to the myophragm along the posterior portion of its length.

OCCURRENCE: *Nucleospira* sp. B is relatively common in the *Trematospira* Zone and the *Acrospirifer kobehana* Zone in the Roberts Mountains and in the Sulphur Spring Range. It is also present, but very rare, in the *Spinoplasia* Zone. Only two valves from the *Spinoplasia* Zone at Rabbit Hill have been found.

COMPARISON: *Nucleospira* sp. B is easily distinguished from *N.* sp. A in having a much thicker shell. In addition, the dorsal interior of *Nucleospira* sp. B has a median ridge supporting the cardinal plate and continuous with the dorsal myophragm in contrast with *Nucleospira* sp. A which lacks any sort of median ridge between the posterior terminus of the myophragm and the cardinal plate.

FIGURED SPECIMENS: USNM 157138-157144.

Nucleospira subsphaerica Johnson, n. sp.
(Pl. 51, figs. 17-24)

DIAGNOSIS: Large, deeply convex *Nucleospira* tending to be slightly transverse.

EXTERIOR: The shells are transversely suboval and are strongly biconvex in lateral profile. Specimens attain a relatively large size, and large specimens are very thick-shelled. Small brachial valves are very nearly exactly suboval, but small pedicle valves are nearly as long as they are wide. Large-sized specimens retain almost the same outline characteristics and length-width ratio as smaller ones. The largest specimens bear a shallow median furrow in the anterior half of the pedicle valve; however, there is no fold on the brachial valve. The umbos on both valves are fairly strongly convex, and the beaks are strongly incurved. The ventral beak almost touches the dorsal umbo. The hinge line is short and curved; there is no interarea developed.

Well-marked concentric growth lines appear on most specimens first at a distance of approximately 12 mm anterior to the umbo, and on larger shells there are several more well-developed concentric growth lines.

INTERIOR OF PEDICLE VALVE: The delthyrium is broad and encompasses an angle of slightly more than 90 degrees. It is closed at its apex by a broad, concave, triangular plate. The hinge teeth are widely set apart laterally and are subelliptical in cross section. They diverge strongly anterolaterally and are unsupported by dental lamellae. The muscle scars are only poorly impressed. The adductor impressions are slightly more distinct than the diductors and form a suboval pair in the posterior portion of the valve, but anterior to the delthyrial cavity. There is a long, low, bladelike median septum originating at the posterior end of the adductor scars and continuing to the anterior commissure.

INTERIOR OF BRACHIAL VALVE: The sockets are relatively deep and narrow and diverge anterolaterally along the inner edges of the valve. The socket plates unite medially, forming a cardinal plate that is directed ventrally and then recurves sharply posteriorly. The anterior face of the cardinal plate bears a strong median ridge. A long, low, V-shaped ridge is present running the length of the valve. The adductor scars are not impressed.

OCCURRENCE: *Nucleospira subsphaerica* is relatively common in the *Eurekaspirifer pinyonensis* Zone in the Sulphur Spring Range. It is also present, but rare, in the *Eurekaspirifer pinyonensis* Zone in the Cortez Range.

COMPARISON: *Nucleospira subsphaerica* is more transverse, thicker shelled, and reaches greater size than other *Nucleospira* present in underlying zones. *Nucleospira* sp. B in its largest specimens may be nearly as large as *N. subsphaerica*, but is elongate rather than transverse. In addition, *Nucleospira* sp. B has a relatively ponderous median ridge supporting the cardinal plate in the posterior portion of the brachial valve.

FIGURED SPECIMENS: USNM 157145-157148. The holotype is USNM 157148.

SUBORDER RETZIOIDEA

Superfamily RETZIACEA Waagen, 1883
Family RETZIIDAE Waagen, 1883
Genus *Trematospira* Hall, 1859
TYPE SPECIES: *Spirifer multistriatus* Hall, 1857, p. 59.
Trematospira perforata (Hall, 1857)
(Pl. 52, figs. 8-25)

Spirifer? perforatus Hall, 1857, p. 60.
Trematospira perforata Hall, 1859, p. 208; 1861, Pl. 28A, figs. 3a-k.
Trematospira perforata Hall and Clarke, 1895, Pl. 49, figs. 5, 6.
Trematospira multistriata Clarke, 1900, p. 43, Pl. 6, figs. 1-4; *not* Hall.
Trematospira perforata var. *atlantica* Clarke, 1909, p. 41, Pl. 7, figs. 21-23.

Trematospira multistriata Schuchert and others, 1913, p. 428, Pl. 73, figs. 3, 4; not Hall.
Trematospira multistriata Stewart, 1923, p. 248, Pl. 65, figs. 4, 5; not Hall.
?*Trematospira* sp. Amsden and Ventress, 1963, p. 127, Pl. 4, fig. 11.
?cf. *Straelenia* sp. Talent, 1964, p. 32, Pl. 15, figs. 1, 2, 5, 7.

DISCUSSION: *Trematospira multistriata* (type species of *Trematospira*) and *T. perforata* are related species from the Helderberg of New York. The syntype collections of both were borrowed from the American Museum of Natural History. *Trematospira multistriata*, represented by 21 syntypes under the number 2459 are from the "lower Helderberg group," Clarksville, New York. The specimens are large and transverse with a broad, blunt beak, not narrowed as in the specimens from Nevada or like the types of *T. perforata*. The radial costae are slightly more rounded than in the specimens of *T. perforata* from New York and from Nevada. The ventral beak of *T. multistriata* is very strongly incurved without an exposed foramen. Shells are preserved in black calcite. *Trematospira perforata* is represented by four syntype specimens under the number 2460 from the "lower Helderberg group," Hudson, New York. The specimens are smaller than those of *T. multistriata* and bear relatively strong bifurcating costae and are characterized by an erect ventral beak with a circular foramen. The specimens are silicified, with a brownish preservation. With respect to the angularity of the costae, a very slight bundling or fasciculation of those costae and of the presence of a circular foramen the Nevada specimens agree closely with the New York specimens of *T. perforata* and differ from *T. multistriata*, furnishing the basis for the present assignment.

EXTERIOR: Very small shells are elongate suboval to pyriform, and large shells are transversely suboval in outline. The valves are unequally biconvex with the pedicle valve about 50 percent more strongly convex than the brachial valve. The ventral umbo is the place of greatest convexity. The beak is incurved, suberect. No interarea is developed. The delthyrium is covered by a deltidium, and the foramen is mesothyrid. In large shells the place of maximum width is at about midlength or slightly posterior to midlength. On the anterior of large shells there is a broad shallow sulcus and there is a corresponding low, poorly defined fold on the brachial valve.

The ornament consists of strong raised costae, subrounded to angular in cross section, which commonly increase in number anteriorly by bifurcation of smaller costae at the lateral edges of larger ones. The commissure is crenulated by the costae.

INTERIOR OF PEDICLE VALVE: The hinge teeth are unsupported by dental lamellae. The pedicle cavity is pentagonal and more or less set off from the diductor muscle field anteriorly. The diductor muscle field is slightly impressed posterolaterally and blends with the interior of the shell anteriorly.

INTERIOR OF BRACHIAL VALVE: The socket plates are attached directly to the posterolateral edges of the interior of the shell. They are strongly curved or coiled, forming cylindroidal sockets. On the posteromedial edges of the socket plates, the thickenings of the crural bases originate as medial lobes that curve strongly posteriorly and then recurve dorsally, largely posterior to the brachial valve interior. A median septum is not developed. The interior of the shell is crenulated by the impress of the costae.

OCCURRENCE: The species is rare in the *Spinoplasia* Zone, occurring in the upper McMonnigal Limestone in the Toquima Range and in the Rabbit Hill Limestone at Rabbit Hill. It is more common in the *Trematospira* Zone in the Roberts Mountains and in the Sulphur Spring Range, and it occurs in that zone in the Andy Hills of the Panamint Range of southeastern California. It is also known in the *Costispirifer* Subzone in the Cortez Mountains. A specimen of *Trematospira*, probably *T. perforata*, occurs in the *Spinoplasia* Zone at Antelope Peak.

FIGURED SPECIMENS: USNM 157149-157154.

Genus *Pseudoparazyga* Johnson, n. gen.
TYPE SPECIES: *Trematospira cooperi* Merriam, 1940, p. 82.

DIAGNOSIS: *Pseudoparazyga* resembles *Parazyga* in shape and in radial ornament. Internally, it is characterized by the presence in large shells of a pedicle collar and the absence of dental lamellae. Socket plates are medially conjunct.

DISCUSSION: The type species is very close to "*T.*" *equistriata* Hall and Clarke (1895, Pl. 49, fig. 47) which also belongs to *Pseudoparazyga*. Both have been assigned to *Trematospira* in the past, but *T. multistriata*, the type species of *Trematospira*, bears a very different ornament of coarse, angular, bifurcating costae as well as strong medial lobes on the inner hinge plates. It seems likely that "*Parazyga*" *deweyi* (Hall; Hall and Clarke, 1895, Pl. 49, figs. 40-46) belongs to *Pseudoparazyga*. It may be no more than the young of *P. equistriata*.

The new genus differs from *Parazyga* by much stronger posterior extension of the socket plates, by the absence of dental lamellae, and by the presence of a complete pedicle collar in large specimens.

Pseudoparazyga cooperi (Merriam, 1940)
(Pl. 52, figs. 1-7; Pl. 53, figs. 1-18)
Trematospira cooperi Merriam, 1940, p. 82, Pl. 6, fig. 12.

EXTERIOR: Shells are transversely suboval in outline and biconvex in lateral profile. The valves are approximately of equal depth or the brachial valve may be slightly more convex. The ventral umbo is small, and the beak is strongly incurved touching the dorsal umbo which rests beneath it. The hinge line is relatively long and gently curved. Interareas are not developed. Maximum width of the shell is commonly about one-third of the distance from posterior to anterior. There is a more or less well-developed ventral sulcus and a low, rounded dorsal fold.

The ornament consists of closely spaced, simple, narrow, rounded costae which increase in size anteriorly. Neither bifurcation nor implantation occur on most shells. The interspaces between the rounded costae are V-shaped and are narrower than the adjoining costae In the midregions of both valves, costae are more tightly set together, with extremely narrow interspaces, a condition only well developed in the anterior half of large specimens. On most shells there are 9 or 10 costae in a space of 5 mm, 10 mm from the beak.

INTERIOR OF PEDICLE VALVE: The hinge teeth are widely spaced, blunt, and triangular in cross section. Apically, there is a pedicle collar formed by the presence of a concave triangular plate beneath the level of the delthyrium. The pedicle cavity is trapezoidal and deeply impressed. Medially, there is a deep groove leading beneath the pedicle collar to join the foramen apically. On large shells the umbonal cavities are deeply filled with secondary shell material which is pustulose. The diductor muscle field is triangular, strongly impressed posteriorly, longitudinally striate, and blends anteriorly with the interior of the shell. Large specimens are thick shelled and are not crenulate along the internal margins of the valve.

INTERIOR OF BRACHIAL VALVE: The socket plates are attached directly to the posterolateral walls of the valve. They curve sharply medially and posteriorly, and join medially to form a single plate. On some shells, extensions of the inner edges of the socket plates are directed smoothly posteriorly and bifurcate again into two lobes. On others there is a swelling on the posterolateral portion facing the pedicle valve and a medial thickening on the dorsal side of the plate connecting the two lobes. The position of the adductor scars is deeply impressed and elongate, but the muscle impressions themselves are poorly defined. There is a more or less well-developed myophragm in the posterior portion of some valves. A median septum is not developed.

OCCURRENCE: *Pseudoparazyga cooperi* is fairly common in the Rabbit Hill Limestone at Rabbit Hill. It is also present in the upper McMonnigal Limestone in the Toquima Range. It is uncommon, but fairly widely distributed in the *Trematospira* Zone in the Roberts Mountains and in the Sulphur Spring Range.

COMPARISON: *P. cooperi* is most like *P. equistriata* Hall and Clarke which is known from New Scotland age beds in Maryland (Schuchert and others, 1913) and from the Bailey Limestone in southeastern Missouri (Tansey, 1923). Specimens of *P. equistriata* from "New Scotland age bed", U.S. Route 40, east bank of Licking Creek, 2.5 miles west of Indian Springs, Maryland, were examined on loan from the U.S. National Museum. They have the same size, convexity, and outline as *P. cooperi,* and have about 40 costae on the brachial valve They have no pedicle collar, possibly due to their thin shells; their cardinal plate is relatively wider, and the transverse groove that accommodates the posterior edge of the brachial valve is deep. All in all the differences are rather small; possibly detailed comparisons, based on larger collections, will show these two species to be synonymous.

FIGURED SPECIMENS: The specimen, USNM 96372, illustrated in figures 1-5 on Plate 53 is the holotype. It is from locality no. 1 of Merriam (1940) in the northern Roberts Mountains. There are approximately 54 costae at the commissure on the brachial valve and approximately 56 costae on the pedicle valve. Bifurcation of costae occurs once or twice, and implantation occurs once. The holotype is 26.0 mm in length, 35.9 mm in width, and 18.4 mm in thickness. Other figured specimens are USNM 157155-157163.

SUBORDER SPIRIFEROIDEA

DISCUSSION: The members of the suborder Spiriferoidea, sometimes referred to as the true spirifers, compose what is probably the most abundant and diverse group of articulate brachiopods in the Siegenian and Emsian of central Nevada. Spiriferids, however, are not very abundant in the Gedinnian rocks in this region. *Howellella* and a faintly ribbed reticulariid resembling Kozlowski's (1929) species *laeviplicata* occur in the Gedinnian along with the oldest *Megakozlowskiella*, *Metaplasia*, and *Cyrtina* in the upper part of the Roberts Mountains Formation.

The *Quadrithyris* Zone, assigned to the lower Siegenian, contains rare, relatively large *Howellella*, fairly common *Megakozlowskiella* (*M.* sp. A), along with *Ambocoelia*, *Cyrtina*, and *Cyrtinaella*, together with a few rare specimens questionably assigned to *Reticulariopsis* and the newly appearing *Quadrithyris* (*Q.* cf. *minuens*). In the succeeding *Spinoplasia* Zone, ribbed spirifers, identified as *Howellella cycloptera*, are abundant and accompanied by the first Applachian Province ambocoeliids represented by *Plicoplasia cooperi*. *Spinoplasia* is abundant; *Cyrtina* and *Megakozlowskiella* continue from below.

In the succeeding *Trematospira* Zone, the first relatively large, coarse-ribbed spirifers assigned to *Acrospirifer* become common although a few specimens have been seen in the upper part of the Rabbit Hill Limestone (Johnson, 1965, p. 374). *Acrospirifer* is accompanied by its close relative *Dyticospirifer* which probably was derived in Nevada directly from the local and still unnamed variety of *Howellella cycloptera* because some specimens of *H. cycloptera* in Nevada bear split plications. *Megakozlowskiella* is represented by the same species (*M. magnapleura*) as in the *Spinoplasia* Zone. The Appalachian Province reticulariid *Elytha* makes a rare appearance in this zone and is represented by fragmentary remains that were judged inadequate for a basis of description. Also rare, but represented by a few identifiable specimens, is what the writer believes to be true *Reticulariopsis* (*R.* sp. B). Ambocoeliids are represented by *Metaplasia* (*M* cf. *paucicostata*), and *Cyrtina* is represented by large multiplicate specimens identified as *Cyrtina* cf. *varia*, but some of these appear to fall within the range of variation of the important Oriskany species *Cyrtina rostrata*. In the higher part of the *Trematospira* Zone, there is a great abundance of an indeterminate species of *Costispirifer*. Insofar as the present evidence can be assessed, the genus appears to be restricted to a narrow interval, the *Costispirifer* Subzone, in Nevada and southeastern California.

The *Acrospirifer kobehana* Zone, which is judged to represent the beginning of the Emsian, is characterized by an abundance of *A. kobehana* which is a species with broad, low, asymmetric plications. *Megakozlowskiella*, *Metaplasia*, and *Cyrtina* are present, as they are in older beds.

Succeeding the *Acrospirifer kobehana* Zone is the *Eurekaspirifer pinyonensis* Zone characterized by a diverse group of newly appearing spiriferids of a multiplicate type including *Hysterolites*, *Brachyspirifer*, *Spinella*, and *Eurekaspirifer*. *Howellella* reappears in this zone, but *Cyrtina* is rare, and *Megakozlowskiella* may be wholly absent. The *pinyonensis* Zone has yielded a few occurrences of *Elythyna* (*E.* sp. A) in the Cortez Range and in the northern Simpson Park Range. That genus becomes abundant in a new fauna called *Elythyna* beds succeeding the true *pinyonensis* Zone in a number of localities in central Nevada (Johnson and others, 1967).

The relatively great diversity and considerable abundance of radially plicate spirifers in the Lower Devonian contrasts with their virtual absence in the Middle Devonian of central Nevada. The latter interval is characterized by abundant smooth spirifers of the genus *Warrenella*, but in the collections studied by the writer less than 10 radially plicate spiriferids have been seen.

Superfamily DELTHYRIDACEA Phillips, 1841
Family DELTHYRIDIDAE Phillips, 1841
Subfamily DELTHYRIDINAE Phillips, 1841
Genus *Howellella* Kozlowski, 1946

TYPE SPECIES: *Delthyris elegans* Muir-Wood, 1925 (=*Terebratula crispa* Hisinger, 1827, Pl. 7, fig. 4); not Linnaeus, 1758.

DISCUSSION: The concept of *Howellella* as a genus has been a difficult one to stabilize along relatively restricted lines. The genus has been too broadly characterized from the beginning (as *Crispella* of Kozlowski, 1929). Kozlowski included both plicate and almost completely smooth species in the genus, but the form called *Spirifer (Crispella) laeviplicatus* Kozlowski (1929, Pl. 10, figs. 22-27) is believed by the writer to be a reticulariid in its form because it generally lacks well-developed plications, and because of its ornament as shown in Kozlowski's figure 27. Boucot further complicated the picture by including a diverse array of spiriferids in the genus (1957, p. 316, 317). Boucot's list of assigned species includes some Silurian forms completely lacking in plications, such as *Spirifer modestus* Hall, as well as the multiplicate *Spirifer gaspensis* Billings from the Emsian of the northern Appalachians. These assignments are based on the reasoning that the presence of crural plates in the brachial valves of the species included is a prime diagnostic feature. Yet such a concept, if rigorously applied, would require the genus *Spirinella* Johnston, 1941, to enter the synonymy of *Howellella* and both to be suppressed in favor of the genus *Reticulariopsis* Frederiks, 1916. The problem, however, is not a real one because now it is known that crural plates are present in many widely separated spiriferid groups. Application of the generic term *Howellella* needs to be restricted to a narrower, less heterogeneous group.

In addition to the removal of the so-called smooth *Howellella* group (many of which will probably find their proper place in *Spirinella*), multiplicate species, such as *Spirifer gaspensis*, should be removed. The Silurian species that are relatively closely allied to the type species are small pauciplicate forms, commonly with only three to five plications on each flank of the valves. A relatively small size goes hand in hand with the general absence of umbonal cavity fillings of secondary shell material. Larger forms have been assigned to *Howellella cycloptera*, but *cycloptera* has thick shell material accumulations in the umbonal cavities, and commonly has seven to nine or even more plications on each flank, and probably is sufficiently removed from the predominantly Silurian group of species close to the type. These differences were pointed out some years ago (Berdan, 1949). On the basis of a relatively large accumulation of information, *Howellella* was ancestral to mid-Lower Devonian species of *Acrospirifer*, probably through such large intermediates as *Howellella cycloptera*, and even though the transition is gradual, it undoubtedly will be desirable to recognize *cycloptera* and its morphologic associates as a separate genus. With these reservations in mind *cycloptera* is retained in *Howellella* for the present, but even when the genus is restricted, it probably will include Emsian and Eifelian forms, such as *textilis* and *aculeata* (Pl. 55).

Howellella cycloptera (Hall, 1859)
(Pl. 54, figs. 1-24)

DISCUSSION: A synonymy is omitted because the available information on the dorsal interiors of most of the forms presently assigned is inadequate.

EXTERIOR: The shells are transversely suboval in outline and biconvex in lateral profile with the pedicle valve more convex than the brachial valve. The cardinal angles are commonly slightly obtuse and are rounded. The interarea is low and triangular and is equal to about half the maximum width of the valves. Maximum width is attained posterior to midlength. The ventral interarea is strongly apsacline and incurved. The delthyrium is low, triangular, and encloses an angle of about 45 degrees. The dorsal interarea is linear and anacline. The pedicle valve bears a relatively broad U-shaped sulcus. On some specimens it may be slightly angular anteriorly. There is a corresponding low, rounded fold on the brachial valve. The ventral sulcus is equal in width to about three of the adjoining plications and their intervening interspaces.

The ornament consists of relatively strong, rounded, radial plications which are crossed by indistinct concentric growth lines. The growth lines are prominent anteriorly on some large specimens. There are commonly seven or eight plications on each flank of the valves. They are separated by U-shaped interspaces of approximately the same width as adjacent plications. The plications bifurcate anteriorly on some specimens (Pl. 54, fig. 18), but this feature is seldom observed on silicified specimens that were etched free, because of poor preservation of the anterior portions of the valves. The fine ornament consists of numerous concentric rows of radially arranged striae on each of the growth lamellae.

INTERIOR OF PEDICLE VALVE: The hinge teeth are relatively small and triangular in cross section. They are supported basally by more or less well-developed dental lamellae which diverge moderately to strongly anterolaterally. The dental lamellae in some specimens are closely set and thickened medially on their posterior portions. In other specimens they are relatively widely set apart. The muscle impressions are elongate subpyriform and vary considerably in width. They are moderately strongly impressed posterolaterally, medially striated, and they blend anteriorly with the interior of the shell. There may be a myophragm dividing them into two parts. The umbonal cavities of most specimens are moderately filled with secondary shell material, but only in rare specimens are the dental lamellae made completely obsolete. In some shells where the secondary shell material is strongly developed, it projects nearly as far anterior as the dental lamellae, but there is a narrow slot between the lateral edges of the dental lamellae separating them from the adjacent secondary shell material. The interior of the valves is moderately to strongly crenulated by the impress of the plications.

INTERIOR OF BRACHIAL VALVE: The sockets diverge and expand anterolaterally. Basally, the socket plates are set well above the interior surface of the shell. Relatively strongly developed triangular crural plates are attached to the inner edges of the socket plates and converge slightly toward the midline basally. The adductor muscle scars are not strongly impressed. The interior of the valves is crenulated by the impress of the plications.

OCCURRENCE: This is a common species in the Rabbit Hill Limestone in the Monitor Range (at Rabbit Hill and Dobbin Summit) and in the northern Simpson Park Range. Poorly preserved specimens, probably assignable to *H. cycloptera*, occur in the *Spinoplasia* Zone at Antelope Peak.

COMPARISON: *Howellella cycloptera* is multiplicate and thus differs from *H.* cf. *textilis* which has fewer plications, separated by deeper, broader interspaces.

DISCUSSION: *Howellella cycloptera* is a difficult species to delineate clearly and decisively on the basis of its morphology. It is reported to range throughout the Helderbergian (summarized by Boucot and Johnson, 1967b, p. 74, 75) and supposedly could be regarded as giving way to *Howellella tribulis* in the Oriskany. Hall (1859, p. 421) noted, in proposing *Howellella tribulis*, that it resembles *Howellella cycloptera* and may represent only a variety in form. However, *Howellella cycloptera* is unknown as high as the type horizon of *H. tribulis*, which Hall noted was in the Oriskany Sandstone of Cumberland, Maryland. Schuchert (*in* Schuchert and others, 1913, p. 408) remarked that either *Howellella tribulis* or *H. cycloptera* gave rise to *Acrospirifer? intermedius* and thought that *Acrospirifer murchisoni* and *Acrospirifer angularis* were derived from *H. tribulis*. He reported the common occurrence of *H. tribulis* to be in the Shriver Chert of Maryland, as well as in the lower beds of the Ridgely Sandstone. The writer has examined the types of both *cycloptera* and *tribulis,* and while it seems that in a general way the two species can be separated on the slight diffrences of ribbing (there seem to be very slightly smallr plicatons and one or two greater in number on the flanks in *H. tribulis*), the two closely resemble one another. The dorsal interiors of the two species are not displayed on any specimens in the two syntype collections; however, the slight differences that can be seen in comparing the syntype lots of *cycloptera* and *tribulis* appear to be too small to base a decisive assignment of the form from Nevada to one species or the other. The specimens studied from the Rabbit Hill Limestone certainly display a variation that includes the specimens in the syntype collections of *H. cycloptera* and *H. tribulis*. Presuming that the interiors of the New York and Maryland specimens prove to be like those from Nevada, then the Nevada specimens might logically be assigned a new subspecific name, based on the presence in some individuals of bifurcating plications. A good deal more information is necessary before the matter can be solved.

FIGURED SPECIMENS: USNM 157164-157175.

Howellella cf. *textilis* Talent, 1963
(Pl. 55, figs. 1-19)
Howellella textilis Talent, 1963, p. 81, Pl. 50, figs. 1-43.

EXTERIOR: The shells are biconvex, varying from almost equally biconvex in small shells to strongly unequally biconvex in large shells. Larger shells have the pedicle valve distinctly more convex in the umbonal region. The outline is transversely suboval. The lateral margins slope off gently and evenly from the midline giving a lens-shaped lateral cross section of the two valves. Maximum width is near midlength. The cardinal angles are obtuse and rounded. The ventral interarea is triangular, apsacline, and moderately incurved. It is relatively short and is equal to only slightly more than half the maximum width of the valves. The delthyrium encompasses an angle about 55 to 60 degrees and is bounded along its lateral margins by deltidial plates that project posteriorly and converge slightly toward the midline. The dorsal interarea is sublinear and short.

The pedicle valve is sulcate, and the brachial valve has a corresponding fold. The fold and sulcus are only slightly more prominent in width than the adjoining plications and interspaces, giving the plications the appearance of being almost evenly distributed across the shell. The fold, however, reaches considerable height above the adjoining plications anteriorly. There are four well-defined plications on each flank of the pedicle valve and there are three on each flank of the brachial valve. The plications are separated by deep U-shaped interspaces. The largest available specimen has a plication that bifurcates anteriorly.

The fine ornament is very poorly preserved on the exfoliated specimens available for study, but one specimen retains a small patch of shell material that appears to have prominent concentric growth lines. Preservation is inadequate to determine if radial striae or spines were present.

INTERIOR OF PEDICLE VALVE: Short dental lamellae are present, originating in the first or second interspaces lateral to the sulcus. The dental lamellae diverge anteriorly, but at an angle less than the adjoining plications. The dental lamellae converge medially where they meet the floor of the valve, but they diverge from the median line away from the floor of the valve. The umbonal cavities contain very little secondary shell material.

The muscle field is impressed posteriorly and merges anteriorly with the shell interior over a very slight transverse ridge. The muscle field is longitudinally striated and occupies the position of the sulcus, and also the first two lateral interspaces in some specimens.

INTERIOR OF BRACHIAL VALVE: The socket plates diverge widely laterally between the posterior shell wall on the outside and crural plates on the inner side. The crural plates are relatively strong and are attached to the floor of the valve posteriorly. They diverge slightly anteriorly. The notothyrial cavity terminates posteriorly at an angle of about 90 degrees and is partially filled with deeply striate shell material at the place of diductor muscle attachment. About 20 such striations are present on one specimen. The resulting plates are straight, thin, and flat-sided (comblike in section). The adductor scars are divided by a prominent, thin median ridge, but the musculature is otherwise not well-defined.

OCCURRENCE: *Howellella* cf. *textilis* is a rare species in Nevada where it is known only from the *Eurekaspirifer pinyonensis* Zone. Less than 20 specimens are known to the writer. They were collected in the Simpson Park Range, Roberts Mountains, Sulphur Spring Range, and at Lone Mountain.

DISCUSSION: *H. textilis* appears to be related to the Eifelian spiriferid *Howellella aculeata* (Schnur). The latter species has resided under the name *Spiriferina*, indicating the presence of a median septum in the pedicle valve. Struve (1964, p. 335) placed *aculeata* tentatively in the septate genus *Ivanothyris*. Investigation of typical specimens of *aculeata* from the German Eifelian by means of preparation of internal molds reveals, however, that there is no median septum, but only a well-developed myophragm in the pedicle valve. Relatively well-developed crural plates are present in the brachial valve and there seem to be no certain criteria that would distinguish the species from *Howellella*. The lone noteworthy feature of *Howellella aculeata* is the slightly swollen nature of its radially arranged spine bases. This fine ornament is illustrated along with molds revealing the internal structure in figures 20-22 of Plate 55. There seem to be no strong reasons at present for the establishment of a new genus which would be distinguished by the presence of swollen spine bases. They are present on the Givetian species *Euryspirifer pseudocheehiel alatus* of the Altai Mountains (Ivanova, 1962, Pl. 7, fig. 4) and may represent modification of more than one stock in Emsian and later Devonian time.

Vandercammen (1963, p. 112) described *aculeata* as a *Delthyris*, suggesting that the pedicle valve bears a median septum (1963, p. 115, Fig. 74). The internal mold prepared by the writer agrees with Vandercammen's sections (Fig. 74), but it is certain that the ridge is a myophragm and not a median septum. Vandercammen compounds an old and often repeated error in failing to recognize the significance of a high, bladelike, median septum in the pedicle valve of spiriferids, as noted by Kozlowski (1929, p. 185, Fig. 60). The writer's own observations bear out the fact that genera with a high, bladelike median septum, such as is present in *Delthyris*, do not develop species or individuals with a low median septum or ridge. Accordingly, there are no grounds on which to base an assignment of *aculeata* to *Delthyris*.

There is a single extenuating circumstance that may have led to the advocacy of the median septum as an evanescent feature. This is the development of a platelike myophragm in the apex of pedicle valves of some species. The structure may appear in serial sections as a true median septum in that it is two or three times higher than thick. Wherever a platelike septum of this sort is developed, however, it is confined to the posterior part of the muscle field, commonly posterior to the adductor scars, in contrast to the true median septum which attains its principal development well into the visceral cavity of the pedicle valve.

FIGURED SPECIMENS: USNM 157176-157179.

Genus *Acrospirifer* Helmbrecht and Wedekind, 1923

TYPE SPECIES: *Spirifera primaeva* Steininger, 1853, p. 72, Pl. 6, fig. 1, by subsequent designation of Wedekind *in* Solomon, 1926, p. 202.

DISCUSSION: *Acrospirifer* is a large, pauciplicate, biconvex genus with a deeply impressed ventral muscle scar set between dental lamellae that tend to obsolescence. No median septum is present. In the brachial valve the cardinalia consist simply of divergent sockets and triangular, platelike crural bases without well-developed crural plates, but in specimens that lack thick shells, short slots can be seen in some dorsal internal molds at the position where the crural bases meet the floor of the valve near the apex (Plate 56, fig. 17). The genus is common only in the *Trematospira* and *kobehana* Zones in the Great Basin. A few specimens have been seen high in the *Spinoplasia* Zone below the *Trematospira* Zone in the northern Simpson Park Range. All of the western forms seen by the writer differ slightly from their counterparts in the typical area of the Appalachian province. The Great Basin forms tend to have a smaller and relatively narrower ventral muscle scar and fewer plications on the valves. This is the case with specimens of *Acrospirifer* aff. *murchisoni* from the *Trematospira* Zone and *Acrospirifer kobehana* from the *kobehana* Zone. In examples of Appalachian Province *Acrospirifer* examined by the writer and even including those from the Oriskany horizons of Chihuahua (Boucot and Johnson, 1964), the ventral muscle scar is much more prominent, large, and broad, even when comparing specimens in the same size range. The closet occurrence to the Great Basin of an Appalachian Province *Acrospirifer* that is well illustrated is in Oklahoma (Amsden and Ventress, 1963, Pl. 6). The Nevada specimens commonly have three to six plications on each flank of a pedicle valve, while the Oklahoma specimens and other eastern-occurring specimens have five to eight plications. Probably the Great Basin form, called *Acrospirifer* aff. *murchisoni* needs a new name, but a better understanding of the named species in the east is desirable before renaming the Nevada form.

Acrospirifer aff. *murchisoni* (Castelneau, 1843)
(Pl. 56, figs. 5-13; Pl. 57, figs. 1-6)

Spirifera murchisoni Castelneau, 1843, p. 41, Pl. 12, figs. 1, 2.
Spirifer murchisoni Williams and Breger, 1916, p. 95 (with an extensive synonymy), Pl. 2, fig. 9.
Acrospirifer murchisoni Cooper, 1944, p. 325, Pl. 123, figs. 15, 16.
Hysterolites (Acrospirifer) murchisoni Amsden and Ventress, 1963, p. 105, Pl. 6, figs. 3-18.

EXTERIOR: Small shells are transversely suboval and large shells are transversely shield-shaped to subrectangular in outline and strongly biconvex in lateral profile. The brachial valve in some specimens is more strongly convex than the pedicle valve. The hinge line is long and straight. Cardinal angles are almost right angles or are obtuse, and they are rounded or angular at their extremities. The pedicle valve bears a strong, broad subangular or rounded sulcus. The brachial valve has a corresponding fold. The breadth and depth of the sulcus gives the pedicle valve a bilobate outline in transverse section, while that of the brachial valve is carinate with the flanks sloping away without appreciable curvature. The ventral interarea is long, low, and triangular. It is apsacline, and the beak is strongly incurved. The dorsal interarea is long and ribbon-like, and on large shells it is apsacline.

The ornament consists of rounded or subangular, coarse plications crossed by concentric growth lines. There are four to seven plications on each flank of the pedicle valves, commonly, there are six on either flank of a pedicle valve and five on each flank of the associated brachial valve. The largest available specimen bears seven plications on the flank of the pedicle valve. The furthest lateral plications originate about two-thirds of the radial distance from the beak to the commissure and are parallel and well anterior to the hinge line. Plications on most large specimens tend to become broader, more angular, and lower anteriorly. The fine ornament consists of concentric rows of radially aligned ridges on the growth lamellae.

INTERIOR OF PEDICLE VALVE: The hinge teeth are supported by short dental lamellae which diverge toward the base of the pedicle valve as well as anteriorly. The dental lamellae bound the posterolateral edges of the muscle impressions and at their anterior edges commonly tend to recurve gently inward. On some specimens there is a thickening of shell material on the inner sides of the dental lamellae below the level of the delthyrium forming a constriction in the pedicle cavity. In small shells the muscle field is relatively elongate and narrow and lies within the first plications lateral to the sulcus. On larger specimens the diductor field becomes relatively broader and is generally moderately to deeply impressed. A low myophragm may be present. Anteriorly the muscle scar merges relatively abruptly with the shell interior, but muscle-bounding ridges are not present.

INTERIOR OF BRACHIAL VALVE: The socket plates lie along the inner edges of the notothyrium and diverge anterolaterally. They are relatively shallow and become strongly drawn out on their inner edges into ventral projections. Short, medially convergent plates are attached to the inner edges of the socket plates and are thickened along their dorsal edges. Posteriorly, the dorsal edges are attached to the inner walls of the valve. Shell material at the apex of the notothyrial cavity is deeply striate longitudinally, serving as a site of diductor muscle attachment. The adductor scars are not impressed.

OCCURRENCE: In Nevada, *Acrospirifer* aff. *murchisoni* is present in the *Trematospira* Zone in the Roberts Mountains and in the Sulphur Spring Range. It occurs rarely in the upper *Spinoplasia* Zone.

FIGURED SPECIMENS: USNM 157180-157182, 157183 A and B.

Acrospirifer kobehana (Merriam, 1940)
(Pl. 57, figs. 7-11; Pl. 58, figs. 1-18; ?Pl. 56, figs. 1-4, 14-17)
Spirifer raricosta Walcott, 1884, p. 135, Pl. 4, figs 2, 2a(?); Pl. 14, fig. 12; not Conrad, 1842.
Spirifer kobehana Merriam, 1940, p. 86, Pl. 6, figs. 8-10; Pl. 11, figs. 12, 13.

EXTERIOR: The shells are transverse and broadly shield-shaped in outline and commonly slightly emarginate anteriorly. The valves are biconvex in lateral profile with both valves about equally convex. There is a broad median sulcus in the pedicle valve and a corresponding fold in the brachial valve. The considerable width of the sulcus gives the pedicle valve a bilobate profile in transverse section. The flanks of the brachial valve commonly slope away from the fold and are gently concave viewed in transverse section. The ventral interarea is broad, low, and triangular, and is the place of maximum width. Horizontal growth lines are well developed on the interareas of some specimens. The interarea is moderately incurved and apsacline and is overhung medially by a fairly strongly incurved beak. The delthyrium is low and triangular, enclosing an angle of about 100 degrees. Short deltidial plates are attached to sides of the delthyrium. The dorsal interarea is long, linear, and apsacline. The place of greatest width is commonly at the hinge line. The lateral margin is nearly parallel for a short distance before incurving sharply anteromedially. The anterior margin is roughly parallel to the hinge line due to sharp curvature of the shell. Large shells tend to become increasingly carinate anteriorly, and the sulcus is then formed into a strongly incurved spathulate extension.

The ornament consists of strong, low, broad, asymmetrically rounded plications which are disposed radially and which are crossed by low concentric growth lines that become more numerous and slightly more prominent at the anterior of the shell. There are from five to eight plications on each ventral flank; most commonly there are six on a pedicle valve and five on each flank of the associated brachial valve. Variation in the number of plications appears to be due to addition of new plications on the posterolateral flanks and in specimens with seven or eight plications on each flank, the lateral pairs commonly originate along the hinge line rather than at the beak. This is best seen on the brachial valve as plications tend to form very slightly anterior to the ventral edge of the ventral interarea. In cross section the plications are very low, broad, and rounded and are separated by relatively narrow U-shaped interspaces. The plications are asymmetric anteriorly with their crests skewed laterally. The ventral sulcus is broad, approximately the width of three plications, and it encompasses the two medial plications on the pedicle valve. On the brachial valve, the plications adjoining the fold are slightly raised above the level of the adjoining plications complementing the disposition of plications in the sulcus of the pedicle valve. The two plications in the ventral sulcus become flatter anteriorly and on large shells may become completely obsolescent.

INTERIOR OF PEDICLE VALVE: The hinge teeth are prominent and triangular in cross section. They diverge slightly laterally and project or curve posteriorly. They bear small circular pits on their inner sides near the distal ends. The hinge teeth are supported by stout dental lamellae which diverge anteriorly and slightly laterally as they approach the base of the valve. They may continue anteriorly as low ridges curving around the posterolateral part of the impressed muscle field. In the apex of the valve, there are two small sharp ridges on the dental lamellae beneath the tracks of the hinge teeth leaving relatively deep grooves separating the ridges from the hinge-teeth tracks. These ridges meet posteriorly at the thickened medial surface of the shell, but do not join to form a discrete plate. The structure is possibly an incipient development of a subdelthyrial plate as is common in Emsian and younger age spiriferids. The umbonal cavities are slightly filled in with shell material and in some specimens, deeply filled in. The shell material may not be deposited exactly adjacent to the dental lamellae, but slightly lateral to it, and the shell thickenings tend to converge medially as ponderous, low, rounded ridges. They meet at the base of the sulcus at the anterior edge of the muscle field. Some shells lack anteriorly convergent ridges of shell material and display exceptionally large and broad diductor impressions. The diductor scars are elongate-oval and are longitudinally striate. They are separated medially by a more or less well-defined broad, low, V-shaped myophragm. The interior of the shell is crenulated by the impress of the plications.

INTERIOR OF BRACHIAL VALVE: The sockets originate at the beak and increase in width anterolaterally as they diverge. The socket plates are free with their bases above the floor of the valve, and on large shells are nearly flat plates which recurve relatively sharply medially to enclose the sockets. The sockets recurve slightly on their inner distal ends and fit into the circular depressions on the hinge teeth to act as pivots.

The notothyrial cavity is thickened posteriorly into a slightly raised triangular pad which is deeply longitudinally striate and which served as a site of diductor muscle attachment. The crura are attached to the anteromedial sides of the socket plates and are more or less triangular in shape, narrowing to simple bandlike structures as they project anteriorly and incurve slightly medially. The dorsal edges of the crura project as free, ribbon-like plates along the dorsomedial edges of the socket plates and connect posteriorly with a pair of more or less rounded ridges along the sides of the notothyrial cavity and join the anterolateral corners of the place of diductor muscle attachment. It seems likely that the ridges on the sides of the notothyrial cavity, which are joined by the dorsal edges of the crura, are actually the crural bases. In this case, since there are no supporting plates dorsal to them, the cardinalia appear to lack true crural plates. The spiralia were not observed.

OCCURRENCE: According to Merriam (1940, p. 52) *Acrospirifer kobehana* is present between 100 and 150 feet above the Lone Mountain Dolomite at Lone Mountain. It is common in the *A. kobehana* Zone in the Sulphur Spring Range and in the Roberts Mountains. The writer has identified a single specimen from the east slope of the Cortez Range. The species has also been reported by McAllister (1952, p. 17) from the Quartz Spring area in southeastern California. *Acrospirifer kobehana* specimens reported at the base of the Piute Formation in the northern Arrow Canyon Range (Waines, 1962, p. 62) were examined by the writer and found not to belong to *A. kobehana*. The pattern of ribbing on the lone well-preserved brachial valve suggests closer affinity to *Acrospirifer* aff. *murchisoni*, as described above.

DISCUSSION: Inasmuch as *A. kobehana* is found to be restricted to a single zone, special attention should be focused on the morphologic features that allow discrimination between it and *Acrospirifer* aff. *murchisoni* which is restricted to the next older *Trematospira* Zone. The distinction is best illustrated by the rib pattern, which on *A. kobehana* consists of broad, low, somewhat asymmetric plications which contrast with the more elevated, rounded or anteriorly subangular plications of *A.* aff. *murchisoni*. In addition, the dorsal fold of *A.* aff. *murchisoni* is markedly higher than adjoining plications, in contrast with the dorsal fold of *A. kobehana* which encompasses the first lateral pair of plications as an integral part (*compare* figs. 7 and 8 of Pl. 56 with figs. 9 and 11 of Pl. 57).

FIGURED SPECIMENS: The holotype is USNM 96387. Other figured specimens are USNM 96388A, 157188-157194. Questionably assigned are USNM 157184-157187.

Genus *Dyticospirifer* Johnson, 1966
TYPE SPECIES: *D. mccolleyensis* Johnson, 1966.
DIAGNOSIS: Acrospiriferoid species with bifurcating lateral plications and with a smooth fold and sulcus.

Dyticospirifer mccolleyensis Johnson, 1966
(Pl. 59, figs. 1-16)

Dyticospirifer mccolleyensis Johnson, 1966b, p. 1044, Pl. 127, figs. 1-16.

DISCUSSION: This is a relatively rare form that appears so far to be restricted to the lower or *Trematospira* Subzone of the *Trematospira* Zone in the Sulphur Spring Range. The species *Howellella* cf. *textilis*, described elsewhere in this paper, has bifurcating plications on one large specimen found at Lone Mountain.

Since the genus *Dyticospirifer* was erected (Johnson, 1966b), the presence of bifurcating plications has been established on specimens of *Howellella cycloptera* from the Rabbit Hill Limestone, *Spinoplasia* Zone, at Rabbit Hill, at Dobbin Summit, and in the Simpson Park Range. The specimens noted by Johnson (1966b, p. 1045) in the northern Simpson Park Range belong to this group. The *Spinoplasia* Zone form of *Howellella cycloptera* is almost certainly the ancestor of *Dyticospirifer mccolleyensis*. The two forms are very close, but *D. mccolleyensis* lacks well-developed crural plates which are present in *Howellella cycloptera*. In addition, *D. mccolleyensis* has slightly coarser plications than the *Spinoplasia* Zone specimens of *Howellella cycloptera*.

FIGURED SPECIMENS: USNM 146512-146520. The holotype is USNM 146512.

Subfamily COSTISPIRIFERINAE H. and G. Termier, 1949

DISCUSSION: This subfamily must center around *Costispirifer* Cooper, but it appears unlikely to the writer that there is any close connection between *Costispirifer* and *Eudoxina, Lazutkinia, Theodossia*, or *Urella*—genera included in a family Costispiriferidae by Pitrat (1965, p. H696). According to Rzhonsnitskaya (1964, p. 45) her genus *Urella* is a junior synonym of *Retzispirifer* Kulkov. The latter genus and *Lazutkinia* seem closer to *Branikia* Havlíček, which Pitrat (1965, p. 719) placed in the Reticulariidae.

The species "*Spirifer*" *cumberlandiae* Hall (1857, p. 421, Pl. 96, fig. 9) is a form very closely allied to *Costispirifer*. Internally, the two are identical. Externally, they differ only in the absence of plications on the fold and in the sulcus of "*Spirifer*" *cumberlandiae*. However, the writer sees considerable significance in the shape of the ventral interarea and deltidium. In relatively large specimens, both of *Costispirifer* and of "*Spirifer*" *cumberlandiae*, the interarea is trapezoidal rather than triangular, and the delthyrium is covered by a thick, externally convex, chevron-shaped deltidium (*see* Pl. 64, fig. 5). No other genus has such a combination of characters, and as the absence of median plications in "*Spirifer*" *cumberlandiae* distinguishes it from *Costispirifer*, the eventual erection of a new genus is in order.

Costispirifer and "*Spirifer*" *cumberlandiae* are Siegenian to Emsian contemporaries that must have been derived from an aseptate delthyridid, possibly a variety of *"Spirifer" concinnus.* Thus, a separate family seems unwarranted. *Costispirifer* and *"Spirifer" cumberlandiae* compose a compact subfamily that may be diagnosed as follows: Delthyrididae developing a trapezoidal ventral interarea and a thick apical deltidium; cardinalia lack crural plates; plications on flanks are simple.

Genus *Costispirifer* Cooper, 1942

TYPE SPECIES: *Spirifer arenosus* var. *planicostatus* Swartz, 1929, p. 56. Pl. 9, figs. 13-15.

Costispirifer sp.
(Pl. 64, figs. 1-9)

EXTERIOR: The shells are transversely suboval to subsemicircular in outline and unequally biconvex in lateral profile. The pedicle valve is more strongly convex than the brachial valve. The hinge line is long and straight and is the place of maximum width. Cardinal angles are commonly acute and rounded. The greatest convexity is at the ventral umbo, and the beak is strongly incurved. The ventral interarea is curved and apsacline. On small shells it is low and triangular. On the largest specimens the interarea is trapezoidal and ribbon-like with the beak ridges parallel or nearly parallel to the dorsal margin of the hinge line. The delthyrium is broad, low, and triangular. It encloses an angle of from 75 to almost 90 degrees. There is a relatively flat, V-shaped deltidium filling the apical part of the delthyrium and joining its lateral edges. The width of the separate limbs of the deltidium narrow approaching the hinge line. There may be a median indentation at the apex of the deltidium. The dorsal interarea is linear and anacline.

The pedicle valve bears a broad, shallow, poorly defined median sulcus. The brachial valve is carinate, and does not develop a separately distinguishable median fold. The ornament consists of numerous, low, rounded, radial plications separated by narrow, shallow, V-shaped interspaces. The plications are moderately rounded on the posterior portion of the valves, but anteriorly on large shells they become very low, flat, and broad relative to the interspaces, which remain narrow. There are commonly about twelve plications on the flank of a medium- to large-sized valve. Fifteen radial plications are present on the flank of the best preserved specimen available. The lateral plications do not bifurcate; however, bifurcation of plications occurs adjacent to the midline of the valves. In the brachial valve the basic pattern consists of a single median plication posteriorly which splits into a pair of plications of equal size. Each of these, in turn, gives rise to a new plication which splits off from its lateral flank. No concentric ornament was observed on the specimens examined, and their preservation is such as to indicate that no strong development of concentric ornament was present. No fine radial ornament is preserved.

INTERIOR OF PEDICLE VALVE: The teeth are relatively long and thin. They are elliptical in transverse section and are inclined about midway between the median plane and the plane of the commissure. They project posteriorly as well as dorsally and are situated within the open part of the delthyrium near the dorsal edge of the hinge line. The hinge teeth are supported by thick dental lamellae that are short adjacent to the palintrope, but which extend relatively strongly anteriorly as they diverge along the base of the valve. In large specimens the umbonal region is deeply filled with shell material as are the umbonal cavities. The diductor muscle scars are deeply impressed posteriorly. They constitute a broadly flaring pyriform impression that blends anteriorly with the interior of the shell. The posterior ends of the diductor impressions are deeply and separately impressed and are divided by a triangular wedge of shell material which is commonly medially grooved. The interior of the shell is not crenulated by the impress of the plications.

INTERIOR OF BRACHIAL VALVE: The sockets expand and diverge moderately anterolaterally. The socket plates attach directly to the posterior margin of the valve beneath the interarea and curve inward sharply above the floor of the valve. They are thickened along their medial edges which diverge relatively strongly anterolaterally. Crural plates are not present on medium- and large-sized valves and apparently the crural bases originated adjacent to the inner edges of the socket plates; however, no small, well-preserved specimens that would give an understanding of the ontogeny of the cardinalia were available for study. There commonly is a very slightly raised arcuate ridge of shell material defining the anterior edge of the notothyrial cavity. The apex of the notothyrial cavity is longitudinally striated at the site of diductor muscle attachment. The adductor muscle scars are not impressed, and the interior of the valves is faintly crenulated by the impress of the plications.

OCCURRENCE: *Costispirifer* sp. is restricted to the *Costispirifer* Subzone in the upper part of the *Trematospira* Zone. It is abundant in the Sulphur Spring Range and in the northern Roberts Mountains. Several specimens have been seen by the writer from the Cortez Mountains. Merriam identified *Costispirifer* at Dobbin Summit in the Monitor Range (written commun., 1967). The genus has not been reported south of the Dobbin Summit area in Nevada, but is present in the Quartz Spring area of southeastern California (McAllister, 1952, p. 17).

FIGURED SPECIMENS: USNM 157225-157227, 157228 A and B.

Subfamily HYSTEROLITINAE H. and G. Termier, 1949
Genus *Hysterolites* Schlotheim, 1820

TYPE SPECIES: *H. hystericus* Schlotheim, 1820, p. 249; 1822, Pl. 29, figs. 1a, 1b.

Hysterolites sp. A
(Pl. 60, figs. 1-18)

EXTERIOR: The shells are transversely subtrigonal to nearly semicircular in outline and unequally biconvex in lateral profile. The brachial valve is very gently convex to moderately convex in large specimens, but the pedicle valve is commonly two or three times as deep. The hinge line is long and straight and is the place of maximum width. The cardinal angles are acute and slightly rounded to moderately auriculate. The ventral interarea is commonly catacline to steeply apsacline and tends to be flat or nearly so. There may be slight incurvature at the beak. The delthyrium is triangular and relatively narrow, enclosing an angle of slightly less than 45 degrees. The dorsal interarea is commonly strongly anacline, but varies through the orthocline position in some shells to become slightly apsacline. The ventral sulcus may be shallow and U-shaped, but is commonly slightly flattened along its base. The dorsal fold is low and rounded to relatively strongly flattened medially. Fold and sulcus have a width equal to about two plications and an intervening interspace.

The ornament consists of numerous rounded radial plications separated by U-shaped interspaces approximately the same width as the adjoining plications. There are commonly 8 to 14 radial plications on medium- to large-sized shells. The higher number results from the presence of 5 or 6 very fine plications on the posterolateral flanks of relatively well-preserved large specimens. The plications are crossed by regularly spaced growth lines that commonly become imbricate in the posterior half to two-thirds of medium- and large-sized shells. The fine ornament consists of numerous, radially arranged, closely set striae on each of the concentric growth lamellae.

INTERIOR OF PEDICLE VALVE: The tracks of the hinge teeth are set beneath the inner edges of the delthyrium and do not project within the delthyrium. They are supported by short divergent dental lamellae which tend to be slightly arcuate and concave laterally. In the larger specimens the dental lamellae bound approximately the posterior half or slightly more of the muscle impressions. At the apex of the valve, there is a small triangular thickening of shell material joining laterally with a pair of narrow elongate ridges on the medial sides of the dental lamellae below the level of the delthyrium. These evidently constitute an incipient development of a transverse subdelthyrial plate, but they do not join in any of the specimens to form a single plate that would compartment or cover the umbonal chamber. The muscle scars are not strongly impressed and commonly they blend with the interior of the valve anteriorly. The adductor impressions consist of a pair of elongate tracks medially which may or may not be bounded laterally by muscle-bounding ridges. The interior of the shell is corrugated in its anterior two-thirds by the impress of the plications.

INTERIOR OF BRACHIAL VALVE: The sockets expand considerably and are divergent anterolaterally. The medial portions of the socket plates diverge slightly from the midline. Crural plates are present extending dorsally from the inner edges of the socket plates and curving very gently medially. They meet the floor of the valve only in the apex. In thick-shelled specimens the crural plates are completely buried by deposition of shell material in the umbonal cavities. The notothyrial cavity bears a thickening of shell material that is raised along the midline and that is set out from the crural plates flanking the cavity by a pair of short, deep radial grooves. The median elevated portion is deeply striate at its posterior end at the place of diductor muscle attachment. Specimens with thin shells show the adductor muscle impressions only faintly impressed and separated by a thin myophragm. In thick-shelled specimens, the adductor muscle field consists of two strongly elongate pairs, the anterior pair lying deeply impressed in the furrow corresponding to the external fold. The shell is crenulated over most of its interior by the impress of the plications.

OCCURRENCE: *Hysterolites* sp. A is present, but not common, in the *Eurekaspirifer pinyonensis* Zone at Lone Mountain and in the Roberts Mountains. It is also present in the *Eurekaspirifer pinyonensis* Zone in the Cortez Mountains where it is the common spiriferoid.

COMPARISON: *Hysterolites* sp. A has slightly narrower and more numerous plications, especially adjoining the fold and sulcus, than does *Hysterolites* sp. B. In addition, the latter species bears much stronger crural plates.

FIGURED SPECIMENS: USNM 157195-157203.

Hysterolites sp. B
(Pl. 61, figs. 1-20)

EXTERIOR: The shells are strongly transverse and subtrigonal in outline. The width is commonly two times or slightly more than two times the length. In lateral profile the valves are strongly biconvex. The pedicle valve is the most strongly convex and bears a high, nearly flat, triangular, catacline or steeply apsacline interarea. The hinge line is long and straight and is the place of maximum width. The cardinal angles are acute and are commonly sharply auriculate. The delthyrium is high, triangular, and open. It encloses an angle of from 30 to 45 degrees. The dorsal interarea is linear and is anacline to orthocline. There is a relatively broad, well-defined ventral sulcus which is U-shaped or which may be slightly flattened on its base. Posteriorly, the sulcus is about equal in width to two adjoining plications and their mutual interspace. It broadens perceptibly anteriorly relative to the adjoining plications and may be equal in width to three or more adjoining plications. The dorsal fold is moderately elevated, rounded posteriorly, and somewhat flattened anteriorly. It curves longitudinally at about the same rate as the lateral portions of the shell and is not strongly elevated above the flanks anteriorly.

The ornament consists of rounded radial plications which commonly number from 7 to 10 on each flank of the valves. On most specimens the first two lateral plications adjoining either side of the fold on the brachial valve and the corresponding interspaces on the pedicle valve are much broader than the remaining lateral plications and interspaces. Numerous relatively well-developed concentric growth lines are developed on some specimens, but the fine ornament is not preserved.

INTERIOR OF PEDICLE VALVE: There is a pair of short, straight, broadly divergent dental lamellae in the posterior portion of the valve. On all of the specimens available, the dental lamellae are situated on the pair of internal ridges which correspond to the interspaces adjoining the sulcus. No delthyrial plate is present.

INTERIOR OF BRACHIAL VALVE: The sockets are relatively narrow and broadly divergent. Crural plates are present attached to the inner edges of the socket plates. The enclosed notothyrial cavity is elevated slightly medially by a deposit of shell material that is striated posteriorly at the site of diductor muscle attachment. In some specimens, deposition of secondary shell material in the umbonal cavities makes the crural plates partly obsolescent, but to a much less greater extent than occurs in *Hysterolites* sp. A. In the notothyrial cavity the elongate thickening of shell material is only faintly defined laterally by grooves adjacent to it and the flanking crural plates. The interior is relatively strongly corrugated by the impress of the plications. The adductor muscle scars are not strongly impressed, but are separated medially by a short narrow myophragm.

OCCURRENCE: *Hysterolites* sp. B is present in only two collections from tne *Eurekaspirifer pinyonensis* Zone, one from the southern Roberts Mountains and one from the Sulphur Spring Range. There is no evidence to suggest the position of the collection horizon at the Sulphur Spring occurrence. However, the specimens from the southern Roberts Mountains may be from the lower part of the zone.

FIGURED SPECIMENS: USNM 157204-157210.

Genus *Brachyspirifer* Wedekind, 1926
TYPE SPECIES: *Spirifer carinatus* Schnur, 1853, p. 202, Pl. 33, figs. 2a-e.

DISCUSSION: In the course of study of *Brachyspirifer* from Nevada, specimens of the type species, *B. carinatus*, from the Wiltzer Schichten, south of Daleiden, Eifel, were examined in order to fully diagnose the genus as well as for comparative purposes.

Brachyspirifer is a multiplicate member of the Delthyrididae. The plications are crossed by numerous, finely lamellose growth lines that bear concentric rows of fine, radially arranged spines. In outline, *B. carinatus* is commonly a little wider than long and does not attain a disproportionately small length-width ratio. The ventral interarea is of moderate height, apsacline, and strongly incurved at the beak. The umbonal cavities generally lack infilling shell material. The dental lamellae are long, relatively straight, and diverge only moderately. Brachial valves bear widely divergent sockets. The inner edges of the socket plates serve as the attachment sites for a pair of crural bases that are inclined only slightly from the plane of the commissure (Pl. 62, figs. 12, 13) in contrast to the disposition of crural bases and crural plates in the more common Silurian delthyrids, such as *Howellella* and *Delthyris*. There is a medial thickening of shell material where the crural bases converge at the apex of the notothyrial cavity, both above and below the level of the crural bases and filling the space between them. The posteroventral extremity of this medial shell thickening is deeply longitudinally striated to serve as the site of diductor muscle attachment.

Brachyspirifer thus bears the basic pattern of cardinalia construction of *Hysterolites*. The cardinalia of *Brachyspirifer*, however, differ by the presence of more widely divergent sockets, flatter and more nearly horizontal crural bases, and a more poorly defined stalk or median column of shell material basal to the site of diductor attachment.

Rzhonsnitskaya's suggestion (1964, p. 45) that *Brachyspirifer* is synonymous with *Spinocyrtia* is indefensible inasmuch as the two genera are similar only in gross external morphology. *Spinocyrtia* bears a completely different external fine ornament than does *Brachyspirifer*, consisting of numerous teardrop-shaped granules more or less aligned on radial threads (Ehlers and Wright, 1955, Pls. 2, 3, 5, 7, 8, and 11). The pedicle valve of *Spinocyrtia* bears a well-developed transverse subdelthyrial plate (Ehlers and Wright, 1955, Pl. 1, fig. 8; Pl. 4, fig. 11; Pl. 6, figs. 10 and 11; Pl. 9, figs. 6

and 7). Other internal structures, including the cardinalia, bear no more than a general structural resemblance.

Cooper's conception of *Brachyspirifer* (1944, p. 323) was erroneous, based as it was on the Appalachian Middle Devonian species *"Spirifer" audaculus*. *"Spirifer" audaculus* belongs to *Mediospirifer* Bublitschenko, 1956, which has *"Spirifer" medialis* Hall as its type species. *"Brachyspirifer" ventroplicatus* Cooper (1945, p. 486, Pl. 64, figs. 29-35), *"B." macronotus*, and *"B." angustus* (*fide* Cooper, 1945, p. 487), and *"Brachyspirifer" palmerae* Caster (1939, p. 146) all must be removed from *Brachyspirifer*.

Brachyspirifer pinyonoides Johnson, n. sp.
(Pl. 62, figs. 1-11; Pl. 63, figs. 1-18)

Spirifer pinyonensis Walcott, 1884 (in part), Pl. 4, fig. 1d; *not* Meek.
Spirifer pinyonensis Merriam, 1940 (in part), Pl. 6, fig. 3; Pl. 7, figs. 3-5, 7, 8; *not* Meek.

EXTERIOR: Small shells are semicircular to transversely shield-shaped in outline and biconvex in lateral profile. The pedicle valve is decidedly more convex than the brachial valve. Large shells are commonly transversely oval to slightly pentagonal in outline. The cardinal angles of small shells are about right angles or are obtuse and rounded, but some shells are moderately auriculate. The ventral interarea is relatively long, low, and triangular. It is apsacline and commonly is strongly incurved. The dorsal interarea is long and anacline. The delthyrium is relatively broad, low, and triangular and encloses an angle of approximately 90 degrees. The lateral margins of the delthyrium, particularly apically, bear narrow deltidial plates which are set nearly normal to the interarea. The pedicle valve bears a relatively broad sulcus equal in width to approximately three plications. The sulcus may be rounded or flat-bottomed and shallow. Some specimens bear a relatively deep U-shaped sulcus, but none with a sharply angular sulcus have been seen. There is a corresponding broad, low, rounded fold on the brachial valve.

The ornament consists of numerous low radial plications separated by relatively narrow shallow U-shaped interspaces. There are commonly 12 plications on each flank of the pedicle valves and on large, well-perserved specimens there may be 14 or, rarely, 15. The plications are crossed by numerous, closely set concentric growth lines which bear rows of fine radial ridges.

INTERIOR OF PEDICLE VALVE: The hinge teeth are supported by strong dental lamellae which diverge laterally and recurve gently near their anterior ends to enclose approximately the posterior half of the muscle field. The diductor scars may be moderately to deeply impressed and are elongate-oval in outline. In most specimens the diductor scars broaden perceptibly anteriorly and are bounded along their anterolateral and anterior margins by muscle-bounding ridges. The adductor muscle tracks are elongate and may be separated by a broad, low, rounded myophragm. Small shells are crenulated across their whole interior by the impress of the plications, but the largest shells are crenulated only on approximately their anterior third.

INTERIOR OF BRACHIAL VALVE: The sockets are narrow and widely divergent. Relatively short subparallel crural plates are present on some shells and absent on others. Small specimens, which invariably have thin shells, display short narrow slots on a mold, indicating the presence of the crural plates. However, on specimens that are thick-shelled, the presence or absence of short slots is variable. In the largest and thickest-shelled specimens, there is no trace of indentation at the place where slots are seen on other specimens; the crural plates evidently were made obsolescent by the deposit of secondary shell material. There is a pad of deeply longitudinally striate shell material in the apex of the notothyrial cavity serving as the site of diductor muscle attachment. This area may be very faintly bilobate, but commonly it is an undivided structure. Small lateral ridges along the base of the notothyrial cavity are poorly developed on some specimens. On thin-shelled specimens, the adductor muscle field is not noticeably impressed, but is commonly divided medially by a relatively long, narrow myophragm. Brachial valves with thick shells have the adductor muscle field strongly impressed and divided into elongate posterior and anterior pairs with the posterior pair lateral to the anterior pair. The interior of the valve is variably crenulated by the impress of the plications.

OCCURRENCE: *Brachyspirifer pinyonoides* is common in the *Eurekaspirifer pinyonensis* Zone at Lone Mountain, in the Roberts Mountains, and in the northern Simpson Park Range.

COMPARISON: *Brachyspirifer carinatus* (Schnur), the type species, has a more angular sulcus. In the pedicle valve, its dental lamellae are longer and are nearly subparallel while those of *B. pinyonoides* are shorter and curved. *B. carinatus* also has more widely spaced socket plates defining the inner edges of the sockets and relatively shorter crural bases than in *B. pinyonoides*. *Brachyspirifer carinatus ignorata* (Maurer) as illustrated by Scupin (1900, Pl. 2, fig. 9) is a large oval form that differs in outline from *B. pinyonoides* and has less divergent dental lamellae. *Brachyspirifer carinatus latissima* (Scupin, 1900, Pl. 2, figs. 12a, 12b) differs from *B. pinyonoides* as does *B. carinatus* in the absence of a strongly impressed muscle scar in the pedicle valve. *Brachyspirifer carinatus crassicosta* (Scupin, 1900, Pl. 2, figs. 13a, 13b) is most like *B. pinyonoides* relative to the pedicle valve interior, but has a more trigonal outline than the latter species. Scupin (1900, p. 323) also associated *"Spirifer" nerei* with the group of *carinatus*. Havlíček (1959, p. 101), in revising the Bohemian spirifers, regarded *nerei* as a *Hysterolites*, but he also relegated *Brachyspirifer* to the synonymy of the latter genus. Since the cardinalia of *"Spirifer" nerei* are still not fully known, it is impossible to be certain of its generic assignment, but should it be assigned to *Brachyspirifer*, as is suggested by its fine ornament (Havlíček, 1959, Pl. 1, fig. 1), it can be distinguished from *B. pinyonoides* by its more angular sulcus and smaller length-width ratio.

DISCUSSION: *Brachyspirifer pinyonoides* is notable in displaying all stages in the development of crural plates—from well developed to completely lacking (Pl. 62, figs. 2, 5; Pl. 63, figs. 10, 14, 18).

FIGURED SPECIMENS: USNM 157212-157224. The holotype is USNM 157218.

Subfamily KOZLOWSKIELLININAE Boucot, 1957
Genus *Kozolowskiellina* Boucot, 1958
Subgenus *Megakozlowskiella* Boucot, 1957
TYPE SPECIES: *Spirifer perlamellosus* Hall, 1857, p. 57, figs. 1-5.
Megakozlowskiella sp. A
(Pl. 72, figs. 21-29)

EXTERIOR: Small shells are semicircular or transversely shield-shaped. Larger shells are transversely suboval in outline and unequally biconvex in lateral profile. The pedicle valve is more strongly convex. The ventral interarea is high, triangular, relatively flat, and strongly apsacline. The delthyrium is relatively narrow and encloses an angle of approximately 45 degrees. No deltidial plates were observed. On large shells the dorsal interarea is narrow, relatively high, and orthocline to apsacline. Laterally, the dorsal interarea blends almost imperceptibly with the posterolateral portion of the shell. The dorsal beak is strongly incurved. On small shells the hinge line may nearly equal the maximum width of the valves, and the cardinal angles are acute or are about right angles. On larger shells the hinge line is more nearly equal to two-thirds of the maximum width, and cardinal angles are decidedly obtuse and rounded. There is a broad, deep, almost V-shaped sulcus on the pedicle valve and a broad, flat, rounded fold on the brachial valve. On large shells the dorsal fold tends to become gently emarginate at the anterior commissure and displays an inverted V-shaped indentation.

The ornament consists of coarse, rounded, radial plications commonly numbering three on each ventral flank and two on each dorsal flank. In general, lateral plications are of considerably less strength than the adjoining plications on their medial sides. The second lateral plications on the dorsal flanks, particularly, tend to be obsolete. Radial plications are crossed by numerous, frilly, concentric growth lamellae which bear rows of fine radial striae.

INTERIOR OF PEDICLE VALVE: The hinge teeth are small and pointed. In small shells there are thin dental lamellae which project only slightly anterior to the inner edge of the delthyrium and adjoin the floor of the valve only apically. Dental lamellae are welded together at their line of junction with the interior of the shell, rather than joining along discrete tracks. There is a short median septum which lies dorsally within the "V" of the united dental lamellae posteriorly and which projects anteriorly beyond the edges of the dental lamellae. On larger shells the dental lamellae become more or less trigonal plates in contrast to their bandlike outline in small shells, and they project relatively farther anteriorly to approximately the anterior end of the median septum where the convergent dental lamellae and the median septum unite forming a single thickened median septal structure slightly anterior to the midlength of the valve.

INTERIOR OF BRACHIAL VALVE: The socket plates join directly to the anterior side of the interarea and are attached to the floor of the valve for part of their distance medially. In most shells the inner edges of the socket plates are free from the base of the valve and recurve strongly ventrally forming anterolaterally divergent sockets. Platelike crural bases are broadly triangular in outline and converge toward the midline of the valve where they join one another on the notothyrial platform. The crural bases tend to become concave ventrally and commonly extend posteriorly, partially covering the sockets. The place of diductor attachment consists of two, strong, posteriorly convergent, elongate rounded ridges which are deeply striate along their posterior portions. There appear to be two pairs of elongate-oval adductor impressions. The posterior pair is lateral to the anterior pair, and they are poorly impressed on the flanks of the furrow that corresponds to the external fold. Medially and somewhat anterior to the socket plates, the adductor scars (anterior adductor scars?) are closely set together, strongly elongate, and deeply impressed. There is a high, almost bladelike myophragm which originates at the posterior end of the adductor impressions and becomes progressively lower anteriorly. The adductor impressions blend anteriorly with the interior of the shell.

OCCURRENCE: *Megakozlowskiella* sp. A is present in the *Quadrithyris* Zone of the Windmill Limestone at Coal Canyon in the northern Simpson Park Range and in the lower McMonnigal Limestone in the Toquima Range.

COMPARISON: *Megakozlowskiella* sp. A is externally comparable to *M.* cf. *raricosta* of the *Acrospirifer kobehana* Zone, although the latter form attains greater size. Internally, however, the two species are easily distinguished. In the pedicle valve of *Megakozlowskiella* sp. A, the dental lamellae are long and thin and join the sides of the median septum. In younger forms, coalescence occurs mainly by the deposition of secondary shell material. In the brachial valve of *Megakozlowskiella* sp. A, there is a strong myophragm separating impressed adductor scars; these features are lacking in younger forms.

FIGURED SPECIMENS: USNM 157284-157288.

Megakozlowskiella magnapleura Johnson n. sp.
(Pl. 71, figs. 1-19)

DIAGNOSIS: Subquadrate *Megakozlowskiella* typically bearing four ventral plications.

EXTERIOR: Small- and medium-sized shells are shield-shaped and transversely quadrate to subsemicircular in outline. They are unequally biconvex in lateral profile. Large shells are strongly convex anteriorly and attain great depth. On these shells the pedicle valves are strongly convex at the umbo, and the beak is prominent and strongly incurved. Anteriorly on the largest specimens, the brachial valve becomes the more convex. The ventral interarea is broad and triangular and is equal to the greatest width of the valves. It is steeply apsacline or cataline and is nearly flat. On larger shells the interarea becomes more incurved and more strongly apsacline. The ventral beak is small and incurved. The beak ridges are strong laterally and project ventrally as flanges, producing an overly large interarea. The delthyrium is high and triangular and encloses an angle of approximately 45 degrees. Ridges are present along the exterior edge of the delthyrium, but none of the specimens have deltidial plates preserved. Lateral to the edges of the delthyrium, there is a pair of triangular areas that are striated both horizontally and vertically, subparallel to the edges of the delthyrium. Laterally, the ventral and lateral portions are deeply striated by horizontal growth lines. The inner two triangular areas are set apart from the lateral pair by presence of a second layer of shell material which fairly effectively covers up the early growth lines in those regions. The dorsal interarea is short, linear, and apsacline.

There is a very deep U-shaped sulcus on the pedicle valve and a high, rounded fold on the brachial valve. Laterally, the flanks bear strong rounded radial plications and deep U-shaped interspaces. On the pedicle valve there is a pair of strong plications bounding the sulcus and an additional, slightly smaller pair between them and the beak ridges, but closer to the adjoining pair of medial plications. There is a single, strong, rounded plication on each flank of the brachial valve. Anterolateral to each flanking plication the shell margin is convex, but does not form a full additional plication. This pattern is seen on dozens of well-preserved specimens however, two specimens from the Rabbit Hill Limestone bear three plications on each flank of the pedicle valve (one is illustrated in Pl. 71, fig. 16). The plications are crossed by numerous regularly spaced frilly growth lamellae. The fine radial ornament is not preserved.

INTERIOR OF PEDICLE VALVE: The hinge teeth are stubby and triangular in cross section. They project dorsally and slightly posteriorly within the inner edges of the delthyrium. A pair of stout arcuate dental lamellae are attached to the hinge teeth and their tracks along the inner edge of the delthyrium. The dental lamellae converge at a decreasing angle toward the ventral midline. A long, high median septum is present and continues about half the distance to the anterior edge of the shell. A thickening of shell material beneath the delthyrium joins the ventral edges of the dental lamellae to the sides of the median septum. The surface of the shell is deeply corrugated by the impress of the plications.

INTERIOR OF BRACHIAL VALVE: The sockets are short, broad, and relatively deep. They expand abruptly and diverge anterolaterally. The socket plates are thickened and raised ventrally along the distal edges of the sockets. Crural plates are attached to the inner edges of the socket plates, but do not join the floor of the valve. Medially, there is a pair of subhorizontal cardinal plates which are made sessile by deposition of shell material in the apex of the valve. Posteriorly, on the cardinal plate there is a pair of ridges which converge toward the apex of the valve. Each lobe is deeply striate at the sites of diductor muscle attachment. The adductor muscle scars are not impressed. The interior of the shell, however, is deeply crenulated by the impress of the plications.

OCCURRENCE: *Megakozlowskiella magnapleura* is common in the Rabbit Hill Limestone in the Monitor Range and in the *Trematospira* Zone in the Roberts Mountains and in the Sulphur Spring Range. It also occurs in the *Spinoplasia* Zone at Antelope Peak.

COMPARISON: The older *Megakozlowskiella* sp. A from the *Quadrithyris* Zone differs by being smaller and having more numerous plications. Internally *M. magnapleura* lacks the well-marked adductor impressions separated by a stout myophragm in the brachial valve. Furthermore, in the interior of the pedicle valve, the dental lamellae of *M. magnapleura* do not themselves join the median septum, but only become coalescent by the addition of secondary shell material. On the contrary, in *Megakozlowskiella* sp. A, the dental lamellae are thin and long, and join the median septum basally along a considerable portion of its length. *Megakozlowskiella magnapleura* differs from *Megakozlowskiella* cf. *raricosta* of the overlying *Acrospirifer kobehana* Zone in being relatively more quadrate in outline and having fewer plications.

FIGURED SPECIMENS: USNM 157277-157283. The holotype is USNM 157283.

Megakozlowskiella cf. *raricosta* (Conrad, 1842)
(Pl. 70, figs. 26-28)

Delthyris raricosta Conrad, 1842, p. 162, Pl. 14, fig. 18.
Spirifera raricosta Hall, 1867, p. 192, Pl. 27, figs. 30-34, Pl. 30, figs. 1-9.
Kozlowskiella (Megakozlowskiella) raricosta Boucot, 1957, Pl. 3, figs. 18, 19.

EXTERIOR: The shells are subsemicircular in outline and strongly biconvex in lateral profile. The pedicle valve is more strongly convex than the brachial valve. The hinge line is long and straight and is the place of maximum width. Cardinal angles are acute on large specimens or are at approximately right angles. The ventral interarea is relatively broad, low, and triangular. It is apsacline and strongly curved. Beak ridges are strongly defined. The beak is incurved. The delthyrium is triangular and encloses an angle of about 60 degrees. No deltidial plates are preserved on any of the specimens examined. Each of the two halves of the interarea is divided into two triangular areas. The lateral areas are deeply striate horizontally by the growth lines and the inner areas are thickened and striate both horizontally and nearly vertically, subparallel to the sides of the delthyrium. The dorsal interarea is linear and orthocline to apsacline.

The pedicle valve bears a relatively broad, moderately deep, rounded sulcus and the brachial valve bears a corresponding rounded fold. The flanks of the valves bear strong rounded radial plications separated by deep U-shaped interspaces. There are commonly three or four plications on each ventral flank. In some specimens the first pair of plications laterally bounding the sulcus appear to be slightly narrower than the next adjoining lateral pair which are the largest. When present the fourth lateral plication is relatively much smaller than the three adjoining lateral plications. The plications are crossed by regularly spaced, frilly, concentric growth lamellae. Numerous fine concentric rows of radially arranged striae are present on each of the growth lamellae.

INTERIOR OF PEDICLE VALVE: The teeth are of moderate size and are triangular in cross section. They project posteriorly and dorsally and in the largest specimen are inclined slightly toward the midline at their distal ends. On their inner edges they connect with a pair of short, stout dental lamellae that converge medially toward the midline, but which do not extend strongly anteriorly. There is a strong median septum which continues slightly beyond half the distance to the anterior commissure. In the smallest specimens the dental lamellae join the inner surface of the valve along two discrete tracks closely flanking the median septum. On larger specimens, infilling secondary shell material may unite the ventral edges of the dental lamellae with the sides of the median septum. Relatively deep crural fossettes may be present on the inner sides of the dental lamellae.

The interior of the shell is moderately crenulated by the impress of the plications.

INTERIOR OF BRACHIAL VALVE: The sockets expand broadly and diverge anterolaterally. Their posterior edges are covered by the inner edge of the interarea. The bases of the sockets are thickened and the socket plates rise well above the base of the valve. The platelike crural bases are curved and join medially forming a transverse concave cadinal plate between the inner edges of the socket plates. There are two lobes of deeply striate shell material at the posterior portion of the cardinal plate serving as the site of diductor muscle attachment. The adductor muscle impressions were not observed. The interior of the shell is deeply corrugated by the impress of the plications.

OCCURRENCE: This species is known in Nevada only from the *Acrospirifer kobehana* Zone in the northern Roberts Mountains and in the Sulphur Spring Range.

FIGURED SPECIMEN: USNM 157276.

Subfamily SPINELLINAE n. subfam.

DIAGNOSIS: Advanced multiplicate Delthyrididae with radial or papillose fine ornament, but lacking a transverse subdelthyrial plate in the pedicle valve.

DISCUSSION: This new subfamily replaces the invalid subfamily Gurichellinae Paeckelmann, 1931.

Genus *Spinella* Talent, 1956

TYPE SPECIES: *S. buchanensis* Talent, 1956, p. 22.

Spinella talenti Johnson, n. sp.
(Pl. 64, figs. 10-16)

Spirifera parryana Walcott, 1884, p. 137, Pl. 14, fig. 10; *not* Hall.
Spinella sp. Johnson, 1966b, Pl. 129, figs. 12-15.

DIAGNOSIS: *Spinella* with low, flat plications and an external ornament of fine radial lirae.

EXTERIOR: The shells are transversely suboval in outline and unequally biconvex in lateral profile. The pedicle valve is more strongly convex than the brachial valve. The ventral interarea is short, low, and triangular. The beak is strongly incurved, and the inclination of the interarea is apsacline. The dorsal interarea is short, low, triangular, and its inclination is anacline. The cardinal angles are rounded and obtuse. Maximum length is attained approximately at midlength. There is a broad, shallow sulcus on the pedicle valve equal in width to about three of the adjoining plications. The brachial valve bears a low, flattened fold.

The ornament consists of numerous low, closely spaced, radial plications crossed by very faintly developed concentric growth lines. The plications are strongly flattened and are separated by very narrow shallow interspaces. As many as twelve flat plications can be seen on one flank on the best preserved specimen. The fine ornament consists of numerous flat, closely spaced, fine radial striae of varying width across any one portion of the shell which are crossed at irregular intervals by low concentric growth lines. In the sulcus anteriorly on one valve, the concentric growth lines become so numerous and well defined that they give the fine ornament a reticulate pattern, but the growth lines are in every case fine and narrow relative to the radial striae which are wide and flat, mimicking the radial macroornament.

INTERIOR OF PEDICLE VALVE: The hinge teeth are relatively strong and are roughly elliptical in transverse section. The tracks of the hinge teeth are lengthened internally by thickenings along the inner sides of the dental lamellae beneath the inner edge of the delthyrium. Dental lamellae are thin and divergent in their posterior portions, but recurve very slightly near their anterior ends. They are not flat plates, but are bowed slightly laterally. There is a slight thickening of shell material in the apex of the valve, but there is no transverse subdelthyrial plate connecting the thickenings of shell material on the medial edges of the dental lamellae near the hinge line. An indistinct myophragm may be present dividing the muscle field. The muscle field is very poorly impressed and blends imperceptibly anteriorly with the interior of the valve. The interior is only faintly crenulated anteriorly by the impress of the external plications.

INTERIOR OF BRACHIAL VALVE: The sockets are moderately divergent and expand anterolaterally. They are relatively flat-bottomed and their bases are elevated strongly above the base of the valve. Relatively strong crural plates are attached to the medial edges of the socket plates and diverge moderately anterolaterally. The crural plates converge only slightly toward the midline and join the inner surface of the valve posteriorly along two separate tracks. The notothyrial cavity is striated in its posterior half at the place of diductor muscle attachment. The adductor scars are not strongly impressed. The interior is only faintly crenulated by the impress of the plications.

OCCURRENCE: At present this species is known only from the *Eurekaspirifer pinyonesis* Zone at Lone Mountain and at Table Mountain on the west slope of the Mahogany Hills. At both localities it occurs as relatively abundant silicified specimens in association with *Parachonetes macrostriatus* in the lowest beds of the zone. However, the specimen illustrated by Walcott (1884, Pl. 14, fig. 10) under the name *Spirifera parryana* is reported (1884, p. 137) to have been collected from "Devonian Limestone, on the divide at the head of the Reese and Berry Canyon, Eureka District, Nevada." Walcott (p. 137) also mentioned a second specimen obtained from the "Pinyon Range" (=Sulphur Spring Range).

COMPARISON: *Spinella talenti* differs from *S. buchanensis* Talent (1956, Pls. 1, 2) in fine ornament and generally in having flatter plications. However, Talent (1956, Pl. 2, figs. 4-7) showed that some specimens of *S. buchanensis* have relatively flat plications.

The species *Spirifer incertus*, assigned to *Spinella* by Vandercammen (1963) has well-differentiated, more rounded plications than *Spinella talenti*. In addition, *S. incertus* appears to have a different cardinalia that lacks crural plates and may prove to be better assigned to some other genus.

DISCUSSION: The assignment to *Spinella* is made with little hesitation. The specimens at hand differ from the Australian species principally in the development of a fine ornament consisting of radial striae rather than teardrop-shaped granules. However, the two types are not regarded by the writer as of generic value. Vandercammen (1957, Pls. 1, 2) showed that similar variation occurs in species of *Adolfia* (=*Gurichella*) from Middle and Upper Devonian beds in western Europe. Similar variations have been observed in external ornament of certain single species from the Middle and Upper Devonian of the midcontinent region of North America (Fenton, 1931, Pls. 18, 19). In its ornament and external form, *Spinella talenti* has much to suggest relation to *Eurekaspirifer*. In the pedicle valve of the Lone Mountain specimens of *Spinella*, the dental lamellae are reminiscent of those of *Eurekaspirifer*, but are relatively more broadly separated. None recurves anteriorly to enclose the diductor muscle impressions. In addition, the place of diductor attachment of the brachial valve is developed on a simple triangular pad of material in the posterior triangular portion of the notothyrial cavity and is not elevated on two free lobes which attach to the sides of the crural plates flanking the notothyrial cavity. Also there is no development seen in the *Spinella* specimens of dorsal adminicula. However, the characters of the Lone Mountain *Spinella* specimens clearly indicate to the writer the possibility that they are a link between some presently unidentified spiriferid genus and the more divergent *Eurekaspirifer* stock.

FIGURED SPECIMENS: USNM 146536, 146537, 146537A, 157229A and B, 157230. The holotype is USNM 146536.

Genus *Eurekaspirifer* Johnson, 1966
TYPE SPECIES: *Spirifer* (*Trigonotreta*) *pinonensis* Meek, 1870, p. 60.

DIAGNOSIS: Pedicle valve lacking a subdelthyrial plate and possessing long, closely spaced dental lamellae that recurve anteriorly toward the midline of the valve. Brachial valve with short to long adminicula variably developed and most prominent in small shells. Cardinal process on large shells advances anteriorly, forming a pair of discrete, deeply striate lobes that attach to platelike crural bases flanking the notothyrial cavity. Fine exterior ornament is radially striate and nonfrilly.

Eurekaspirifer pinyonensis (Meek, 1870)
(Pl. 65, figs. 1-15; Pl. 66, figs. 1-12)
Spirifer (*Trigonotreta*) *pinonensis* Meek, 1870, p. 60.
Spirifer (*Trigonotreta*) *pinonensis* Meek, 1877, p. 45, Pl. 1, figs. 9, 9a, 9b.
Spirifera pinyonensis Walcott, 1884, p. 138, in part, Pl. 4, figs. 1c, 1e, 1f; *not* 1, 1a, 1b, 1d.
Spirifer pinyonensis Merriam, 1940, p. 54, in part, Pl. 7, fig. 6; *not* figs. 3-5, 7, 8; *not* Pl. 6, fig. 3.
Acrospirifer pinyonensis Cooper, 1944, p. 325, Pl. 123, fig. 14.
Eurekaspirifer pinyonensis Johnson, 1966b, p. 1045, Pl. 128, figs. 1-15; Pl. 129, figs. 1-11.

DISCUSSION: This species is very abundant, represented by silicified shells in the upper part of the McColley Canyon Formation in the Sulphur Spring Range and so far has only been found beyond the Sulphur Spring Range at Willow Creek in the Roberts Mountains. At that locality it occurs in the middle of the interval between the *kobehana* Zone and the *Leptathyris circula* Zone (A. R. Ormiston, written commun., 1966). This species is characterized by a subtriangular brachial valve, a rhomboidal pedicle valve, and a relatively low, incurved, apsacline ventral interarea. Internally, the characteristic features are ones that appear to be diagnostic for the genus, that is, long, anteriorly converging dental lamellae and variably developed dorsal adminicula. At McColley Canyon, the most common associates of *E. pinyonensis* are *Atrypa nevadana* and *Phragmostrophia merriami*.

FIGURED SPECIMENS: USNM 146521-146535.

Family RETICULARIIDAE Waagen, 1883
Genus *Reticulariopsis* Frederiks, 1916

TYPE SPECIES: *Spirifer (Reticularia) dereimsi* Oehlert, 1901, p. 236, Pl. 6, figs. 2-16.

DISCUSSION: There are a number of species of spiriferoid brachiopods that lack plications and probably belong to *Reticulariopsis* or to *Spirinella*. However, no thoroughgoing examination of the nonplicate topological group has been accomplished to date. In Nevada, the species *"Howellella" pauciplicata* Waite (1956, p. 17, Pl. 4, figs. 6-10; Johnson and Reso, 1964, p. 82, Pl. 20, figs. 8, 10-12) is fairly common in beds correlated with the Upper Silurian. It is believed to belong to *Spirinella*. The collections from Lower Devonian strata studied in this paper have yielded a total of less than a dozen fragmentary specimens from two horizons, and these specimens are discussed below.

Reticulariopsis? sp. A
(Pl. 67, fgs. 16, 17)

EXTERIOR: The shells are smooth on the exterior with a short hinge line and a poorly developed, rounded, narrow interarea. The beak is moderately strongly incurved. The pedicle valve is more strongly convex than the brachial valve.

INTERIOR OF PEDICLE VALVE: The hinge teeth are thin and pointed. The tracks of the teeth form relatively prominent ridges within the inner edges of the delthyrium. Apically, the tracks of the hinge teeth are continuous with thin, slightly divergent dental lamellae.

INTERIOR OF BRACHIAL VALVE: Small socket plates define narrow, widely divergent sockets. The place of diductor muscle attachment bears a small protuberance of shell material. The crural plates are relatively stout and are attached to the inner edges of the socket plates. They converge toward the midline of the valve and join with the interior of the valve at the apex of the shell.

OCCURRENCE: This species is represented by three fragmentary specimens from the "lower breccia" of the *Quadrithyris* Zone at Coal Canyon.

FIGURED SPECIMENS: USNM 157236, 157237.

Reticulariopsis sp. B
(Pl. 67, figs. 18-24)

EXTERIOR: The shells are transversely suboval in outline and strongly subequally biconvex in lateral profile. The hinge line is relatively short and is less than the maximum width of the valves which is attained slightly posterior to midlength. Cardinal angles are acute and rounded. The ventral interarea is poorly developed, and the beak is strongly incurved. A single brachial valve with well-preserved growth lines shows that the species has a strongly transverse subrhomboidal outline and is deeply convex. There is a relatively strong, narrow, rounded fold set off laterally by a pair of U-shaped furrows. The dorsal interarea is long and triangular. It is relatively strong and flat and is anacline to nearly orthocline.

The ornament consists of very fine, numerous, closely spaced growth lines which bear concentric rows of fine radially arranged striae.

INTERIOR OF PEDICLE VALVE: The hinge teeth are relatively ponderous, blunt, and triangular in transverse section. They are supported by stout slightly divergent dental lamellae which bound an elongate, impressed central muscle scar. The diductor tracks are divided medially by a prominent rounded myophragm which extends anteriorly slightly beyond the place where the diductor impressions fade into the shell anteriorly.

INTERIOR OF BRACHIAL VALVE: The sockets are shallow and diverge moderately anterolaterally and are not closely subparallel to the hinge line. They are elevated strongly above the base of the valve and bear long, thin, triangular platelike crural bases which are well developed anterior to the distal ends of the sockets. The crural bases are strongly elevated above the base of the valve, except in the apex where they are supported by a pair of crural plates that diverge slightly from the midline and then recurve very slightly to be subparallel at their anterior edges. The site of diductor attachment consists of a pair of lobes attached to the sides of the crural bases near the apex of the notothyrial cavity. The adductor muscle impressions are elongate, narrow, and moderately well defined. They are divided medially by a thin long myophragm.

OCCURRENCE: Two brachial valves and a single pedicle valve of this species have been collected from the *Trematospira* Zone at McColley Canyon in the Sulphur Spring Range. Several, probably conspecific, specimens have been found in the *kobehana* Zone at Lone Mountain.

COMPARISON: *Reticulariopsis* sp. B somewhat resembles "*Eomartiniopsis*" *rhomboidalis* (Khalfin) as illustrated by Kulkov, (1963, Pl. 9, fig. 2).

FIGURED SPECIMENS: USNM 157238, 157239.

Genus *Quadrithyris* Havlíček, 1957

TYPE SPECIES: *Spirifer robustus* Barrande, 1848, p. 162, Pl. 15, fig. 1.

Quadrithyris cf. *minuens* (Barrande, 1879)

(Pl. 67, figs. 1-15)

Spirifer minuens Barrande, 1879, Pl. 138, case 9, figs. 4, 5.
Delthyris (Quadrithyris) minuens Havlíček, 1959, p. 122, Pl. 11, fig. 1; Pl. 23, fig. 1.

EXTERIOR: The shells are transversely subpentagonal in outline and strongly biconvex in lateral profile. The ventral interarea is relatively short, rounded rather than straight, and is triangular. It is apsacline, moderately curved, and equal to approximately one-third the maximum width of the shell. The delthyrium is relatively long and narrow, enclosing an angle of about 30 degrees. The cardinal angles are obtuse and well rounded. Maximum width of the valves is attained slightly anterior to the hinge line. The lateral margins of the shell converge anteromedially from the rounded cardinal extremities. The pedicle valve bears a relatively well-defined, shallow, median sulcus. It may be U-shaped or somewhat flat at its base. It is bounded laterally by a pair of low, rounded plications. The flanks are unplicated. The sulcus is anterodorsally prolonged as a tonguelike extension that is accommodated by the dorsal fold. The dorsal fold is fairly well defined and is accentuated by the presence of broad, shallow furrows adjacent to it laterally.

The ornament consists of fine, closely spaced concentric growth lines which bear rows of fine, radially aligned ridges.

INTERIOR OF PEDICLE VALVE: The hinge teeth are poorly preserved on the specimens at hand. They are supported by long, thin, only slightly divergent dental lamellae. There is a median septum beginning at the apex of the valve and continuing about three-quarters of the distance to the anterior of the shell. The median septum is high and triangular, and it slopes away precipitously toward the anterior commissure from its highest point.

INTERIOR OF BRACHIAL VALVE: The sockets diverge relatively strongly anterolaterally from the hinge line. The socket plates medially define the sockets well above the base of the valve. They are supported by strong, triangular, platelike crural bases which are free at their dorsal edges along most of their length, but which are attached to the inner surface of the valve posteriorly. The site of the diductor muscle attachment is too poorly preserved on the available specimens for an accurate description. There is a long narrow myophragm present on some specimens, dividing an elongate adductor muscle field.

OCCURRENCE: *Quadrithyris* cf. *minuens* is a relatively rare species in the *Quadrithyris* Zone at Coal Canyon in the northern Simpson Park Mountains and in the Toquima Range.

DISCUSSION: The Nevada specimens differ slightly from the Czechoslovakian ones in having a flatter, more U-shaped sulcus and in being relatively less transverse.

FIGURED SPECIMENS: USNM 157231-157235.

Genus *Elythyna* Rzhonsnitskaya, 1952

TYPE SPECIES: *E. salairica* Rzhonsnitskaya, 1952, p. 62, Text-Fig. 3, nos. 6a-c, p. 82, Pl. 5, figs. 1-3, Pl. 6, figs. 1-5.

Elythyna sp. A
(Pl. 68, figs. 1-13)

EXTERIOR: The shells are transversely suboval in outline and moderately biconvex in lateral profile. In most specimens the pedicle valve is a little more convex than the brachial valve. The cardinal angles are obtuse and strongly rounded. Maximum width is slightly posterior to midlength. The beak is short and incurved. The ventral interarea is relatively short and triangular and is equal to approximately half the maximum width of the valves. It is apsacline and moderately incurved. The dorsal interarea is low and narrow. Its inclination is anacline. The pedicle valve bears a more or less well-developed rounded sulcus, and the brachial valve bears a corresponding low, rounded fold.

The flanks of medium- and large-sized individuals bear a few low, rounded plications. There are commonly about five plications developed on each flank of the best preserved larger specimens, but the plications are very low and indistinct and affect only the anterior half of the shell. They are separated by equally faintly developed U-shaped interspaces. The larger specimens are more or less exfoliated so that the fine external ornament is not preserved. The silicified ones are fragmentary.

INTERIOR OF PEDICLE VALVE: The hinge teeth are suboval in cross section. They are supported basally by relatively thin dental lamellae which diverge anteriorly and extend anterolaterally along about half the length of the muscle impressions. The diductor muscle scars are moderately strongly impressed and are elongate suboval in outline. There is a more or less well-defined myophragm dividing the muscle impressions in some specimens. A pair of vascular tracks may be impressed at the anteromedial edges of the muscle impressions. The umbonal cavities are moderately infilled with secondary shell material, and their surfaces are pustulose.

INTERIOR OF BRACHIAL VALVE: The sockets expand and diverge moderately from the hinge line anterolaterally. Their inner edges are thick and are attached to a pair of medially concave crural plates which adjoin the inner surface of the valve. In large specimens the spaces beneath the sockets are almost completely filled in by secondary shell material, making the crural plates largely obsolescent. In the nothothyrial cavity of one specimen, there is a strong build-up of shell material forming a moderately bilobed platform. The apex of the notothyrial cavity is longitudinally striate in a small, transverse, triangular area serving as a site of diductor muscle attachment. The adductor muscle scars are strongly elongate and impressed along their lateral edges. Anteriorly, they blend with the interior of the valve. There is a moderately well-developed narrow myophragm dividing the adductor field medially. On specimens bearing plications, the interior of the valve is crenulated by the impress of the plications.

OCCURRENCE: The only available specimens are from the *Eurekaspirifer pinyonensis* Zone in the Cortez Range and in the northern Simpson Park Range.

Two specimens are illustrated from the *Leptocoelia infrequens* Subzone at Coal Canyon. They are decidedly more lenticular than *Elythyna* sp. B from the *Elythyna* beds. Specimens closely comparable to those from the *Leptocoelia infrequens* Subzone are present in collections (GSC 25853, 25850, 25835) examined by the writer from the Stuart Bay Formation on Bathurst Island in the Canadian Arctic.

FIGURED SPECIMENS: USNM 157240-157245.

Elythyna sp. B
(Pl. 68, figs. 14-19)

Spirifera (Martinia) undifera Walcott, 1884, p. 143, Pl. 3, figs. 6, 6a; *not* 3, 3a, 3b (=?*Emanuella* sp.); Pl. 14, fig. 11, 11a, 11b, 13b?, 13c?; *not* Roemer, 1844.

Martinia "undifra Roemer" Merriam, 1940, Pl. 8, fig. 23; Pl. 11, fig. 11; *not* Roemer, 1944.

DISCUSSION: Since the present study was begun, it has become evident that the Lower Devonian of Nevada is characterized by a fauna with abundant *Elythyna* above the *Eurekaspirifer pinyonensis* Zone. This higher horizon, in this paper called "*Elythyna* beds," has been included as a part of the *pinyonensis* Zone in the past, but certainly will rank as a fully distinct zone when its fauna is studied. The abundant species of *Elythyna* of these higher beds is illustrated on Plate 68 as *Elythyna* sp. B for comparison with *Elythyna* sp. A of the *pinyonensis* Zone. The chief distinction between the two forms is the deeply biconvex lateral profile of *Elythyna* sp. B compared to the relatively lenticular lateral profile of *Elythyna* sp. A.

OCCURRENCE: In the *Elythyna* beds above the *Eurekaspirifer pinyonensis* Zone in the J-D Window of the Simpson Park Range, the Roberts Mountains, and at Table Mountain in the Mahogany Hills (Johnson and others, 1968). Walcott (1884, p. 146) recorded what probably is this species in the Lower Devonian of Atrypa Peak in the Eureka district and at Lone Mountain.

FIGURED SPECIMENS: USNM 157246, 157247.

Family AMBOCOELIIDAE George, 1931
Genus *Ambocoelia* Hall, 1860

TYPE SPECIES: *Orthis umbonata* Conrad, 1842, p. 264, Pl. 14, fig. 4.

Ambocoelia sp.
(Pl. 70, figs. 1-11)

EXTERIOR: The shells are transversely suboval in outline and unequally biconvex in lateral profile. The brachial valve is gently convex but the pedicle valve is strongly arched, especially posteriorly, and is several times as deep as the brachial valve. The cardinal angles are rounded and obtuse. Maximum width of the shell is posterior to midlength. The ventral interarea is high, narrow, and triangular; it is somewhat rounded laterally. Its inclination is strongly apsacline, nearly cataclive. The beak is blunt and is slightly to moderately incurved. There is a triangular delthyrium which encloses an angle of approximately 45 degrees. The dorsal interarea is short and linear. It is anacline and almost orthocline. Anteriorly on the pedicle valve, there is a low fold with an extremely faint median indentation, and on the brachial valve there is a low anteromedial raised area defined by broad, shallow, widely divergent bounding furrows.

The fine ornament consists of sparse, indistinct, concentric growth lines. No radial ornament was observed.

INTERIOR OF PEDICLE VALVE: The hinge teeth are unsupported by dental lamellae, but the inner edge of the delthyrium is thickened by the tracks of the hinge teeth. The interior of the valve is smooth.

INTERIOR OF BRACHIAL VALVE: The sockets are narrow and broadly divergent and are bounded posteriorly by ridges on the inner edge of the shell. Medially, the socket plates are attached to triangular crural plates which converge toward the midline, but do not join, and which merge with the inner edge of the shell only posteriorly along two discrete tracks. The cardinal process is poorly preserved in specimens at hand and is only distinguishable as a knoblike structure in the apex of the valve. The crura are relatively thick and project almost straight anteriorly from the crural bases before beginning the first volution. The interior of the valve is smooth.

OCCURRENCE: This species of *Ambocoelia* is present in the *Quadrithyris* Zone at Coal Canyon and in the lower McMonnigal Limestone in the Toquima Range.

FIGURED SPECIMENS: USNM 157264-157268.

Genus *Spinoplasia* Boucot, 1959
TYPE SPECIES: *S. gaspensis* Boucot, 1959, p. 18, Pl. 2, figs. 14-16.
Spinoplasia roeni Johnson, n. sp.
(Pl. 69, figs. 1-13)
?"*Spirifer*" sp. cf. "*S.*" *modestus* Merriam, 1963, p. 43; not Hall.

DIAGNOSIS: *Spinoplasia* lacking any form of plication and virtually without interarea. Ventral posterior thickened internally.

DISCUSSION: *Spinoplasia roeni* is the most abundant brachiopod species in the Rabbit Hill Limestone, to which it is restricted. There are hundreds of specimens present in the collection, but over 99 percent of these are the posterior portions of pedicle valves which are strongly thickened. A few specimens larger than the average, which are known, are comparatively fragile around their anterior and anterolateral edges. The fragmentary nature of the Rabbit Hill fossils together with imperfect silicification of the disarticulated valves is responsible. Not one articulated specimen is known among the hundreds examined.

EXTERIOR: The valves appear to be approximately suboval and variable in length-width ratio, as estimated from the fragmentary remains. They are small shells and the preserved portions are commonly somewhat less than 5 mm in maximum dimension. The ventral interarea is relatively small, low, triangular, and rounded on its lateral edges. The beak is relatively blunt, but strongly incurved over a low triangular delthyrium. The exterior lacks macro-ornament except for a faint median furrow on the pedicle valve. The fine ornament consists of numerous fine concentric growth lines and radially arranged concentric rows of fine striae.

INTERIOR OF PEDICLE VALVE: The hinge teeth are relatively long and pointed. They are subtriangular in cross section and leave moderately well-defined tracks forming ridges along the inner edges of the delthyrium. Dental lamellae are absent. Thin-shelled specimens bear a pair of strongly divergent diductor muscle impressions with relatively straight lateral edges. They are separated medially by a broad myophragm anteriorly. There is commonly a plug of shell material in the umbonal region which partially closes the delthyrium of thicker shells with a pair of lobes deposited between the inner edges of the tracks of the hinge teeth. The two lobes connect anteriorly with thickenings of shell material at the position which normally would be filled by dental lamellae in other spiriferids. These lobes continue anteriorly to join shell material along the posterior edge of the diductor impressions where the thickened ridges of some shells splay very strongly laterally, forming two elevated lobes somewhat similar to those present in *Dayia*. The simple myophragm commonly is cleft medially by a narrow but well-defined groove, giving it the appearance of a pair of closely set elevated ridges.

INTERIOR OF BRACHIAL VALVE: The sockets expand and diverge anterolaterally. Triangular crural plates are attached to the inner edges of the socket plates and converge toward the midline to join the valve along discrete tracks. A somewhat elongate protuberance of shell material in the notothyrial cavity serves as the site of diductor attachment. The shells are very thin.

OCCURRENCE: *Spinoplasia roeni* is present in the *Spinoplasia* Zone in the Monitor Range, northern Simpson Park Mountains, Cortez Range, and at Antelope Peak.

COMPARISON: No other species of *Spinoplasia* combines the features of *S. roeni* noted above in the diagnosis.

FIGURED SPECIMENS: USNM 157248-157257. The holotype is USNM 157252.

Genus *Metaplasia* Hall and Clarke, 1893
TYPE SPECIES: *Spirifer pyxidatus* Hall, 1859, p. 428, Pl. 100, figs. 9-12.
Metaplasia cf. *paucicostata* (Schuchert, 1913)
(Pl. 70, figs. 12-25)
Spirifer paucicostatus Schuchert, *in* Schuchert and others, 1913, p. 402, Pl. 68, figs. 30, 31.
Metaplasia paucicostata Boucot, 1959, Pl. 2, figs. 12, 13.

EXTERIOR: The shells are small and commonly slightly transverse and subpentagonal, but vary to an outline with the width approximately equal to the length. They are strongly, unequally biconvex in lateral profile with the pedicle valve much more convex than the brachial valve. The hinge line is relatively long and equals approximately three-quarters the maximum width. Cardinal angles are commonly rounded and vary from acute to obtuse; in most specimens they are rounded and acute. The lateral margins tend to be moderately straight, not curving around smoothly to the anterior. The ventral interarea is moderately high and triangular and rounded at its lateral edges. Its inclination is apsacline, approaching the catacline position, and it is slightly curved. The ventral beak is small and moderately incurved. The delthyrium is triangular and encloses an angle of about 35 to 40 degrees. The dorsal interarea is low, triangular, flat, and orthocline.

The pedicle valve bears a more or less well-defined median fold composed of a pair of rounded plications separated by a shallow, U-shaped medial furrow. Laterally on the largest specimens, the fold is bounded by a pair of faint furrows. The brachial valve bears a shallow sulcus that flares widely in its anterior portion. The sulcus bears an indistinct, low medial plication near its anterior edge.

INTERIOR OF PEDICLE VALVE: The hinge teeth are small and pointed and project dorsally. The tracks of the hinge teeth form strong ridges on the interior edges of the delthyrium. There is a small plug of shell material at the apex of the delthyrium. Dental lamellae are absent. The shell substance is not thickened in the umbonal regions, and the interior is uniformly smooth.

INTERIOR OF BRACHIAL VALVE: The sockets are relatively narrow and expand anterolaterally. They are defined basally by socket plates lying well above the inner surface of the valve. The inner edges of the socket plates are joined by a pair of stout crural plates which are slightly concave medially and which, in some specimens, join the valve along two subparallel tracks, or converge more strongly medially and unite at about the midline of the valve. The cardinal process consists of a medial rodlike elevation of shell material which is slightly thickened at its posterior end. The interior of the shell is smooth. The adductor scars are not impressed.

OCCURRENCE: *Metaplasia* cf. *paucicostata* is at present known in Nevada only from the *Trematospira* Zone in the Sulphur Spring Range and from the *Acrospirifer kobehana* Zone in the Sulphur Spring Range and in the southern Roberts Mountains.

DISCUSSION: A closely comparable species is *Metaplasia nitidula* (Clarke) from the Grande Greve Limestone of Gaspé, Quebec. Clarke was aware of the ambocoeliid affinities of his species, but unaccountably (or perhaps because of the form of the valves) described it as *Spirifer modestus* var. *nitidulus* (Clarke, 1908, p. 182, Pl. 31). A similar uncertainty about possible relation of "*Spirifer*" *modestus* with the ambocoeliids may be the reason for Merriam's (1963) citation of a species comparable to "*S.*" *modestus* from the Rabbit Hill Limestone. The latter form is believed to be *Spinoplasia roeni* as noted in the *S. roeni* synonymy above. The true Keyser Limestone species bears well-developed dental lamellae and probably belongs to the genus *Spirinella* (cf. *Howellella modesta* as figured by Bowen, 1967).

FIGURED SPECIMENS: USNM 157269-157275.

Genus *Plicoplasia* Boucot, 1959
TYPE SPECIES: *P. cooperi* Boucot, 1959.
Plicoplasia cooperi Boucot, 1959
(Pl. 69, figs. 14-23)
Metaplasia pyxidata Cooper, 1944, Pl. 126, figs. 40-43; not Hall.
Plicoplasia cooperi Boucot, 1959, p. 20, Pl. 1, figs. 13, 14; Pl. 2, figs. 1-5.

DISCUSSION: *Plicoplasia cooperi* is notable for the presence of its closely set, subparallel, rounded plications forming the ventral fold and by the relatively shallow furrow which divides them medially.

EXTERIOR: The shells are transversely suboval to subsemicircular in outline and unequally biconvex in lateral profile. The brachial valve is only gently convex, and the pedicle valve is relatively strongly convex. The hinge line is long and straight and maximum width is at the hinge line or only slightly anterior to it. The cardinal angles are rounded. The ventral interarea is relatively long and low, almost ribbon-like, but slightly higher at the midline. It is curved and nearly cataclinal. The beak is moderately incurved. The interarea is cleft medially by a V-shaped delthyrium with lateral edges only slightly divergent dorsally. The apex is partially filled by a plug of material on the interior. The dorsal interarea is about half the maximum width and is linear and anacline.

The ornament consists of a few strong rounded radial plications on each valve. There is a pair of strong plications medially on the pedicle valve which are elevated above the general level of the flanks to form a medial fold. The two medial plications are closely set together and are separated by a shallow and relatively narrow U-shaped furrow. Laterally, on each flank there are two rounded plications separated by relatively deep U-shaped interspaces. Each lateral plication originates slightly further anterior to the beak than the adjoining plication medial to it. The brachial valve bears a sulcus with a medial plication and two lateral plications on each flank of the valve.

INTERIOR OF PEDICLE VALVE: The hinge teeth are blunt and triangular. They leave relatively strong tracks forming ridges beneath the inner edges of the delthyrium. Dental lamellae are absent. The interior is relatively strongly crenulated by the impress of the plications.

INTERIOR OF BRACHIAL VALVE: The sockets diverge relatively strongly from the hinge line and expand anterolaterally. They are defined basally by relatively flat socket plates which project above the interior surface of the shell. There is a pair of strong, triangular crural plates attached to the inner edges of the socket plates. They converge slightly basally and diverge relatively strongly from one another anteriorly. They join the inner surface of the valve within the notothyrium at about the midline and diverge slightly along the floor of the valve. Medially, there is a rodlike lobe of shell material expanding posteriorly to serve as a site of diductor muscle attachment. The adductor muscle scars are not impressed and the interior is relatively strongly corrugated by the impress of the plications.

OCCURRENCE: *Plicoplasia cooperi* is a relatively rare species in Nevada, presently known only from the *Spinoplasia* Zone at Rabbit Hill.

FIGURED SPECIMENS: USNM 157258-157263.

Superfamily CYRTINACEA Frederiks, 1912
[*nom. transl.* Johnson, 1966, *ex* Cyrtininae Frederiks, 1912]
Family CYRTINIDAE Frederiks, 1912

DISCUSSION: The writer (Johnson, 1966a) suggested that *Cyrtina* belongs to a superfamily not directly connected with any of the younger punctate spiriferids, such as are common in the Carboniferous and Permian, and is relatively sharply differentiated from other spirifers that range through the Silurian and Lower Devonian. No pre-Gedinnian occurrence of *Cyrtina* is known with certainty although a single specimen has been found in collections obtained within an interval 50 feet below where the base of the Gedinnian as drawn in the Roberts Mountains Formation. The Silurian occurrence reported by Bowen (1967) turned out to be erroneous because the specimen was misidentified (Johnson and Boucot *in* Berdan and others, 1969). *Cyrtina* is a common genus occurring widely in the Devonian, and its arrival on the scene in the Gedinnian, in real abundance in some areas, is apparently unheralded.

Cyrtina, because of its peculiar morphology exemplified by the presence of endopunctae and a tichorhinum, is distinct from other Devonian spiriferids. No really satisfactory precursor is known, but it is suggested here that the most likely possibilities lie among the Silurian kozlowskiellinids. Shell shape and ribbing are quite similar and some early forms of *Cyrtina* bear an irregular development of concentric, lamellose ornament.

Moreover, the cardinalia of *Cyrtina* and *Kozlowskiellina* (*Megakozlowskiella*) are quite similar in over-all plan. In the pedicle valve, early representatives of *Megakozlowskiella* have long thin dental plates firmly attached to the median septum to form a spondylium, morphologically similar to that of *Cyrtina* although there is no question of a tichorhinum being present in any species of *Megakozlowskiella*. At the same time it may be noted that the tichorhinum of Gedinnian specimens of *Cyrtina*, such as the abundant occurrence in the lower part of the Coeymans Formation of New York (silicified specimens studied by the writer), is relatively rudimentary. No kozlowskiellinid is known to be endopunctate, so comparisons must end there, at least until some new evidence is available.

Genus *Cyrtinaella* Frederiks, 1916
TYPE SPECIES: *Cyrtina biplicata* Hall, 1857, p. 165.
Cyrtinaella causa Johnson, n. sp.
(Pl. 72, figs. 1-9, 14, 15)

DIAGNOSIS: *Cyrtinaella* with prominent U-shaped furrows on the pedicle valve lateral to the plications bounding the sulcus.

EXTERIOR: The outline is transverse and varies from acutely triangular to subquadrate. In lateral profile the valves are unequally biconvex with a high, subpyramidal pedicle valve, about twice as deep as the brachial valve. The latter is relatively prominently arched from beak to anterior margin. The ventral interarea is catacline to steeply apsacline, nearly flat, but commonly gently curved in its apical half. Horizontal growth lines are present on some specimens. The delthyrium is high and triangular and encompasses an angle between 30 and 45 degrees. It is covered by an outwardly convex deltidium over most of its length, except at the apex which is pierced by an oval foramen. There is no dorsal interarea.

The pedicle valve bears a rounded sulcus bounded by a pair of rounded plications that tend to be accentuated by a pair of bounding furrows lateral to them. The brachial valve has a prominent rounded fold bounded by a pair of lateral furrows. A few irregularly disposed concentric growth lines are present. No fine radial ornament was seen.

INTERIOR OF PEDICLE VALVE: The hinge teeth are triangular in cross section and are rather delicate. They taper and tend to project posteromedially as they extend dorsally. They are supported basally by a pair of thick, bladelike dental lamellae that converge and join with a thin median septum about midway between the palintrope and the anterior of the shell. The median septum bears an oval tichorhinum whose long axis it bisects. The median septum may terminate without extending to the posterior surface of the tichorhinum, or it may form a ridge along the posteromedial surface of the tichorhinum. The median septum commonly is about two-thirds the length of the valve.

INTERIOR OF BRACHIAL VALVE: The sockets open widely toward their anterolaterally diverging extremities and are incised into the flat area along the dorsal hinge line. They are bounded medially by prominently developed socket ridges, of which the crural bases are a part. The notothyrial area is completely filled by a thickened, bilobate cardinal process which is longitudinally striated on its posteroventrally facing myophore. There may or may not be a pair of deep grooves that separate the lateral edges of the cardinal process from the adjacent socket ridges. In some specimens the bases of the sockets and the anterior portions of the socket ridges slightly overhang the base of the valve. There may or may not be a short myophragm extending from the base of the cardinal process, and, where present, it tapers sharply to a termination in what is apparently the posterior part of the adductor muscle field. There commonly is a pair of laterally diverging muscle-bounding ridges that lie atop the interior elevations which correspond to the furrows that bound the dorsal fold. These are evident in Plate 72, figure 9, where they could be mistaken as laterally bent crura. In specimens that lack development of a myophragm and muscle-bounding ridges, the adductor muscle field is deep and suboval (Pl. 72, fig. 3).

SHELL STRUCTURE: The shell substance is endopunctate.

OCCURRENCE: This species is known only from the *Quadrithyris* Zone. It is rare at Coal Canyon and relatively common in the lower McMonnigal Limestone of the Toquima Range.

COMPARISON: *Cyrtinaella causa* has more rounded flanks on both valves and has more prominent furrows bounding the medial pair of plications on the pedicle valve than does the type, *C. biplicata*.

FIGURED SPECIMENS: USNM 157289-157293. The holotype is USNM 157293.

Genus *Cyrtina* Davidson, 1858
TYPE SPECIES: *Calceole heteroclite* Defrance, 1824, p. 306, Pl. 80, figs. 3, 3a.
Cyrtina sp. A
(Pl. 72, figs. 10-13)

EXTERIOR: The shells are transverse and triangular to subsemicircular in outline and strongly unequally biconvex in lateral profile. The brachial valve is only gently convex, but the pedicle valve is high and subpyramidal. The hinge line is the place of maximum width. The cardinal angles are acute and pointed. The anterolateral commissure is straight to only slightly rounded. The ventral interarea is high, triangular, and apsacline to cataline. It is only gently incurved. The delthyrium is long and triangular, enclosing an angle of about 20 degrees. Near the hinge line there is a convex deltidium. It is discontinuous apically leaving an elliptical foramen. The dorsal interarea is relatively well defined, short, linear, and apsacline. There is a strong ventral sulcus and a complementary fold on the brachial valve.

The ornament consists of sparse radial plications. A pair of strong, rounded plications borders the sulcus, and there commonly are two smaller radial plications on each flank of the pedicle valve. On each flank of the brachial valve, there commonly are two plications which are separated from the median fold by deep U-shaped furrows. Concentric ornament is prominently developed on some specimens, forming lamellose growth lines which may be at regular or irregular intervals, but are more commonly well developed near the anterior.

INTERIOR OF PEDICLE VALVE: There is a spondylium developed medially beneath the delthyrium. It is supported by a high, thin median septum which penetrates the trough of the spondylium and bears a tichorhinum in the form of an elongate tube of elliptical cross section.

INTERIOR OF BRACHIAL VALVE: The sockets are shallow and broad and are set in thickened shell material along the inner edge of the interarea. The sockets are bounded medially by thickened, triangular socket plates which are slightly attenuated anteriorly, evidently at the point of attachment of the crura. The notothyrial cavity bears a thickening of shell material which is deeply cleft longitudinally in its posterior half to form a bilobate cardinal process. Each lobe of the cardinal process is longitudinally striated. The interior is strongly thickened and the adductor scars are deeply impressed into the shell material medially. They are situated anterior to the edge of the notothyrial cavity and are narrow and elongate. They are separated by a stout myophragm that tapers measurably in a short distance toward the anterior.

OCCURRENCE: *Cyrtina* sp. A is common in the *Quadrithyris* Zone at Coal Canyon and in the lower McMonnigal Limestone in the Toquima Range.

COMPARISON: *Cyrtina* sp. A appears to differ from an undescribed *Cyrtina* sp. of the upper Roberts Mountains Formation by the presence of strong plications bounding the sulcus on the pedicle valve and in obsolescence of plications toward the anterolateral margins.

FIGURED SPECIMENS: USNM 157294-157296.

Cyrtina cf. *varia* Clarke, 1900
(Pl. 73, figs. 1-14)
Cyrtina varia Clarke, 1900, p. 49, Pl. 6, figs. 15-22.

EXTERIOR: The shells are transversely subtrigonal in outline and strongly unequally biconvex in lateral profile. The cardinal angles are acute and pointed. The hinge line is long and straight and is the place of maximum width. The pedicle valve is deeply convex and subpyramidal. In a number of specimens the length of the valves is equal to or slightly less than the height, and of these the width is approximately twice the length. In other specimens the width-length ratio is less than 2 to 1. The ventral interarea is flat, catacline, and triangular. The delthyrium is high, triangular, and narrow, encompassing an angle of about 15 to 20 degrees. The dorsal half of the delthyrium is covered by a convex deltidium leaving an elongate elliptical opening at the apex and a low, crescentic opening at its base. The dorsal interarea is low and linear and varies from orthocline to apsacline.

There is a rounded sulcus on the pedicle valve and a corresponding low, rounded fold on the brachial valve. The ornament consists of rounded radial plications crossed at irregular intervals by more or less well-developed concentric growth lines. There are four or five plications on one flank of the valve in most specimens. The fifth lateral plication is generally faint, and the posterolateral extremities are unplicated. Plications adjoining the ventral sulcus are decidedly larger than the adjoining lateral plications which may be as little as half the width of the medial pair. The corresponding interspaces adjacent to the dorsal fold are also considerably broader than the interspaces on the remainder of the flanks.

INTERIOR OF PEDICLE VALVE: The hinge teeth are supported by a pair of thick dental lamellae that converge medially and join the median septum.

The spondylium thus formed is relatively deep and the supporting median septum is long and thick. The tichorhinum has a median blade, but the posterior edge of the structure is not well preserved on the available specimens. Large specimens are not crenulated by the impress of the plications.

INTERIOR OF BRACHIAL VALVE: The sockets are deep and expand relatively broadly in a lateral direction. Their posterior edges are defined by the inner edges of the interarea and are parallel to the hinge line. The lateral edges of the socket plates curve up sharply medially and recurve posteriorly on larger specimens to partly enclose the proximal ends of the sockets. The inner edges of the socket plates are united by a transverse crescentic plate which bears a ponderous posteriorly bilobed cardinal process. The cardinal plate is supported medially by a thickening of shell material which joins the base of the valve. The interior is commonly crenulated by the impress of the plications.

OCCURRENCE: In central Nevada *Cyrtina* cf. *varia* is present in the Rabbit Hill Limestone at Rabbit Hill and in the *Trematospira* Zone in the Roberts Mountains and in the Sulphur Spring Range. It also occurs in the *Spinoplasia* Zone at Antelope Peak. The species is more common in the *Trematospira* Zone.

FIGURED SPECIMENS: USNM 157299-157303.

Cyrtina sp. C
(Pl. 73, figs. 15-20, 21-26?)

EXTERIOR: The shells are subsemicircular to transversely subtrigonal in outline and unequally biconvex in lateral profile. The pedicle valve is strongly convex and subpyramidal. The hinge line is long and straight and is the place of maximum width. Cardinal angles are commonly acute and sharp or are approximately right angles. The ventral interarea is long, slightly curved, and apsacline to catacline. The beak ridges are sharp on some specimens, slightly rounded on others. The delthyrium is long, narrow, and triangular, enclosing an angle of about 25 degrees. It is covered along its basal half by a very gently convex deltidium.

The pedicle valve bears a broad, shallow, V-shaped sulcus. The brachial valve bears a corresponding, broad, low, rounded fold. The ventral sulcus is bounded laterally by a pair of well-defined rounded plications which are more strongly developed than plications on the flanks. There are commonly three or four low, rounded plications on each ventral flank, but only the median pair originates at the beak. The second, third, and fourth pairs of plications originate, respectively, further anterior to the umbo. The posterolateral slopes are relatively smooth. The plications are separated by shallow U-shaped interspaces and are crossed at irregular intervals by well-developed concentric growth lines.

INTERIOR OF PEDICLE VALVE: The hinge teeth are long, thinly elliptical, and pointed. They are supported basally by a pair of thin dental lamellae that converge medially to join a high triangular median septum. The median septum is long and may be slightly greater in length than the height of the interarea. Its posterior edge projects within the spondylium and pierces a well developed tichorhinum of suboval cross-section. There is a median ridge or septum along the outer (posterior) edge of the tichorhinum. The interior of the shell is lightly corrugated by the impress of the plications.

It is interesting to note in one specimen which bears a moderately twisted ventral beak, as is relatively common in the genus, that internally the median septum remains essentially perpendicular to the interarea so that its base traverses from a position equivalent to the center of the external sulcus, near the umbo, off the side of the sulcus nearly into the furrow corresponding to one of the external bounding plications at its anterior end.

INTERIOR OF BRACHIAL VALVE: The sockets diverge and expand anterolaterally. Their inner edges bear a pair of nearly horizontal plates which are pointed anteriorly forming the crural bases. There is a strong thickening of shell material at the apex of the valve divided into two strong lobes which project ventrally, serving as a site of diductor muscle attachment. The interior is lightly crenulated by the impress of the plications.

OCCURRENCE: This species is present in the *Acrospirifer kobehana* Zone in the Sulphur Spring Range. Probably also assignable are specimens from the *Acrospirifer kobehana* Zone on the north side of the Roberts Mountains, but these are smaller and more lamellose. In the absence of a growth series showing the small specimens of *Cyrtina* sp. C, whose description is based on the specimens from the Sulphur Spring Range, species from the northern Roberts Mountains are questionably assigned.

FIGURED SPECIMENS: USNM 157304-157307.

Cyrtina sp. D
(Pl. 72, figs. 16-20)

DISCUSSION: A few fragmentary pedicle valves of *Cyrtina* were recovered from the *Eurekaspirifer pinyonensis* Zone in the Cortez Mountains and are the basis for the short description given below.

EXTERIOR: Pedicle valves are roughly trigonal with a long hinge line and sharply acute cardinal angles. They are strongly convex, bear a long straight hinge line and a well-defined, nearly flat, triangular interarea. The delthyrium is long and narrow, enclosing an angle of about 20 degrees. There is a relatively deep sulcus medially and five or six narrow, well-defined, radial plications on each flank. The plications are crossed by concentric growth lines.

INTERIOR OF PEDICLE VALVE: The hinge teeth are very small and are supported only in the apex of the valve by dental lamellae. The dental lamellae are short and converge in a relatively short distance beneath the palintrope to join a short median septum. The median septum pierces the base of the spondylium formed by the dental lamellae, but it was not possible to be certain of the formation of a tichorhinum. The interior is moderately crenulated by the impress of the plications.

OCCURRENCE: In the *Eurekaspirifer pinyonensis* Zone in the Cortez Range and at Lone Mountain.

FIGURED SPECIMENS: USNM 157297, 157298.

Order Terebratulida

SUBORDER TEREBRATULOIDEA

DISCUSSION: Terebratulid brachiopods are less common than any of the other articulate brachiopod suborders present in the Lower Devonian of central Nevada, with the exception of the Productoidea. Considering worldwide distribution, both in the Old World and in eastern North America, terebratulids first appeared at the base of the Gedinnian.

In central Nevada no terebratulid brachiopod is known below the Siegenian *Spinoplasia* Zone where fragmentary remains of a centronellid, probably assignable to *Rensselaerina*, occur. A few occurrences of *Rensselaeria* sp. have been noted in the succeeding *Trematospira* and *Acrospirifer kobehana* Zones. *Rensselaerina*? sp. and *Rensselaeria* sp. of the *Spinoplasia* and *Trematospira* Zones are terebratulids that are typical of the Appalachian Province Lower Devonian (Boucot and others, 1967), and occur in brachiopod faunas of Appalachian Province affinity. The occurrence in the *Acrospirifer kobehana* Zone is a situation of an Appalachian Province genus occurring with other shells of a different provincial affinity. However, the *kobehana* Zone representatives of *Rensselaeria* may be regarded as holdovers like *Megakozlowskiella*, *Leptocoelia*, and *Anoplia*, which are uncommon and which apparently persist from older faunas that by their composition were an essential part of the Appalachian Province. The *Acrospirifer kobehana* Zone is believed to be lower Emsian and, therefore, younger than the highest (Oriskany) occurrence of *Rensselaeria* in eastern North America, and by position, to be the approximate equivalent of the *Etymothyris* Zone of eastern North America. No such correlation can be made or refuted at present on the basis of the brachiopods because brachiopod genera generally present as newly appearing elements in the *Etymothyris* and *Amphigenia* Zones of the Appalachian Province Emsian do not occur in Nevada. A single specimen, probably assignable to *Mutationella*, has been found in the *Eurekaspirifer pinyonensis* Zone of the Sulphur Spring Range. As *Mutationella* is typical of Old World faunas rather than Appalachian Province faunas, its occurrence in that zone instead of a typical Appalachian Province genus, such as *Amphigenia*, is consistent. Most of the other *pinyonensis* Zone elements, either are endemic or have their affinities with Old World faunas.

Superfamily TEREBRATULACEA Waagen 1883
Family DIELASMATIDAE Schuchert, 1913
Subfamily MUTATIONELLINAE Cloud, 1942
Genus *Mutationella* Kozlowski, 1929

TYPE SPECIES: *Waldheimia podolica* Siemiradzki 1906, p. 177, Pl. 7, fig. 10.

Mutationella? sp.
(Pl. 74, figs. 15-19)

EXTERIOR: A single specimen is available for study. It is elongate suboval in outline and biconvex in lateral profile. The beak is prominent, blunt, and nearly straight. The foramen is large, circular, and mesothyrid. A well-developed deltidium completely fills the delthyrium anterior to the foramen. Neither fold nor sulcus is developed; however, there is a gentle deflection of the commissure medially toward the brachial valve. The commissure is crenulated by the radial costae. Approximately the posterior half of the shell is smooth, crossed only by very fine concentric growth lines. Anteriorly, there are well-developed radial costae, about 33 in number on the available specimen. The costae are rounded in cross section and are separated by U-shaped interspaces of approximately the same width as the adjoining costae.

SHELL SUBSTANCE: The shell material is finely endopunctate on both the smooth and ribbed portions of the valves and on the deltidium.

FIGURED SPECIMEN: USNM 157308.

Family CENTRONELLIDAE Waagen, 1882
Subfamily RENSSELAERIINAE Raymond, 1923
Genus *Rensselaerina* Dunbar, 1917

TYPE SPECIES: *Rensselaerina medioplicata* Dunbar, 1917, p. 469.

Rensselaerina? sp.
(Pl. 74, figs. 1-7)

DISCUSSION: It is impossible to be certain of the generic assignment of the species described here since no specimens showing the anterior of the shell have been found. The anteromedial portions of *Rensselaerina* are characterized by relatively prominent costae, which are not developed on the lateral portions of the shell. The assignment here is based on shell shape, pedicle valve interior, and on the structure of the cardinalia of the brachial valve. As discussed by Cloud (1942, p. 51), the cardinal plate of *Rensselaerina* is variable, commonly thickened, and nearly sessile. These are characters found in the specimens from the Rabbit Hill Limestone.

EXTERIOR: No available shell is preserved well enough to allow adequate description of the shape; however, the one pedicle valve at hand tends to be elongate oval judging from its convexity. The shells are unequally biconvex with the pedicle valve decidedly more convex than the brachial valve. The beak is small and nearly straight. The delthyrium encloses an angle of approximately 45 degrees anterior to an apical foramen. No deltidial plates are present. Only smooth posterior sections of the shell are available for study.

INTERIOR OF PEDICLE VALVE: The teeth are small and blunt and project slightly dorsomedially. They are supported by very small, stout, nearly obsolescent dental lamellae which are set closely against the thickened inner lateral walls of the shell. The pedicle cavity is bounded laterally by the dental lamellae, and medially is continuous with the diductor muscle impressions. The diductor muscle scars are deeply impressed along their posterolateral edges and their lateral edges diverge only slightly anteriorly. Medially, there is a sharp myophragm continuing from the anterior portion of the pedicle cavity into the posterior portion of the diductor impression. The diductor impressions blend imperceptibly anteriorly with the interior of the valve.

INTERIOR OF BRACHIAL VALVE: The socket plates originate against the posterolateral portions of the shell interior and, at the bases of the sockets, lie almost on the base of the valve. Medially, they rise slightly above the floor of the valve and join the posterolateral edges of the cardinal plate. The cardinal plate comprises three triangular divisions, the lateral pair has its long sides medially and the medial portion is equilateral in outline. Its anterior edge is commonly concave; apically there is an elongate-oval foramen. The cardinal plate is somewhat ponderous and on the thicker shells may be strongly thickened and raised along its anteromedial edge. Short crural plates support the cardinal plate posteriorly, but the cardinal plate is situated only slightly above the inner surface of the valve, and upon the deposition of a small amount of shell material, it appears to be sessile. The adductor muscle scars are impressed anterior to the distal edge of the cardinal plate and form a pair of elongate impressed scars.

OCCURRENCE: *Rensselaerina*? sp. is known only from the *Spinoplasia* Zone in the Monitor Range and at Antelope Peak.

FIGURED SPECIMENS: USNM 157309-157313.

Genus *Rensselaeria* Hall, 1859

TYPE SPECIES: *Atrypa elongata* Conrad, 1839, p. 65 (synonym of *Terebratula ovoides* Eaton, 1832, which was named as type species by Miller, 1889, p. 366, but which is a junior homonym of *Terebratula ovoides* Sowerby, 1812).

Rensselaeria sp.
(Pl. 74, figs. 8-14)

EXTERIOR: The shells are of only moderate size for the genus. They are elongate suboval in outline and gently subequally biconvex in lateral profile. The ventral beak is prominent and nearly straight rather than strongly incurved. The delthyrium is relatively broad and encloses an angle of approximately 70 to 75 degrees. The deltidial plates were not observed. There is an apical (mesothyrid?) foramen.

The umbonal regions are smooth, and the anterior portion of the shells is ornamented with low, rounded, radial costellae, the same strength on the flanks as on the medial regions. On one brachial valve there are 12 costellae in a space of 5 mm, 15 mm anterior to the beak. There appears to be no development of a fold or sulcus, and the anterior commissure is rectimarginate.

INTERIOR OF PEDICLE VALVE: The hinge teeth are supported by short but well-defined dental lamellae which converge slightly toward the midline of the pedicle valve and define the pedicle cavity. The diductor muscle field broadens slightly at the anterior of the pedicle cavity and is lightly impressed for some distance anteriorly. Its sides are rectilinear, diverging only slightly anteriorly. There as a well-developed myophragm beginning at the anterior of the pedicle cavity and extending through most of the muscle impression.

INTERIOR OF BRACHIAL VALVE: The socket plates are attached directly to the posterolateral portions of the shell interior and curve up gently toward the pedicle valve to define the sockets. A tripartite cardinal plate of modified pentagonal outline is situated between the inner edges of the socket plates. Laterally there are two triangular divisions with their long sides medial. The middle portion of the cardinal plate is equilaterally triangular with the base of the triangle concave and facing anteriorly. The anterior edge of the medial portion of the cardinal plate is commonly slightly arched toward the pedicle valve. The cardinal plate is supported by stout crural plates which lie beneath the sutures dividing the three portions at angles approximately normal to the inner surface of the valves. There generally is a circular or elongate foramen developed at the posterior end of the cardinal plate. The adductor scars are more or less well impressed somewhat anterior to the cardinalia and consist of a pair of elongate oval impressions separated by a stout myophragm.

OCCURRENCE: *Rensselaeria* sp. it at present known from the *Trematospira* Subzone of the *Trematospira* Zone at the base of the McColley Canyon Formation at McColley Canyon, Sulphur Spring Range, and from the *Acrospirifer kobehana* Zone at Lone Mountain. Thick-shelled individuals probably belonging to *Rensselaeria,* are known from a collection of reworked shells from the basal sandy member of the Sultan Limestone in the northern Nopah Range, southeastern California (Boucot and others, 1969).

DISCUSSION: Cloud (1942, p. 55) noted that *Rensselaeria* comprises "two externally dissimilar groups; one of forms with subcircular outline and thinly elliptical cross-section, typified by *R. cayuga;* a second of forms with elongately subrectangular or subelliptical outline and subcircular to subquadrate cross-section, typified by the genotype." The Nevada form belongs to the former group along with *R. cayuga* and possibly *R. ovulum.*

The species from Nevada has smooth umbones and thus is not conspecific with either *R. cayuga* or *R. ovulum.*

FIGURED SPECIMENS: USNM 157314-157318.

PLATE SECTION

PLATE 1. "DOLERORTHIS" AND ORTHOSTROPHELLA

Figures 1-11 *"Dolerorthis"* sp.
  Quadrithyris Zone, Coal Canyon, loc. 10758.
  1, 2 Exterior and interior of brachial valve × 1.5, USNM 156749.
  3, 4 Exterior and interior of brachial valve × 1.5, USNM 156750.
  5, 6 Interior and exterior of pedicle valve × 1.5, USNM 156751.
  7 Exterior of brachial valve × 2, USNM 156752.
  8 Posterior view of pedicle valve × 3, USNM 156753.
  9 Posterior view of pedicle valve × 3, USNM 156754.
  10, 11 Interior and exterior views of brachial valve × 2, USNM 156755.

Figures 12-22 *Orthostrophella monitorensis* Johnson, n. sp.
  Spinoplasia Zone, figures 12-21, loc. 10713, Rabbit Hill; figure 22, loc. 10766, Toquima Range.
  12, 13 Exterior and interior of pedicle valve × 3, paratype, USNM 156757.
  14, 15 Exterior and interior of pedicle valve × 2, paratype, USNM 156758.
  16 Interior of brachial valve × 2, paratype, USNM 156759.
  17 Interior of brachial valve × 2, paratype, USNM 156760.
  18, 19 Interior and exterior of brachial valve × 2, paratype, USNM 156761.
  20, 21 Interior and exterior of pedicle valve × 2, holotype, USNM 156756.
  22 Internal mold of pedicle valve × 1.5, USNM 156762.

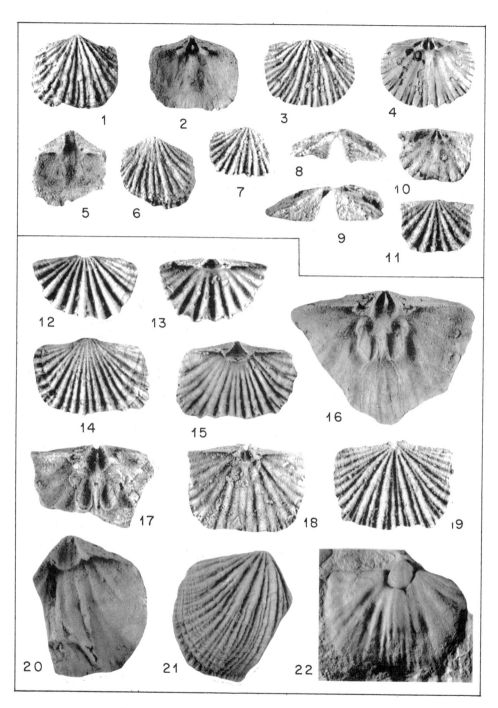

"DOLERORTHIS" AND ORTHOSTROPHELLA

JOHNSON, PLATE 1
Geological Society of America Memoir 121

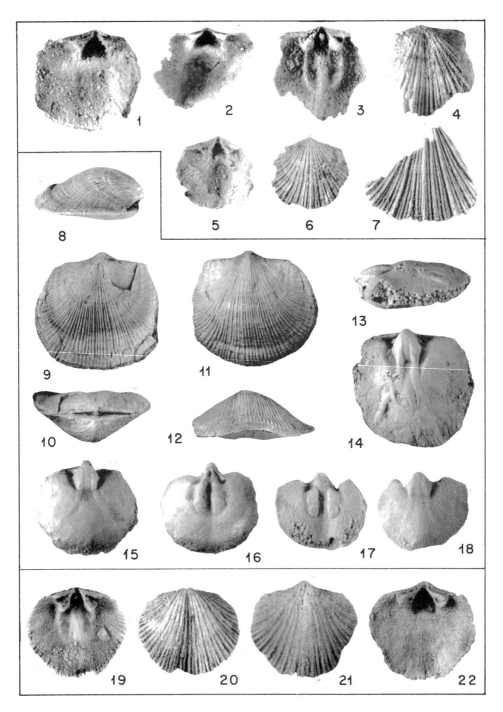

LEVENEA

JOHNSON, PLATE 2
Geological Society of America Memoir 121

PLATE 2. LEVENEA

Figures 1-7 *Levenea* sp. A.
  Quadrithyris Zone, Coal Canyon.
  1 Interior of pedicle valve × 4, USNM 156763, loc. 10757.
  2 Interior of pedicle valve × 3, USNM 156764, loc. 10758.
  3, 4 Interior and exterior of brachial valve × 3, USNM 156765, loc. 10758.
  5, 6 Interior and exterior of pedicle valve × 2, USNM 156766, loc. 10758.
  7 Ventral view of part of exterior × 3, USNM 156767, loc. 10758.

Figures 8-18 *Levenea fagerholmi* Johnson, n. sp.
  Eurekaspirifer pinyonensis Zone, Coal Canyon.
  8-12 Side, dorsal, posterior, ventral, and anterior views × 1.25, holotype, USNM 156768, loc. 10762.
  13, 14 Side and ventral views of internal mold × 1.5, paratype, USNM 156769, loc. 10762.
  15, 16 Ventral and dorsal views of internal mold × 1.5, USNM 156770, loc. 10761, *Leptocoelia infrequens* Subzone.
  17, 18 Dorsal and ventral views of internal mold × 2, USNM 156771, loc. 10761, *Leptocoelia infrequens* Subzone.

Figures 19-22 *Levenea navicula* Johnson, n. sp.
  Spinoplasia Zone, Rabbit Hill.
  19, 20 Interior and exterior of brachial valve × 3, USNM 156773, loc. 10713.
  21, 22 Exterior and interior of pedicle valve × 4, USNM 156774, loc. 10713.

PLATE 3. LEVENEA

Figures 1-19 *Levenea navicula* Johnson, n. sp.
  1 Posterior view of cardinalia of brachial valve × 3, USNM 156779, loc. 10713, *Spinoplasia* Zone, Rabbit Hill. Note lamellose bilobate cardinal process.
  2, 3 Interior and exterior of brachial valve × 1.5, USNM 156778, loc. 10730, *Trematospira* Zone, Sulphur Spring Range.
  4 Interior of brachial valve × 3, USNM 156780, loc. 10715, *Trematospira* Zone, northern Roberts Mountains.
  5-7 Interior, exterior, and posterior views of brachial valve × 2, USNM 156775, loc. 10713, *Spinoplasia* Zone, Rabbit Hill.
  8, 9 Interior and exterior of a small pedicle valve × 5, USNM 156776, loc. 10713, *Spinoplasia* Zone, Rabbit Hill.
  10 Rubber impression of specimen shown in figure 5 × 2.
  11, 12 Interior and posterior of pedicle valve × 2, USNM 156777, loc. 10713, *Spinoplasia* Zone, Rabbit Hill.
  13, 14 Ventral and side views of rubber impression of specimen in figure 11 × 2.
  15-19 Dorsal, ventral, side, anterior, and posterior views × 2, holotype, USNM 156772, loc. 10715, *Trematospira* Zone, northern Roberts Mountains.

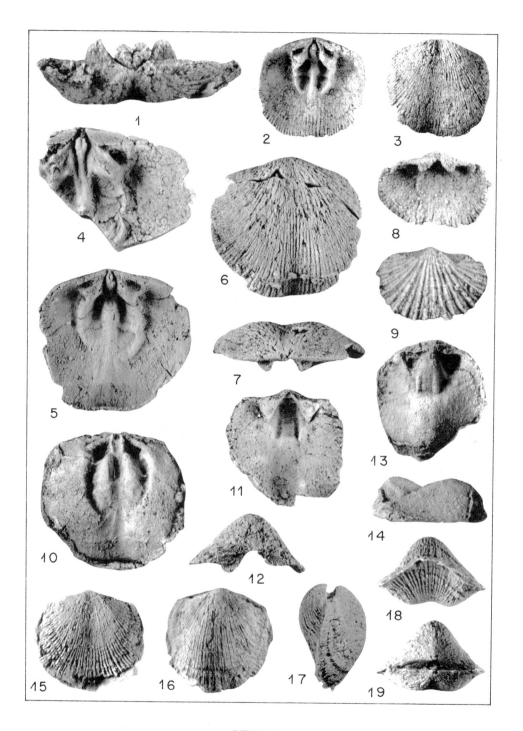

LEVENEA

JOHNSON, PLATE 3
Geological Society of America Memoir 121

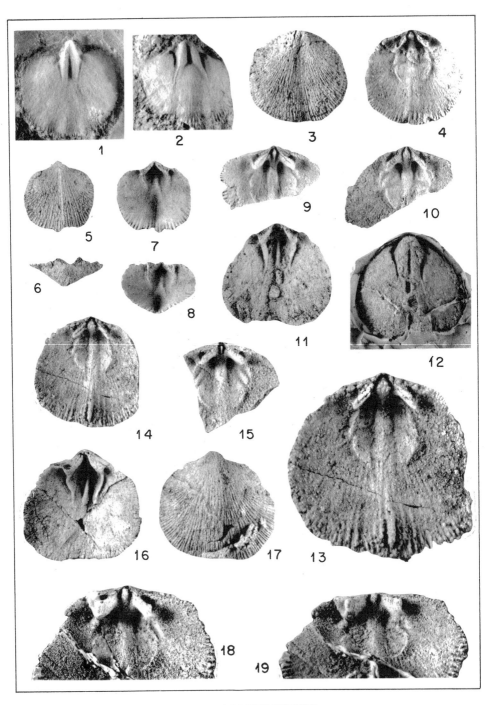

PROTOCORTEZORTHIS

PLATE 4. PROTOCORTEZORTHIS

Figures 1, 2 *Protocortezorthis* sp.
"Tor" Limestone, Toquima Range.
1 Internal mold of pedicle valve × 2, USNM 156783, loc. 10763.
2 Internal mold of pedicle valve × 2, USNM 156784, loc. 10763.

Figures 3-19 *Protocortezorthis windmillensis* Johnson and Talent.
Quadrithyris Zone, Coal Canyon.
3, 4 Exterior and interior of brachial valve × 1.5, paratype, USNM 147343, loc. 10758.
5-8 Exterior, posterior, interior, and anterior views of pedicle valve × 2, paratype, USNM 147344, loc. 10758.
9 Interior of brachial valve × 2, paratype, USNM 156785, loc. 10758.
10 Interior of brachial valve × 1.5, paratype, USNM 147340, loc. 10758.
11, 12 Interior of pedicle valve and rubber impression of interior of pedicle valve × 1.5, paratype, USNM 147342, loc. 10758.
13, 14 Interior of brachial valve × 2.5 and 1.5, holotype, USNM 147339, loc. 10758.
15 Interior of brachial valve × 2, paratype, USNM 147341, loc. 10758.
16, 17 Interior and exterior of pedicle valve × 1.5, USNM 147345, loc. 10757.
18, 19 Two views of interior of brachial valve × 3, USNM 147338, loc. 10758.

PLATE 5. CORTEZORTHIS

Figures 1-12 *Cortezorthis cortezensis* Johnson and Talent.
Eurekaspirifer pinyonesis Zone.
1-5 Ventral, dorsal, side, anterior, and posterior views × 1.75, USNM 156786, loc. 13270, Lone Mountain.
6-10, Side, dorsal, anterior, ventral, and posterior views of internal mold × 1.25, holotype, USNM 159069, loc. 10762, northern Simpson Park Range.
11 Rubber impression of specimen in figure 7 × 1.25.
12 Interior of brachial valve × 2, USNM 159066, loc. 10752, Cortez Range.

Figures 13-16 *Cortezorthis* aff. *bathurstensis* Johnson and Talent.
Beds of Emsian age, 100 feet above the base of the Blue Fiord Formation, south bank of Sutherland River, lat. 76°19'; long. 92°51', Prince Alfred Bay area, Devon Island, USNM loc. 12346. Two other specimens from this locality were illustrated by Johnson and Talent (1967b, Pl. 19, figs. 21-23).
13 Side view of dorsal interior × 4, GSC 19614.
14-16 Anterodorsal, anterior, and interior views of brachial valve × 4, GSC 19615.

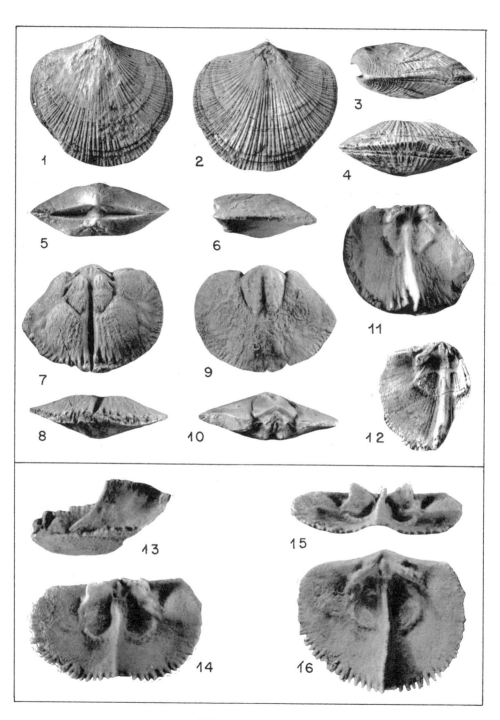

CORTEZORTHIS

JOHNSON, PLATE 5
Geological Society of America Memoir 121

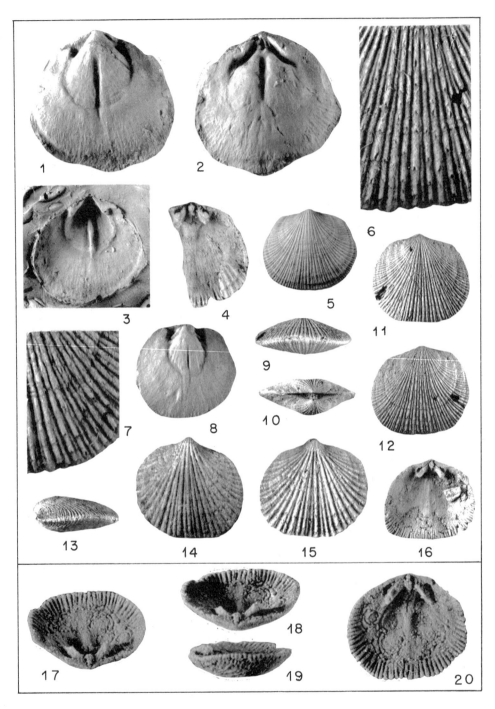

DALEJINA AND RHIPIDOMELLA

PLATE 6. DALEJINA AND RHIPIDOMELLA

Figures 1-16 *Dalejina* sp. C
  Eurekaspirifer pinyonensis Zone.
  1-3 Ventral and dorsal views of internal mold × 2, and rubber impression of internal mold of pedicle valve × 1.5, USNM 156788, loc. 10705, northern Roberts Mountains.
  4 Interior of brachial valve × 4, USNM 156789, loc. 12771, Roberts Mountains.
  5 Ventral view × 1.5, USNM 156790, loc. 12393, northern Simpson Park Range.
  6, 7 Ventral and dorsal views of portions of exterior × 5, USNM 156791, loc. 12771, Roberts Mountains.
  8 Internal mold of pedicle valve × 2, USNM 156792, loc. 10761. *Leptocoelia infrequens* Subzone, Coal Canyon.
  9-13 Anterior, posterior, dorsal, ventral, and side views × 1.5, USNM 156791, loc. 12771, Roberts Mountains.
  14, 15 Ventral and dorsal views × 4, USNM 156793, loc. 12771, Roberts Mountains.
  16 Interior of brachial valve × 2, USNM 156794, loc. 13267, Sulphur Spring Range.

Figures 17-20 *Rhipidomella* sp.
  Eurekaspirifer pinyonensis Zone, loc. 10729, Sulphur Spring Range.
  17-20 Posterodorsal (two views), side, and interior views of brachial valve × 2, USNM 156787.

PLATE 7. DALEJINA AND DISCOMYORTHIS

Figures 1-22 *Dalejina* sp. B.
    1, 2 Exterior and interior of pedicle valve × 2, USNM 156795, loc. 10713, *Spinoplasia* Zone, Rabbit Hill.
    3, 4 Rubber impression of interior of pedicle valve and interior of pedicle valve × 1.5, USNM 156796, loc. 10713, *Spinoplasia* Zone, Rabbit Hill.
    5 Interior of brachial valve × 2, USNM 156797, loc. 10713, *Spinoplasia* Zone, Rabbit Hill.
    6-9 Dorsal, ventral, side, and anterior views × 2, USNM 156798, loc. 10713, *Spinoplasia* Zone, Rabbit Hill.
    10 Interior of pedicle valve × 1.5, USNM 156799, loc. 10713, *Spinoplasia* Zone, Rabbit Hill.
    11 Interior of pedicle valve × 1.5, USNM 156800, loc. 12793, *Trematospira* Zone, McColley Canyon.
    12, 13 Dorsal and anterior views of interior of brachial valve × 3, USNM 156801, loc. 12793, *Trematospira* Zone, McColley Canyon.
    14-18, Anterior, ventral, dorsal, side, and posterior views × 1.5, USNM 156802, loc. 10726, *Trematospira* Subzone, Sulphur Spring Range.
    19, 20 Dorsal and anterior views of interior of brachial valve × 4, USNM 156803, loc. 12793, *Trematospira* Zone, McColley Canyon.
    21, 22 Dorsal and ventral views × 3, USNM 156804, loc. 10744, *Trematospira* Zone, Sulphur Spring Range.

Figures 23-26 *Discomyorthis musculosa* (Hall).
    Trematospira Subzone, McColley Canyon, loc. 12793.
    23-26 Posterior, side, exterior, and interior of pedicle valve × 1, USNM 156805.

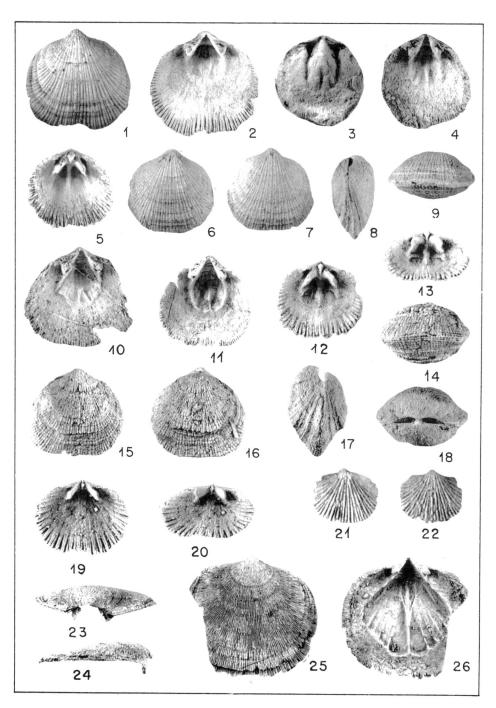

DALEJINA AND DISCOMYORTHIS

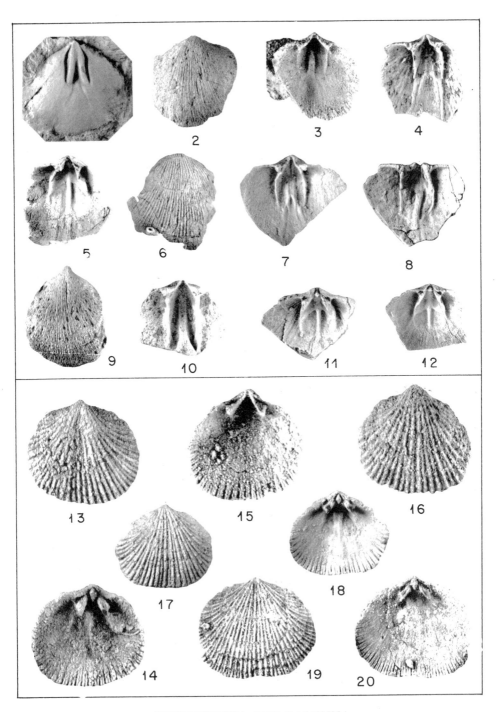

SCHIZOPHORIA AND DALEJINA

JOHNSON, PLATE 8
Geological Society of America Memoir 121

PLATE 8. SCHIZOPHORIA AND DALEJINA

Figures 1-12 *Schizophoria parafragilis* Johnson, n. sp.
Fig. 1, "Tor" Limestone, Sawlog Ridge; figs. 3, 5, 6, *Quadrithyris* Zone, Coal Canyon; figs. 2, 4, 7-12, *Quadrithyris* Zone, Ikes Canyon. "Tor" Limestone, Sawlog Ridge.
1 Internal mold of pedicle valve × 1.5, USNM 156810, loc. 10763.
2 Exterior of brachial valve × 1.5, USNM 156811, paratype, loc. 10808.
3 Interior of pedicle valve × 1.5, USNM 156812, loc. 10758.
4 Interior of pedicle valve × 2, USNM 156813, paratype, loc. 10808.
5, 6 Interior and exterior of brachial valve × 2, USNM 156814, loc. 10757.
7 Interior of pedicle valve × 2, USNM 156815, paratype, loc. 10808.
8 Interior of pedicle valve × 2.5, USNM 156816, paratype, loc. 10808.
9 Exterior of pedicle valve × 1.5, USNM 156817, paratype, loc. 10808.
10 Interior of pedicle valve × 2, USNM 156818, paratype, loc. 10808.
11 Interior of brachial valve × 2, USNM 156819, paratype, loc. 10808.
12 Interior of brachial valve × 2, USNM 156820, holotype, loc. 10808.

Figures 13-20 *Dalejina* sp. A.
Quadrithyris Zone, Coal Canyon.
13, 14 Exterior and interior of brachial valve × 4, USNM 156806, loc. 10757.
15, 16 Interior and exterior of pedicle valve × 4, USNM 156807, loc. 10757.
17, 18 Exterior and interior of brachial valve × 3, USNM 156808, loc. 10758.
19, 20 Exterior and interior of brachial valve × 3, USNM 156809, loc. 10757.

PLATE 9. SCHIZOPHORIA

Figures 1-18 *Schizophoria nevadaensis* **Merriam**.
Eurekaspirifer pinyonensis Zone.
1-3, 5, 8 Dorsal, ventral, side, anterior, and posterior views of paratype × 1, USNM 96376, Lone Mountain.
4, 6, 7, 9, 10 Side, anterior, posterior, dorsal, and ventral views of holotype × 1, USNM 96375, Lone Mountain.
11 Internal mold of brachial valve × 1.5, USNM 156821, loc. 10788, Roberts Mountains.
12 Internal mold of pedicle valve × 2, USNM 156822, loc. 10788, Roberts Mountains.
13, 14 Posterior and dorsal views of internal mold of brachial valve × 1.5, USNM 156823, loc. 10788, Roberts Mountains.
15 Internal mold of pedicle valve × 1, USNM 156824, loc. 10775, Roberts Mountains.
16 Internal mold of pedicle valve × 1, USNM 156825, loc. 10779, Roberts Mountains.
17, 18 Internal mold of pedicle valve and rubber impression of internal mold of pedicle valve × 1.5, USNM 156826, loc. 10762, Coal Canyon.

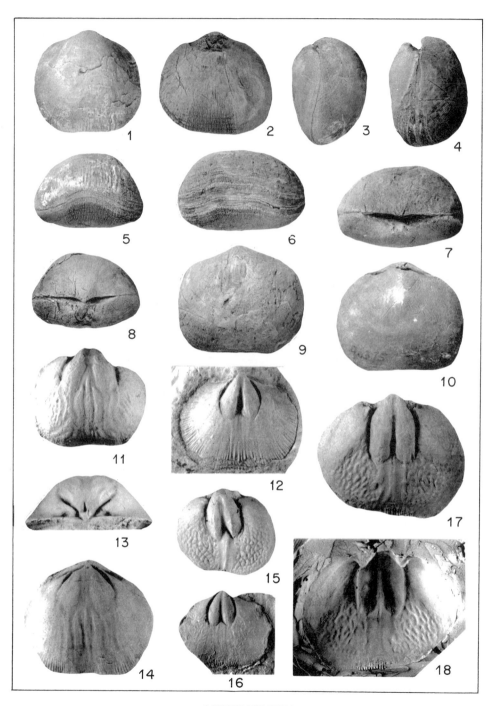

SCHIZOPHORIA

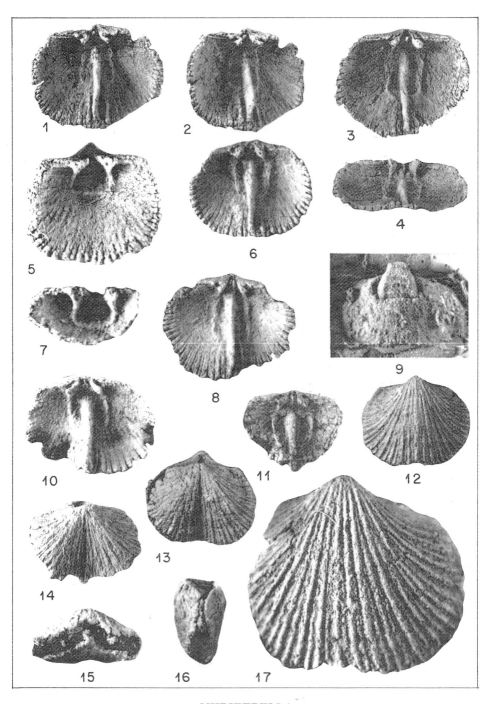

MURIFERELLA

JOHNSON, PLATE 10
Geological Society of America Memoir 121

PLATE 10. MURIFERELLA

Figures 1-17 *Muriferella masurskyi* Johnson and Talent.
  Eurekaspirifer pinyonensis Zone, Cortez Mountains.
  1, 2 Plan and oblique views of dorsal interior × 6, USNM 162009, loc. 10752.
  3, 4 Plan and anterior views of dorsal interior × 6, USNM 162010, loc. 10752.
  5 Interior of pedicle valve × 7, USNM 162011, loc. 10754.
  6 Interior of brachial valve × 6, USNM 162014, loc. 10752.
  7 Anterior view of ventral interior × 6, USNM 156827, loc. 10754.
  8 Interior of brachial valve × 6, USNM 162013, loc. 10752.
  9 Rubber mold of interior of pedicle valve × 5, impression of USNM 162012, loc. 10752.
  10 Interior of brachial valve × 6, USNM 156828, loc. 10754.
  11 Interior of brachial valve × 4, USNM 156829, loc. 10754.
  12, 13 Ventral and dorsal views × 4, USNM 162007, loc. 10752.
  14-16 Dorsal, anterior, and side views × 6, USNM 156830, loc. 10754.
  17 Exterior of pedicle valve × 10, USNM 156831, loc. 10752.

PLATE 11. ANASTROPHIA AND "CAMERELLA"

Figures 1-8 *Anastrophia* cf. *magnifica* Kozlowski.
    Quadrithyris Zone.
  1 Exterior of pedicle valve × 3, USNM 156832, loc. 10758, Coal Canyon.
  2, 3 Interior × 3 and exterior × 2 of pedicle valve, USNM 156833, loc. 10758, Coal Canyon.
  4, 5 Two views of interior of pedicle valve × 3, USNM 156834, loc. 10758, Coal Canyon.
  6 Internal mold of pedicle valve × 1.5, USNM 156835, loc. 10765, Toquima Range.
  7, 8 Posterior and dorsal views of internal mold of brachial valve × 2, USNM 156836, loc. 10765, Toquima Range. Note pair of subparallel slots of the outer plates.

Figures 9-18 "*Camerella*" sp.
    Spinoplasia Zone, Monitor Range.
  9, 10 Exterior and interior of pedicle valve × 2, USNM 156837, loc. 10713.
  11 Interior of pedicle valve × 4, USNM 156838, loc. 10713.
  12 Interior of brachial valve × 3, USNM 156839, loc. 10712.
  13, 14 Exterior and interior of pedicle valve × 3, USNM 156840, loc. 10713.
  15 Interior of brachial valve × 5, USNM 156841, loc. 10713.
  16, 17 Exterior and interior of brachial valve × 4, USNM 156842, loc. 10713.
  18 Interior of pedicle valve × 4, USNM 156843, loc. 10713.

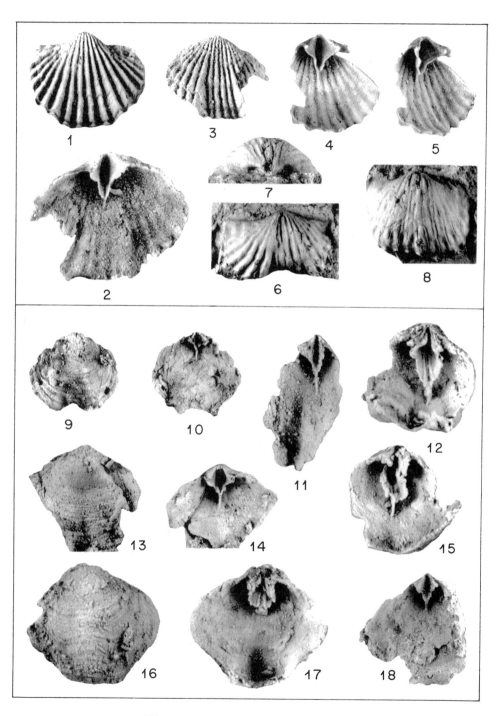

ANASTROPHIA AND "CAMERELLA"

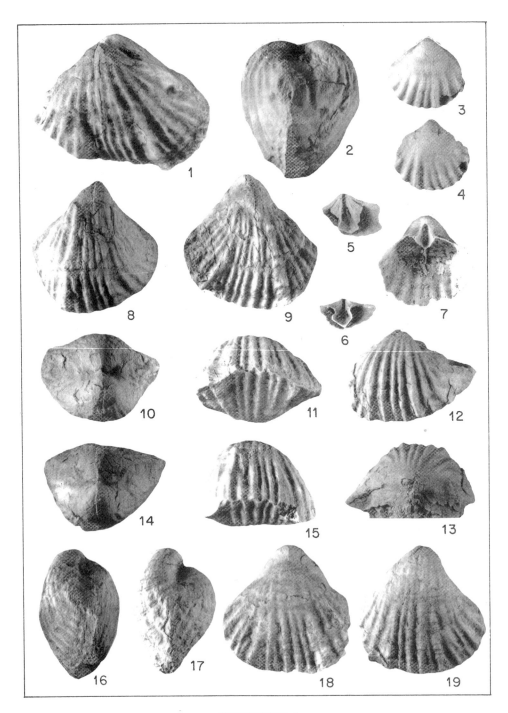

SIEBERELLA

PLATE 12. SIEBERELLA

Figures 1-19 *Sieberella pyriforma* Johnson n. sp.
  Trematospira Zone, Sulphur Spring Range.
  1, 2 Dorsal and side views × 1.5, USNM 156844, loc. 10707.
  3, 4 Ventral and dorsal views of a small specimen × 2, USNM 156845, loc. 10745.
  5, 6 Two views of interior of brachial valve × 1.5, USNM 156846, loc. 10744.
  7 Interior of pedicle valve × 1.5, USNM 156847, loc. 10744.
  8, 9, 14, 15, 16 Ventral, dorsal, posterior, anterior, and side views × 1.5, USNM 156848, loc. 10707.
  10, 11, 17-19 Posterior, anterior, side, dorsal, and ventral views × 1.5, holotype, USNM 156850, loc. 10707.
  12, 13 Ventral and posterior views of pedicle valve × 1.5, USNM 156849, loc. 10707.

PLATE 13. SIEBERELLA, "CAMERELLA"?, AND GYPIDULA

Figures 1-5 *Sieberella* cf. *problematica* (Barrande).
   Quadrithyris Zone, Coal Canyon, loc. 10758.
   1, 2 Exterior and interior of pedicle valve × 1.5, USNM 156851.
   3 Ventral view × 1.5, USNM 156852.
   4 Ventral view × 1.5, USNM 156853.
   5 Interior of brachial valve × 2, USNM 156854.

Figures 6, 7 *"Camerella"?* sp.
   Quadrithyris Zone, Coal Canyon, loc. 10758.
   6, 7 Exterior and interior of pedicle valve × 3, USNM 156855.

Figures 8-17 *Gypidula loweryi* Merriam.
   Eurekaspirifer pinyonensis Zone.
   8, 9 Anterior and ventral views × 1, USNM 156856, loc. 10762, northern Simpson Park Range. Note the very faint plications on the fold anteriorly.
   10-13 Anterior, posterior, dorsal, and ventral views × 1.5, USNM 156857, loc. 10762, northern Simpson Park Range.
   14 Posterior view of the spondylium × 5, USNM 156858, loc. 10754, Cortez Range. Note fine longitudinal grooves at the base of the spondylium.
   15 Interior of brachial valve × 1.5, USNM 156859, loc. 10754, Cortez Range.
   16, 17 Side and interior views of brachial valve × 1.5, USNM 156860, loc. 10754, Cortez Range.

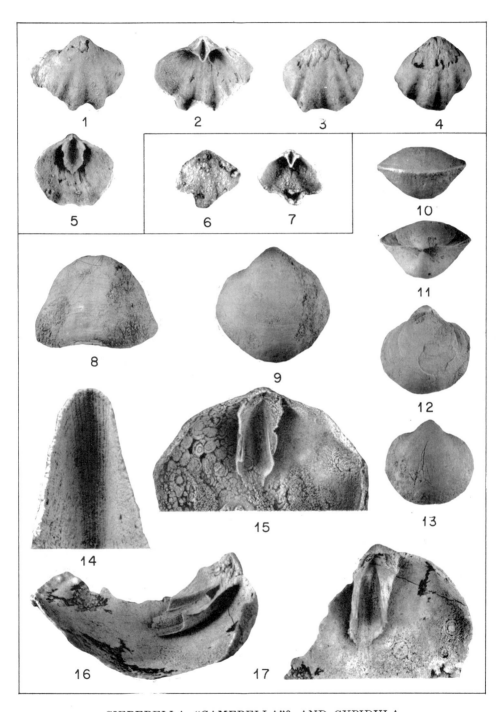

SIEBERELLA, "CAMERELLA"?, AND GYPIDULA

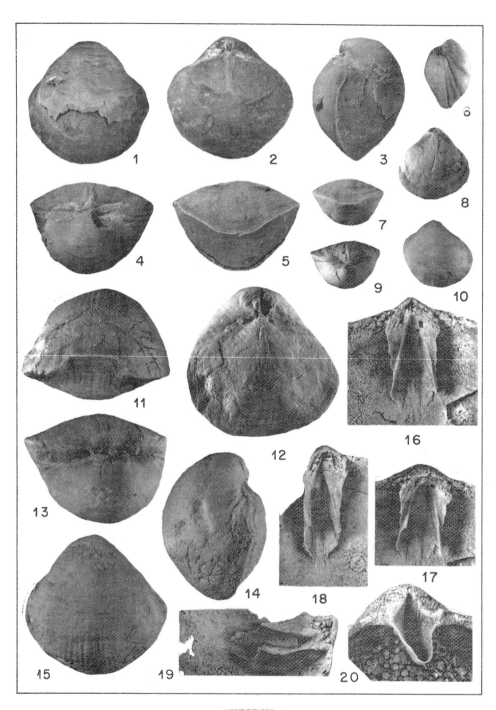

GYPIDULA

PLATE 14. GYPIDULA

Figures 1-20 *Gypidula loweryi* Merriam.
Eurekaspirifer pinyonensis Zone.
1-5 Ventral, dorsal, side, posterior, and anterior views × 1 of the holotype, USNM 96368, Lone Mountain. Shell material remains only on ventral posterior half of specimen.
6-10 Side, anterior, dorsal, posterior, and ventral views × 1, USNM 156861, loc. 10769, northern Roberts Mountains.
11-15 Anterior, dorsal, posterior, side, and ventral views × 1.5, USNM 156862, loc. 10769, northern Roberts Mountains. Note faint plications on the anteromedial portions of the shell.
16, 17 Two views of interior of brachial valve × 1.5, USNM 156863, loc. 10754, Cortez Range. Note well developed carinae.
18, 19 Two views of interior of brachial valve × 1.5, USNM 156860, loc. 10754, Cortez Range. Note pair of divergent ridges (cardinal process lobes) in the apex of the valve.
20 Spondylium of pedicle valve × 1.5, USNM 156858, loc. 10754, Cortez Range. This specimen is also illustrated in figure 14, Plate 13.

PLATE 15. GYPIDULA

Figures 1-8 *Gypidula praeloweryi* Johnson n. sp.
Acrospirifer kobehana Zone.
1-7 northern Roberts Mountains; 8, Sulphur Spring Range.
1-5 Ventral, dorsal, side, posterior, and anterior views × 1, holotype, USNM 156867, loc. 10772.
6 Anterior view of internal mold × 1, USNM 156865, loc. 10705.
7 Dorsal view of internal mold × 1, USNM 156866, loc. 10705.
8 Interior of articulated valves × 1.5, USNM 156864, loc. 10710.

Figures 9-11 *Gypidula* cf. *pseudogaleata* (Hall).
Quadrithyris Zone, Toquima Range.
9-11 Ventral, anterior, and side views × 1.25, USNM 157015, loc. 10808.

Figures 12-15 *Gypidula pelagica* (Barrande).
Gedinnian age beds, upper Roberts Mountains Formation, northern Roberts Mountains.
12-15 Anterior, dorsal, ventral, and side views × 1.25, USNM 157016, loc. 12328.

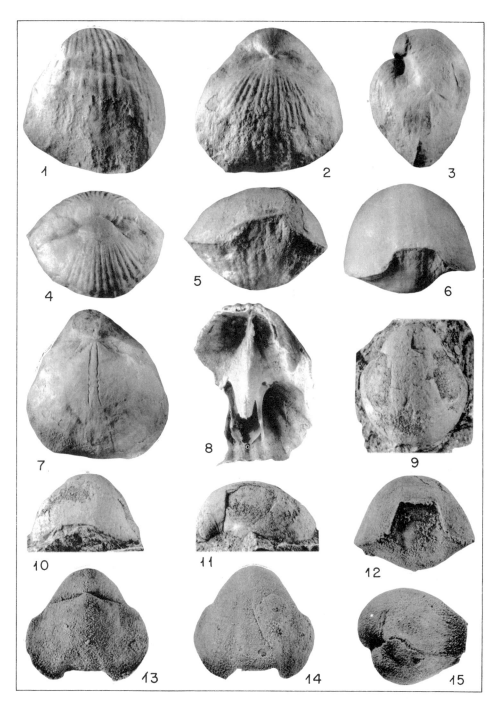

GYPIDULA

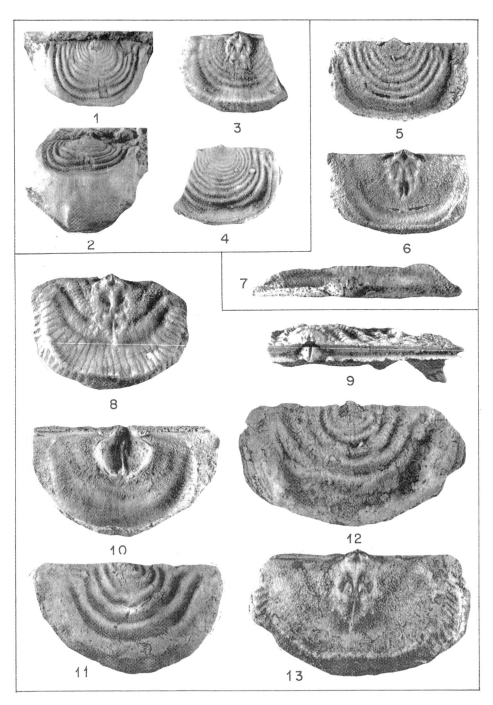

LEPTAENA

PLATE 16. LEPTAENA

Figures 1-4 *Leptaena* sp. A.
    1, 2 Ventral and anterior views × 1.5, USNM 156868, loc. 10708 *Quadrithyris* Zone, Toquima Range.
    3, 4 Interior and exterior of brachial valve × 2, USNM 156869, loc. 10758, *Quadrithyris* Zone, Coal Canyon.

Figures 5-7 *Leptaena* sp. C
    Acrospirifer kobehana Zone, Sulphur Spring Range.
    5-7 Exterior and interior views × 1 and posterior view × 1.5 of brachial valve, USNM 156870, loc. 10739.

Figures 8-13 *Leptaena* cf. *acuticuspidata* Amsden.
    8 Interior of brachial valve × 1.5, USNM 156871, loc. 10713, *Spinoplasia* Zone, Rabbit Hill.
    9 Posterior view × 2, USNM 156872, loc. 10726, *Trematospira* Zone, Sulphur Spring Range. Note the medially grooved chilidium.
    10, 11 Interior and exterior views of pedicle valve × 1.5, USNM 156873, loc. 10712, *Spinoplasia* Zone, Monitor Range.
    12, 13 Exterior and interior of brachial valve × 1.5, USNM 156874, loc. 10713, *Spinoplasia* Zone, Rabbit Hill.

PLATE 17. AESOPOMUM AND LEPTAENISCA

Figures 1-8 *Aesopomum* cf. *varistriatus* Johnson n. sp.
    1-3 Anterior view × 4, dorsal view × 3, and interior view × 2 of brachial valve, USNM 156880, loc. 10758, *Quadrithyris* Zone, Coal Canyon.
    4 Anterior view of interior of brachial valve × 3, USNM 156881, loc. 10758, *Quadrithyris* Zone, Coal Canyon.
    5, 6 Exterior and interior views of brachial valve × 3, USNM 156882, loc. 10758, *Quadrithyris* Zone, Coal Canyon.
    7, 8 Interior and exterior views of brachial valve × 3, USNM 156883, upper Roberts Mountains Formation, Lynn Window.

Figures 9-14 *Aesopomum varistriatus* Johnson n. sp.
    9, 10 Anterior and posterior views of brachial valve × 3, USNM 156884, paratype, upper Roberts Mountains Formation, northern Roberts Mountains. Same collection as specimen in figures 11-14.
    11-14 Ventral, interior, side, and posterior views of pedicle valve × 2, USNM 156885, holotype, upper Roberts Mountains Formation, northern Roberts Mountains, loc. 12857.

Figures 15-21 *Leptaenisca* sp.
    Quadrithyris Zone, Coal Canyon.
    15 Interior of pedicle valve × 3, USNM 156875, loc. 10758.
    16, 17 Interior and exterior of pedicle valve × 5, USNM 156876, loc. 10757.
    18, 19 Exterior and interior of brachial valve × 3, USNM 156877, loc. 10757.
    20 Interior of pedicle valve × 4, USNM 156878, loc. 10757.
    21 Interior of pedicle valve × 4, USNM 156879, loc. 10757.

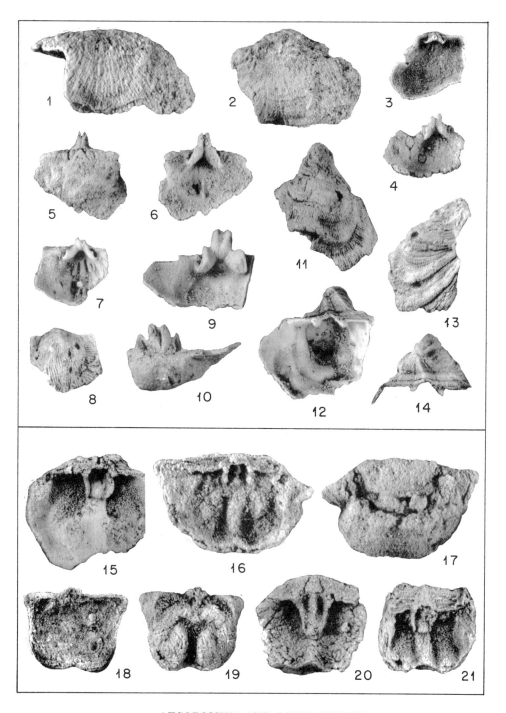

AESOPOMUM AND LEPTAENISCA

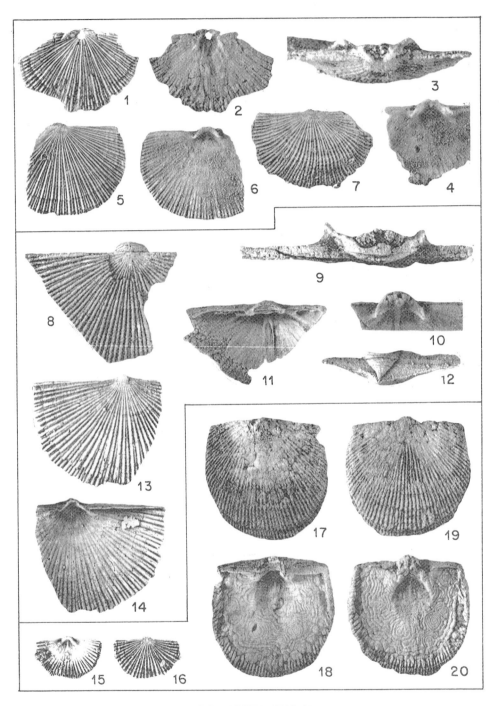

"SCHUCHERTELLA"

PLATE 18. "SCHUCHERTELLA"

Figures 1-7 *"Schuchertella"* sp. A.
  Quadrithyris Zone, loc. 10758, Coal Canyon.
  1, 2 Exterior and interior of brachial valve × 2, USNM 156886.
  3, 4 Posterior × 5 and interior × 2 of brachial valve, USNM 156887.
  5, 6 Exterior and interior of brachial valve × 2, USNM 156888.
  7 Exterior of brachial valve × 2, USNM 156889.

Figures 8-14 *"Schuchertella"* sp. C.
  Acrospirifer kobehana Zone, Sulphur Spring Range.
  8-10 Exterior of brachial valve × 1.5, posterior of brachial valve × 3, interior of brachial valve showing cardinalia × 1.5, USNM 156893, loc. 10739.
  11, 12 Interior and posterior views of pedicle valve × 1.5, USNM 156894, loc. 10739. Note dental lamellae in fig. 11.
  13, 14 Exterior and interior of pedicle valve × 1.5, USNM 156895, loc. 10739.

Figures 15-20 *"Schuchertella"* sp. B.
  Trematospira Zone, Sulphur Spring Range.
  15, 16 Interior and exterior of brachial valve × 2, USNM 156890, loc. 10743.
  17, 18 Exterior and interior of pedicle valve × 1, USNM 156891, loc. 10737.
  19, 20 Exterior and interior of brachial valve × 1, USNM 156892, loc. 10737.

PLATE 19. "SCHUCHERTELLA"

Figures 1-13 *"Schuchertella" nevadaensis* Merriam.
  Eurekaspirifer pinyonensis Zone, figs. 1-4, Lone Mountain; figs. 5-13, Roberts Mountains.
  1-4 Dorsal, side, anterior, and posterior views of holotype × 1, USNM 96366.
  5 Internal mold of brachial valve × 1, USNM 156896, loc. 10705.
  6, 7 Rubber impression of internal mold and internal mold of brachial valve × 1.5, USNM 156897, loc. 10776.
  8 Internal mold of pedicle valve × 1.5, USNM 156898, loc. 10782.
  9 Iternal mold of pedicle valve × 1.25, USNM 156904, loc. 10765.
  10 Posterior view of internal mold × 1.5, USNM 156900, loc. 12772.
  11-13 Posterior view of brachial valve × 2.5, interior and rubber impression of interior × 1, USNM 156901, loc. 10784.

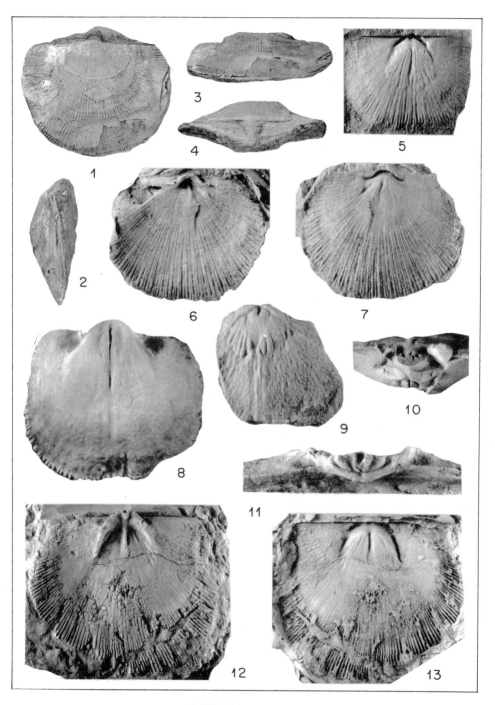

"SCHUCHERTELLA"

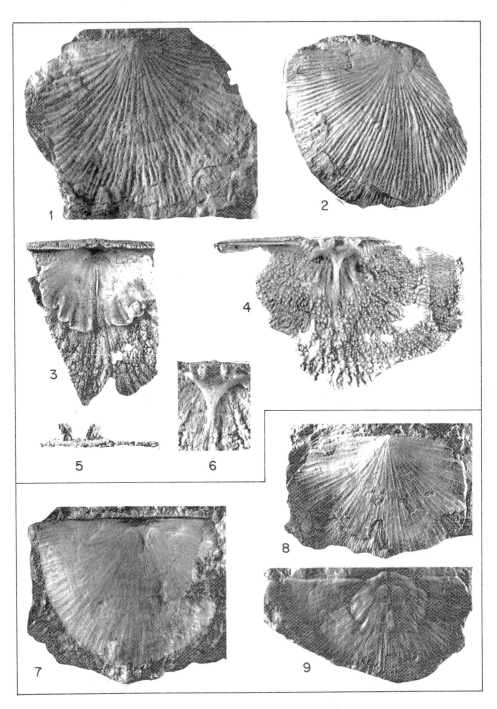

STROPHONELLA

JOHNSON, PLATE 20
Geological Society of America Memoir 121

PLATE 20. STROPHONELLA

Figures 1-6 *Strophonella* cf. *punctulifera* (Conrad).
   1 Ventral view × 1, USNM 156905, loc. 10706, *Trematospira* Zone, Sulphur Spring Range.
   2 Dorsal view × 1, USNM 156906, loc. 10706, *Trematospira* Zone, Sulphur Spring Range.
   3 Interior of pedicle valve × 1.5, USNM 156907, loc. 10739, *Acrospirifer kobehana* Zone, Sulphur Spring Range.
   4 Interior of brachial valve × 1.5, USNM 156908, loc. 10739, *Acrospirifer kobehana* Zone, Sulphur Spring Range.
   5, 6 Posterior and interior views of brachial valve × 2, USNM 156909, loc. 10739, *Acrospirifer kobehana* Zone, Sulphur Spring Range.

Figures 7-9 *Strophonella* cf. *bohemica* (Barrande).
   Lower McMonnigal Formation (*Quadrithyris* Zone), Toquima Range.
   7 Internal mold of brachial valve × 1.25, USNM 156902, loc. 10808.
   8 Ventral view × 1.25, USNM 156903, loc. 10765.
   9 Internal mold of pedicle valve × 1.25, USNM 156904, loc. 10765.

PLATE 21. STROPHEODONTA AND STROPHONELLA

Figures 1-12 *Stropheodonta magnacosta* Johnson, n. sp.
   1-5 Ventral, dorsal, side, anterior, and posterior views × 2, USNM 156911, loc. 10726 *Trematospira* Zone, Sulphur Spring Range.
   6 Interior of pedicle valve × 1.5, USNM 156912, loc. 10712, *Spinoplasia* Zone, Monitor Range.
   7 Interior of pedicle valve × 1.5, USNM 156913, loc. 10726, *Trematospira* Zone, Sulphur Spring Range.
   8, 9 Dorsal and ventral views × 2, USNM 156914, loc. 10777, *Acrospirifer kobehana* Zone, northern Roberts Mountains.
   10, 11 Posterior and interior views of brachial valve × 1.5, holotype, USNM 156915, loc. 10730, *Trematospira* Zone, Sulphur Spring Range.
   12 Dorsal view × 1, holotype, USNM 156915, loc. 10730, *Trematospira* Zone, Sulphur Spring Range.

Figures 13, 14 *Strophonella* cf. *punctulifera* (Conrad).
   13, 14 Interior and exterior views of brachial valve × 1.5, USNM 156910, loc. 10713, *Spinoplasia* Zone, Rabbit Hill.

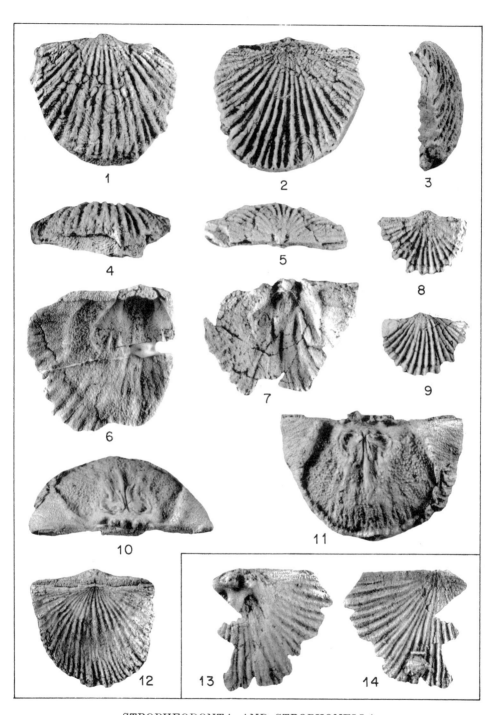

STROPHEODONTA AND STROPHONELLA

JOHNSON, PLATE 21
Geological Society of America Memoir 121

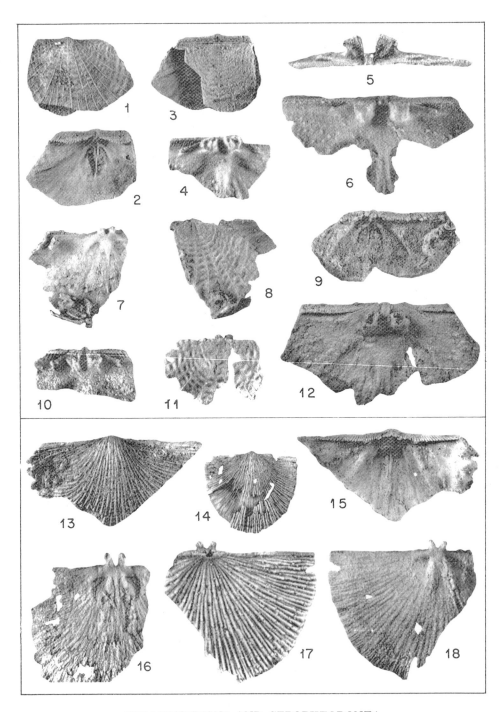

"BRACHYPRION" AND STROPHEODONTA

PLATE 22. "BRACHYPRION" AND STROPHEODONTA

Figures 1-12 *"Brachyprion" mirabilis* Johnson, n. sp.
  Quadrithyris Zone, Coal Canyon.
  1, 3 Ventral and dorsal views × 2, holotype, USNM 156920 A and B, loc. 10758.
  2, 4 Interior of pedicle valve × 2, USNM 156920A, and interior of brachial valve × 3, USNM 156920B, of specimens illustrated in figures 1 and 3 after disarticulation, loc. 10758.
  5, 6 Posterior and interior of brachial valve × 3, USNM 156921, loc. 10758. Note long, slender, widely divergent socket plates.
  7, 8 Interior and exterior of brachial valve × 2, USNM 156922, loc. 10758.
  9 Interior of posterior portion of pedicle valve × 1.5, USNM 156923, loc. 10758.
  10 Interior of part of brachial valve × 4, USNM 156924, loc. 10758.
  11 Exterior of brachial valve × 2, USNM 156925, loc. 10758.
  12 Interior of brachial valve × 3, USNM 156926, loc. 10757.

Figures 13-18 *Stropheodonta filicosta* Johnson, n. sp.
  Trematospira Zone, Sulphur Spring Range.
  13, 15 Exterior and interior of pedicle valve × 3, USNM 156917, loc. 10748.
  14 Ventral view × 2, holotype, USNM 156916, loc. 10749.
  16 Interior of brachial valve × 4, USNM 156918, loc. 10749.
  17, 18 Exterior and interior of brachial valve × 5, USNM 156919, loc. 10749.

PLATE 23. MESODOUVILLINA AND LEPTOSTROPHIA

Figures 1-15 *Mesodouvillina* cf. *varistriata* (Conrad).
Quadrithyris Zone, Coal Canyon.
1-5 Ventral, interior, anterior, posterior, and side views of pedicle valve × 2, USNM 156927, loc. 10758.
6 Interior of pedicle valve × 3, USNM 156928, loc. 10758.
7 Exterior of pedicle valve × 3, USNM 156929, loc. 10758.
8 Part of interior of brachial valve × 4, USNM 156930, loc. 10757.
9, 10 Interior and exterior of pedicle valve × 2, USNM 156931, loc. 10758.
11 Exterior of pedicle valve × 2, USNM 156936, loc. 10758.
12 Interior of brachial valve × 3, USNM 156932, loc. 10758.
13 Interior of pedicle valve × 2, USNM 156933, loc. 10757.
14 Interior of pedicle valve × 3, USNM 156934, loc. 10758.
15 Interior of part of brachial valve × 2, USNM 156935, loc. 10758.

Figure 16 *Mesodouvillina* cf. *varistriata arata* (Hall).
16 Exterior of pedicle valve × 2, USNM 156937, loc. 10758, *Quadrithyris* Zone, Coal Canyon.

Figures 17, 18 *Leptostrophia* sp. A.
Quadrithyris Zone, Coal Canyon.
17 Interior of brachial valve × 3, USNM 156975, loc. 10757.
18 Interior of brachial valve × 3, USNM 156976, loc. 10757.

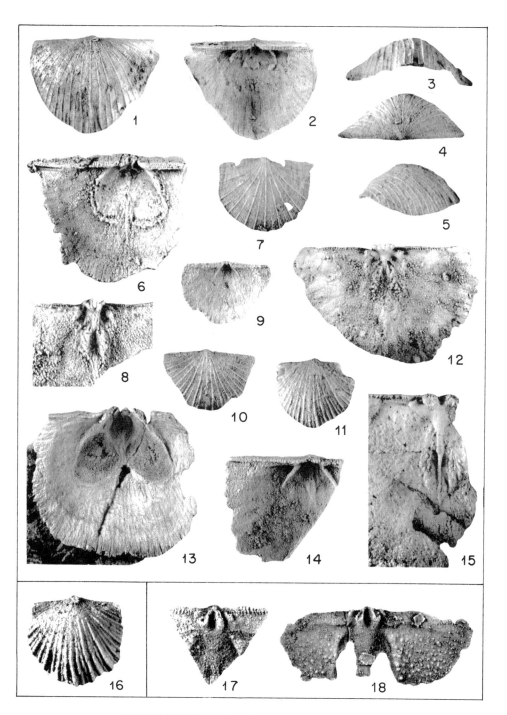

MESODOUVILLINA AND LEPTOSTROPHIA

MCLEARNITES

JOHNSON, PLATE 24
Geological Society of America Memoir 121

PLATE 24. MCLEARNITES

Figures 1-8 *Mclearnites invasor* Johnson, n. sp.
    Acrospirifer kobehana Zone, Sulphur Spring Range.
  1 Interior of pedicle valve × 1.5, USNM 156939, loc. 10739.
  2, 3 Interior and exterior of pedicle valve × 1.5, holotype, USNM 156938, loc. 10740.
  4 Posterior view × 1.5, USNM 156940, loc. 10740.
  5 Interior of brachial valve × 1.5, USNM 156941, loc. 10739.
  6 Interior of brachial valve × 2, USNM 156942, loc. 10740.
  7 Ventral view × 3, USNM 156943, loc. 10739.
  8 Interior of part of brachial valve × 4, USNM 156944, loc. 10739.

Figures 9-13 *Mclearnites* sp. B.
    Eurekaspirifer pinyonensis Zone, Cortez Range.
  9 Ventral view of internal mold × 1.5, USNM 156945, loc. 10756. Note deeply impressed muscle scars and ponderous muscle-bounding ridges laterally.
  10 Ventral view × 2, USNM 156946, loc. 10755. Note finely parvicostellate ornament.
  11, 13 Ventral and dorsal views of internal mold × 2, USNM 156947, loc. 10753. In figure 13 note the relatively strongly impressed anterior adductor ridges.
  12 Interior of posteromedial part of brachial valve × 2, USNM 156948, loc. 10754. Note strongly elevated muscle-bounding ridges lateral to the adductor muscle impressions.

PLATE 25. MEGASTROPHIA

Figures 1-19 *Megastrophia transitans* Johnson n. sp.
  1-3 Interior, exterior, and posterior views of pedicle valve × 2, USNM 156951, loc. 10733, *Trematospira* Zone, Sulphur Spring Range.
  4, 5 Interior and exterior of brachial valve × 3, USNM 156950, loc. 10777, *Acrospirifer kobehana* Zone, northern Roberts Mountains.
  6-8 Posterior, exterior, and interior views of brachial valve × 3, holotype, USNM 156949, loc. 10748 *Trematospira* Zone, Sulphur Spring Range.
  9 Interior of brachial valve × 4, USNM 156952, loc. 10749. *Trematospira* Zone, Sulphur Spring Range.
  10 Interior of pedicle valve × 1.5, USNM 156953, loc. 10726, *Trematospira* Zone, Sulphur Spring Range.
  11 Interior of brachial valve × 4, USNM 156954, loc. 10749, *Trematospira* Zone, Sulphur Spring Range.
  12, 13 Ventral and dorsal views × 4, USNM 156955, loc. 10748, *Trematospira* Zone, Sulphur Spring Range.
  14 Interior of brachial valve × 4, USNM 156956, loc. 10749, *Trematospira* Zone, Sulphur Spring Range.
  15-19 Ventral, dorsal, posterior, side, and anterior views × 2, USNM 156957, loc. 10714, *Trematospira* Zone, northern Roberts Mountains.

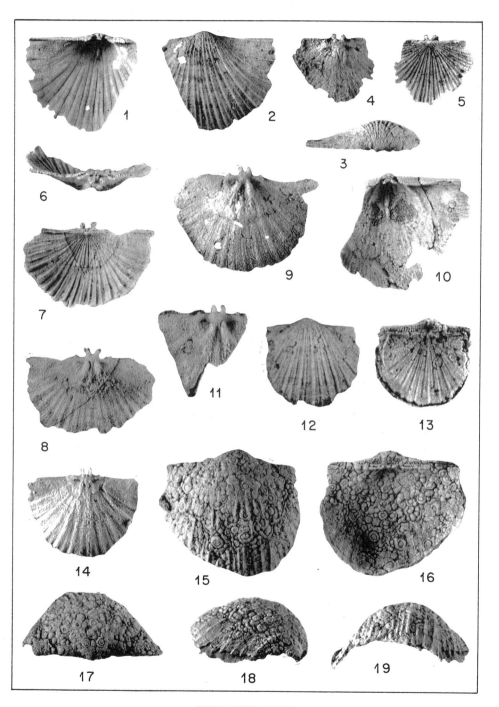

MEGASTROPHIA

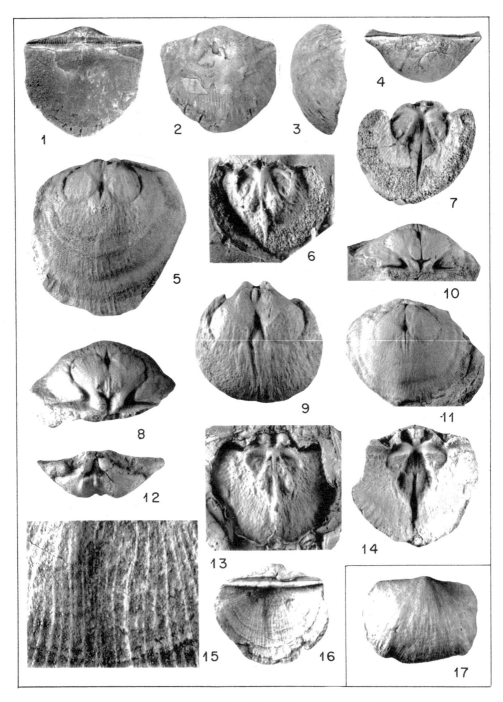

MEGASTROPHIA AND PHRAGMOSTROPHIA

PLATE 26. MEGASTROPHIA AND PHRAGMOSTROPHIA

Figures 1-16 *Megastrophia iddingsi* (Merriam).
Eurekaspirifer pinyonensis Zone.
1-4 Dorsal, ventral, side, and posterior views × 1.5 of the holotype, USNM 96370, Lone Mountain.
5, 8 Ventral and posterior views of internal mold × 2, USNM 156958, loc. 10781, Roberts Mountains.
6, 7 Rubber impression of brachial valve internal mold and brachial valve internal mold × 2, USNM 156959, loc. 10775, Roberts Mountains.
9, 12 Ventral and posterior views of internal mold of pedicle valve × 2, USNM 156960, loc. 10782, Roberts Mountains.
10, 11 Posterior and ventral views of internal mold of pedicle valve × 1.5, USNM 156961, loc. 12775, Lone Mountain.
13, 14 Rubber impression of brachial valve internal mold and brachial valve internal mold × 2, USNM 156962, loc. 10781, Roberts Mountains.
15, 16 Exterior of brachial valve × 5, and dorsal view × 1, USNM 156963, loc. 10762, Coal Canyon.

Figure 17 *Phragmostrophia merriami* Harper, Johnson, and Boucot.
17 Ventral view of exterior × 2, USNM 140420, loc. 10762, Coal Canyon.

PLATE 27. PHRAGMOSTROPHIA

Figures 1-20 *Phragmostrophia merriami* Harper, Johnson, and Boucot.
Eurekaspirifer pinyonensis Zone.

1, 2 Rubber impression of internal mold and internal mold of brachial valve × 2, holotype, USNM 140409, loc. 10711, Sulphur Spring Range.

3 Exterior of pedicle valve × 5, USNM 140414, loc. 12774, northern Roberts Mountains.

4 Exterior of brachial valve × 4, USNM 140415, Lone Mountain (G. A. Cooper coll.).

5, 6 Rubber impression of internal mold and internal mold of pedicle valve × 2, USNM 140408, loc, 10751, Cortez Range.

7 Internal mold of brachial valve × 3, paratype, USNM 140416, loc. 10711, Sulphur Spring Range.

8 Internal mold of brachial valve × 2, USNM 140417, loc. 10782, Roberts Mountains.

9, 10 Dorsal and ventral views of exterior of an unusually convex individual × 1, USNM 140413, Lone Mountain (G. A. Cooper coll.).

11-14, 17 Ventral, dorsal, anterior, posterior, and side views × 1.5, USNM 140410, loc. 10774, Roberts Mountains.

15, 16 Dorsal view × 1.5 and posterior view × 2, USNM 140418, loc. 10782, Roberts Mountains.

18, 19 Rubber impression of internal mold and internal mold of pedicle valve × 2, paratype, USNM 140411, loc. 10771, Sulphur Spring Range.

20 Interior of brachial valve × 2, USNM 140419, loc. 10754, Cortez Range.

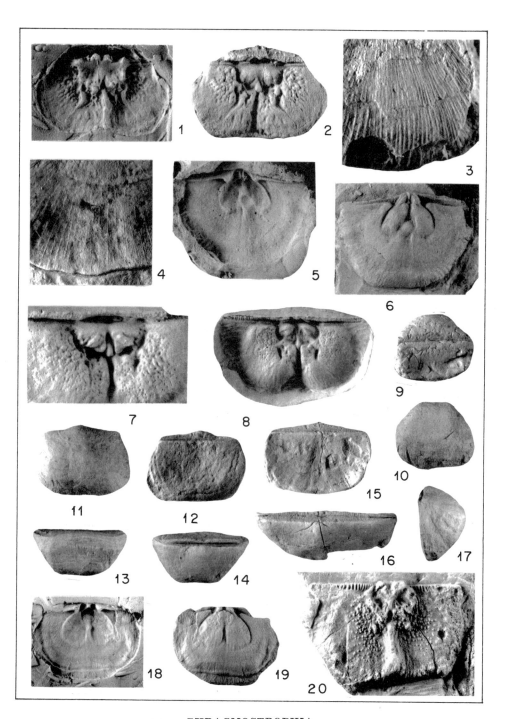

PHRAGMOSTROPHIA

LEPTOSTROPHIA AND PHOLIDOSTROPHIA

PLATE 28. LEPTOSTROPHIA AND PHOLIDOSTROPHIA

Figures 1-10 *Leptostrophia* sp. D.
  Acrospirifer kobehana Zone, loc. 12778, McColley Canyon, Sulphur Spring Range.
  1, 2 Posterior and interior views of brachial valve × 3, USNM 156968.
  3, 4 Exterior and interior views of pedicle valve × 4, USNM 156969.
  5 Interior of pedicle valve × 2, USNM 156970.
  6 Exterior view of pedicle valve × 2, USNM 156971.
  7, 8 Interior and exterior of pedicle valve × 5, USNM 156972.
  9 Interior of brachial valve × 4, USNM 156973.
  10 Interior of brachial valve × 3, USNM 156974.

Figures 11-17 *Pholidostrophia* (*Pholidostrophia*) sp.
  Acrospirifer kobehana Zone, loc. 12778, McColley Canyon, Sulphur Spring Range.
  11-13 Exterior, interior, and posterior views of brachial valve × 4, USNM 156964.
  14, 15 Exterior and interior of pedicle valve × 5, USNM 156965.
  16 Exterior of pedicle valve × 6, USNM 156966.
  17 Interior of brachial valve × 4, USNM 156967.

PLATE 29. LEPTOSTROPHIA

Figures 1-8 *Leptostrophia inequicostella* Johnson, n. sp.
    Spinoplasia Zone, loc. 10713, Rabbit Hill.
    1, 2 Interior and exterior of pedicle valve × 3, holotype, USNM 156977.
    3 Interior of part of pedicle valve × 2, USNM 156978.
    4 Interior of pedicle valve × 2, USNM 156979.
    5 Interior of brachial valve × 4, USNM 156980.
    6 Interior of brachial valve × 3, USNM 156981.
    7, 8 Exterior and interior of brachial valve × 2, USNM 156982.

Figure 9 *Leptostrophia?* sp D?
    Eurekaspirifer pinyonensis Zone, northern Roberts Mountains.
    9 Internal mold of pedicle valve × 1, USNM 156983, loc. 10781.

Figures 10, 11 *Leptostrophia* cf. *beckii* (Hall).
    Trematospira Zone, Sulphur Spring Range.
    10, 11 Exterior and interior of pedicle valve × 1, USNM 156984, loc. 10730. Note absence of diductor muscle impression on the left side of the valve in figure 11.

LEPTOSTROPHIA

STROPHOCHONETES, "STROPHOCHONETES," AND CHONETES

JOHNSON, PLATE 30
Geological Society of America Memoir 121

PLATE 30. STROPHOCHONETES, "STROPHOCHONETES", AND CHONETES

Figures 1-5 *Strophochonetes cingulatus* (Lindström, 1861).
Silurian, Fröjel, Gotland.
1 Internal mold of brachial valve × 5, *BR5398b. Note the pair of anderidia and the double depressions of the ventral face of the cardinal process.
2 Internal mold of brachial valve × 6, *BR5398c. Note pair of anderidia and impression of the socket ridges posteriorly.
3 Exterior of pedicle valve × 4, *BR102296. Note the unsymmetrically developed pair of enlarged median capillae.
4 Exterior of pedicle valve × 5, *BR5400. Note faint development of enlarged median capilla.
5 Exterior of pedicle valve × 4, *BR5398a.

Figure 6 *"Strophochonetes"* sp.
Spinoplasia Zone.
6 Exterior of pedicle valve × 3, USNM 156985, loc. 10759, Rabbit Hill Limestone, Coal Canyon. Note pair of very long spines projecting approximately normal to the hinge line.

Figures 7-10 *Chonetes* sp.
Eurekaspirifer pinyonensis Zone.
7 Internal mold of pedicle valve × 1.5, USNM 156987, loc. 10782, Roberts Mountains.
8 Interior of brachial valve × 2, USNM 156986, Lone Mountain (G. A. Cooper coll.)
9 Mold of exteriors of both valves × 3, USNM 156988, loc. 12775, Lone Mountain.
10 Exterior of pedicle valve × 3, USNM 156989, loc. 12775, Lone Mountain.

---

*Dept. Palaeozool., Swedish Mus. Nat. Hist., Stockholm.

PLATE 31. "STROPHOCHONETES"

Figures 1-17 *"Strophochonetes" filistriata* (Walcott).
 Eurekaspirifer pinyonensis Zone, Figs. 1-3, Combs Peak; fig. 4, locality unknown; figs. 5, 6, 10-14, 16, 17, northern Roberts Mountains; fig. 7, Lone Mountain; figs. 8, 15, Simpson Park Range; fig. 9, Garden Valley quadrangle.
1 Ventral view of the holotype × 1.5, USNM 13811.
2 Ventral view × 1.5 of a paratype, USNM 13811A.
3 Dorsal view of internal mold × 2 of a paratype, USNM 13811C.
4 Ventral view of internal mold × 3, USNM 156990, loc. unknown.
5, 12 Ventral views × 2 and × 3, USNM 156991, loc. 12773.
6 Posterior view of internal mold of pedicle valve × 4, USNM 156992, loc. 10775.
7 Dorsal view × 2, USNM 156993, loc. 12775.
8 Ventral view × 3, USNM 156994, loc. 10796.
9 Ventral views of two specimens × 2, USNM 156995, loc. 10724.
10 Ventral view × 4, USNM 156996, loc. 12773.
11 Dorsal view of internal mold × 4, USNM 156997, loc. 10775.
13 Dorsal view of internal mold × 4, USNM 156998, loc, 10775.
14 Dorsal view × 4, USNM 156999, loc. 12773.
15 Ventral view × 3, USNM 157000, loc. 10796.
16, 17 Dorsal view of internal mold and rubber impression of internal mold of brachial valve × 4, USNM 157001, loc. 10775.

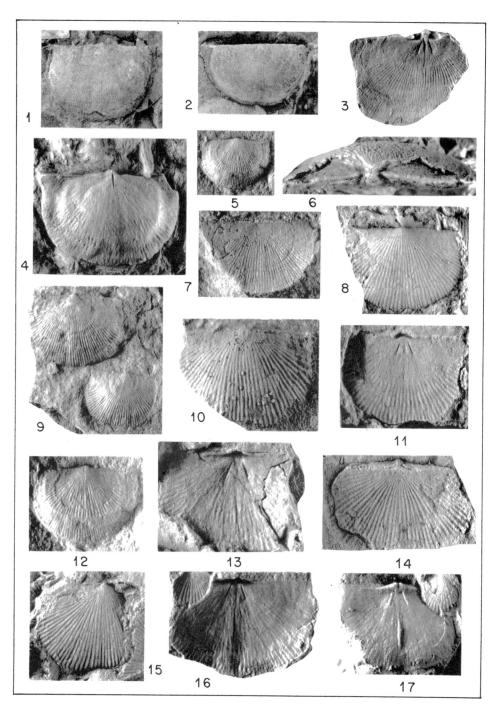

"STROPHOCHONETES"

JOHNSON, PLATE 31
Geological Society of America Memoir 121

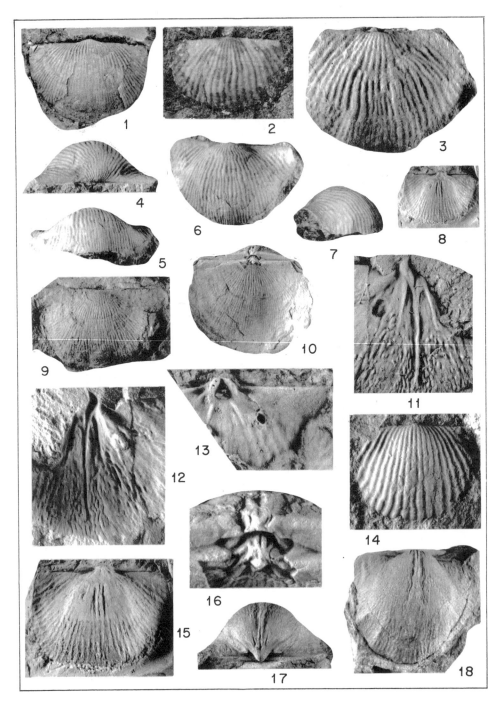

PARACHONETES

PLATE 32. PARACHONETES

Figures 1-18 *Parachonetes macrostriatus* (Walcott).
Eurekaspirifer pinyonensis Zone.
1 Exterior of pedicle valve × 1, USNM 13809A, paratype, Lone Mountain.
2 Exterior of pedicle valve × 2, USNM 13809, holotype, Lone Mountain.
3 Exterior of pedicle valve × 2, USNM 140436, loc. 10709, Sulphur Spring Range.
4-7 Posterior, anterior, ventral, and side views of pedicle valve × 1, USNM 140446, loc. 10705, Roberts Mountains.
8, 15 Internal mold of pedicle valve × 1 and × 2, USNM 140444, loc. 13267, Sulphur Spring Range.
9 Exterior of pedicle valve × 1, USNM 140439, loc. 10788, Roberts Mountains.
10, 16 Exterior of brachial valve × 1 and posterior view at center of hinge line × 5, USNM 140438, loc. 12775, Lone Mountain.
11, 12 Rubber impression of dorsal internal mold and dorsal internal mold × 3, USNM 140441, loc. 13267, Sulphur Spring Range.
13 Rubber impression of interior of brachial valve of small specimen × 5, USNM 140440, loc. 13267, Sulphur Spring Range.
14 Exterior of pedicle valve × 3, USNM 140435, loc. 10769, northern Roberts Mountains.
17, 18 Posterior and ventral views of internal mold of pedicle valve × 1.5, USNM 140445, loc. 13267, Sulphur Spring Range.

PLATE 33. PARACHONETES AND ECCENTRICOSTA

Figures 1-5 *Parachonetes macrostriatus* (Walcott).
Eurekaspirifer pinyonensis Zone.
1, 2 Mold of interior of brachial valve and rubber impression × 2, USNM 140437, loc. 10762, Coal Canyon.
3 Posterior view of internal mold of pedicle valve × 2, USNM 140443, loc. 10705, Roberts Mountains.
4, 5 Rubber impression of internal mold and internal mold of brachial valve × 2, USNM 140442, loc. 13267, Sulphur Spring Range.

Figures 6-10 *Parachonetes verneuili* (Barrande).
Upper Koneprus Limestone, Czechoslovakia.
6 Posteromedial part of brachial internal mold × 3, MCZ 9440. Note mold of the alveolus and of the pair of lateral septa and median septum.
7 Mold of interior of brachial valve × 1.5, MCZ 9440.
8, 9 Ventral view × 1 and posterior view × 2 of internal mold of pedicle valve, MCZ 9441.
10 Posterior view of internal mold of pedicle valve × 2, USNM 140447.

Figures 11, 12 *Eccentricosta jerseyensis* (Weller).
Lower part of Keyser Limestone (Bowen, 1967).
11, 12 Internal mold of brachial valve and rubber impression × 3, MCZ 9445b. Note the relatively small cardinal process composed of a pair of posteriorly disjunct lobes separated anteriorly by a V-shaped alveolus.

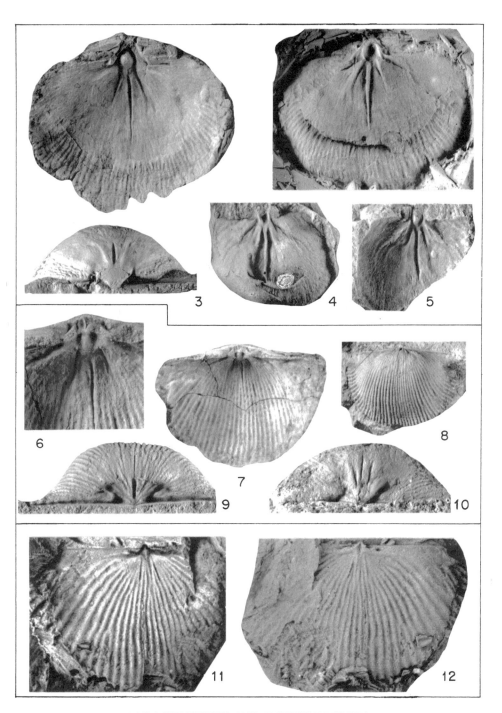

PARACHONETES AND ECCENTRICOSTA

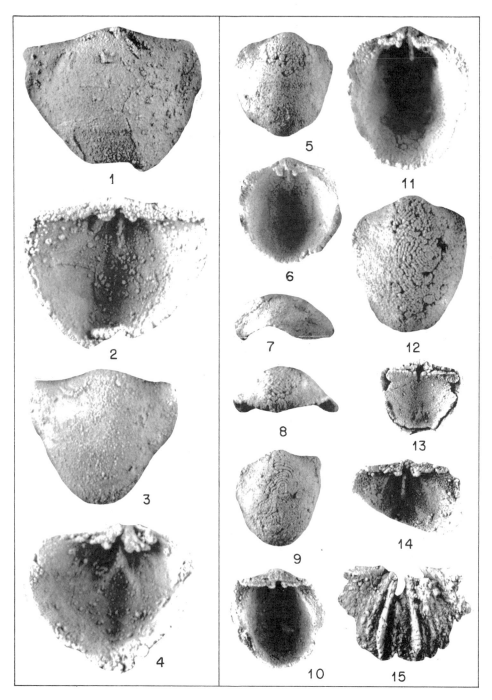

ANOPLIA

JOHNSON, PLATE 34
Geological Society of America Memoir 121

PLATE 34. ANOPLIA

Figures 1-4 *Anoplia* sp. A.
    Spinoplasia Zone, loc. 10713, Rabbit Hill.
    1, 2 Exterior and interior of pedicle valve × 10, USNM 157002.
    3 Exterior of pedicle valve × 10, USNM 157003.
    4 Interior of pedicle valve × 10, USNM 157004.

Figures 5-15 *Anoplia elongata* Johnson, n. sp.
    Acrospirifer kobehana Zone, figures 5-8, 11, 12, southern Roberts Mountains; figs. 9, 10, 13-15, Sulphur Spring Range.
    5-8 Exterior, interior, side, and posterior views of pedicle valve × 7, paratype, USNM 157006, loc. 10791.
    9, 10 Exterior and interior of pedicle valve × 7, USNM 157007, loc. 10739.
    11, 12 Interior and exterior of the pedicle valve, holotype, × 10, USNM 157005, loc. 10791.
    13 Dorsal view × 7, USNM 157008, loc. 10739.
    14 Interior of pedicle valve × 10, USNM 157009, loc. 10740.
    15 Interior of brachial valve × 10, USNM 157010, loc. 10739. Cardinal process is broken away.

PLATE 35. CYMOSTROPHIA AND SPINULICOSTA?

Figures 1, 2 *Cymostrophia* sp.
Eurekaspirifer pinyonensis Zone, northern Roberts Mountains
1 Posterior view of internal mold of pedicle valve × 1, USNM 156781, loc. 10774.
2 Posterior view of external mold of brachial valve × 2, USNM 156782, loc. 10781.

Figures 3, 4 *Productella?* sp., or *Spinulicosta?* sp.
Eurekaspirifer pinyonensis Zone, loc. 12775, Lone Mountain.
3, 4 Ventral and posterior views of pedicle valve × 2, USNM 157011.

Figures 5-13 *Spinulicosta?* sp.
Eurekaspirifer pinyonensis Zone.
5-7 Ventral, side, and anterior views of pedicle valve × 3, USNM 157012, loc. 13269, Lone Mountain.
8-10 Ventral and two side views of pedicle valve × 4, USNM 157013, loc. 12775, Lone Mountain.
11-13 Ventral, posterior, and side views of pedicle valve × 3, USNM 157014, loc. 13267, McColley Canyon, Sulphur Spring Range.

CYMOSTROPHIA AND SPINULICOSTA?

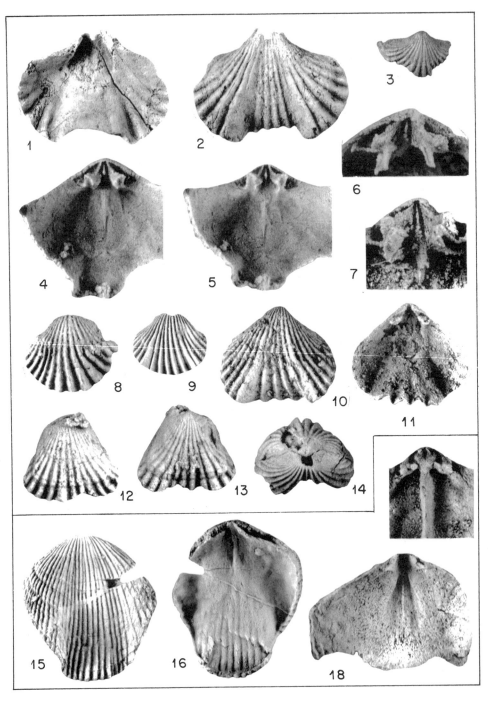

MACHAERARIA AND PLEIOPLEURINA

JOHNSON, PLATE 36
Geological Society of America Memoir 121

PLATE 36. MACHAERARIA AND PLEIOPLEURINA

Figures 1-14 *Machaeraria* sp.
  Quadrithyris Zone, Coal Canyon.
  1, 2 Interior and exterior of pedicle valve × 1.5, USNM 157017, loc. 10758.
  3-5 Exterior × 1.5, and two views of interior of brachial valve × 4, USNM 157018, loc. 10758.
  6, 7 Two views of cardinalia of brachial valve × 5, USNM 157019, loc 10757.
  8 Dorsal view × 3, USNM 157020, loc. 10758.
  9 Ventral view × 1.5, USNM 157021, loc. 10758.
  10, 11 Exterior and interior views of pedicle valve × 2, USNM 157022, loc. 10758.
  12, 13 Dorsal and ventral views × 4, USNM 157023, loc. 10757.
  14 Posterior view × 1.5, USNM 157024, loc. 10758.

Figures 15-18 *Pleiopleurina anticlastica* Johnson, n. sp.
  Spinoplasia Zone, loc. 10713, Rabbit Hill.
  15, 16 Exterior and interior views of pedicle valve × 1.5, USNM 157025.
  17 Part of interior of brachial valve × 3, USNM 157026.
  18 Interior of pedicle valve × 2, USNM 157027.

PLATE 37. PLEIOPLEURINA

Figures 1-16 *Pleiopleurina anticlastica* Johnson, n. sp.

1, 2 Interior and exterior of pedicle valve × 1.5, USNM 157028, paratype, loc. 10726, *Trematospira* Zone, Sulphur Spring Range.

3 Interior of brachial valve × 5, USNM 157029, loc. 10713, *Spinoplasia* Zone, Rabbit Hill.

4 Interior of posterior part of brachial valve × 4, USNM 157030, loc. 10713, *Spinoplasia* Zone, Rabbit Hill.

5 Interior of brachial valve × 5, USNM 157031, loc. 10713, *Spinoplasia* Zone, Rabbit Hill.

6-10 Ventral, dorsal, side, anterior, and posterior views of the holotype × 1.5, USNM 157032, loc. 10726, *Trematospira* Zone, Sulphur Spring Range.

11 Cardinalia of brachial valve × 5, USNM 157033, loc. 10713, *Spinoplasia* Zone, Rabbit Hill.

12-16 Posterior, dorsal, ventral, side, and anterior views × 1, paratype, USNM 157034, loc. 10726, *Trematospira* Zone, Sulphur Spring Range.

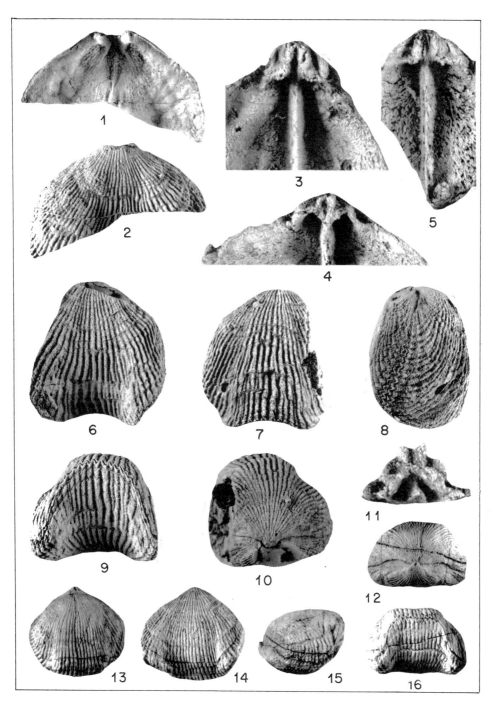

PLEIOPLEURINA

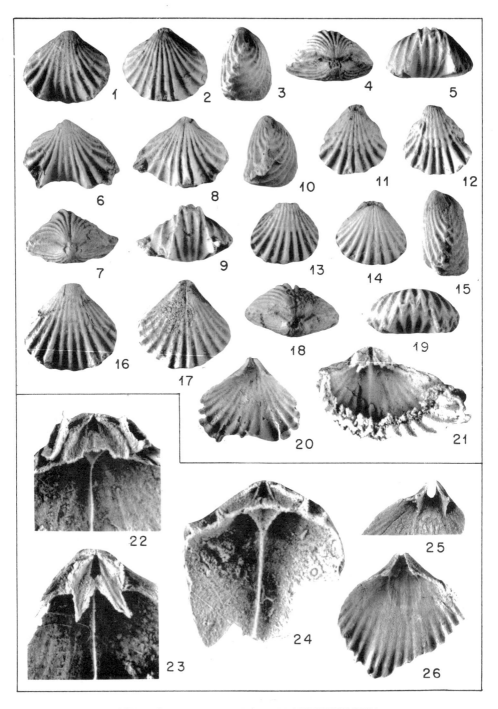

ANCILLOTOECHIA AND TRIGONIRHYNCHIA

JOHNSON, PLATE 38
Geological Society of America Memoir 121

PLATE 38. ANCILLOTOECHIA AND TRIGONIRHYNCHIA

Figures 1-21 *Ancillotoechia aptata* Johnson, n. sp.
Trematospira Zone, Sulphur Spring Range.
1-5 Ventral, dorsal, side, posterior, and anterior views of the holotype × 2, USNM 157035, loc. 10708.
6-10 Ventral, posterior, dorsal, anterior, and side views × 2, paratype, USNM 157036, loc. 10708.
11, 12 Dorsal and ventral views of a small specimen × 3, paratype, USNM 157037, loc. 10708.
13, 14 Ventral and dorsal views of a small specimen × 3, paratype, USNM 157038, loc. 10708.
15-19 Side, ventral, dorsal, posterior, and anterior views × 3, paratype, USNM 157039, loc. 10708.
20 Interior of pedicle valve × 2, USNM 157040, loc. 10726.
21 Interior of brachial valve × 4, USNM 157041, loc. 10726.

Figures 22-26 *Trigonirhynchia occidens* (Walcott)
Eurekaspirifer pinyonensis Zone, loc. 10754, Cortez Range.
22 Interior of brachial valve × 5, USNM 157042. Note that the hinge teeth are still articulated.
23 Interior of brachial valve × 5, USNM 157043. Note crural flanges covering the septalium anteriorly.
24 Interior of brachial valve × 5, USNM 157044.
25 Interior of posterior portion of pedicle valve × 2, USNM 157045. Note dental lamellae; note beak is broken.
26 Interior of pedicle valve × 1.5, USNM 157046.

PLATE 39. CORVINOPUGNAX AND TRIGONIRHYNCHIA

Figures 1-5 *Corvinopugnax* sp.
?Eurekaspirifer pinyonensis Zone, Garden Valley quadrangle.
1-5 Posterior, anterior, ventral, dorsal, and side views of internal mold × 1.5, USNM 157053, loc. 10725. Note the transverse internal ridge at the anterior of the fold in figures 2 and 4.

Figures 6-28 *Trigonirhynchia occidens* (Walcott).
Eurekaspirifer pinyonensis Zone.
6-10 Dorsal, ventral, side, anterior, and posterior views of the holotype × 3, USNM 13855, Combs Peak.
11-15 Anterior, ventral, side, posterior, and dorsal views × 1.5, USNM 157052, loc. 10716, Garden Valley quadrangle.
16-20 Ventral, dorsal, anterior, posterior, and side views × 1.5, USNM 157051, loc. 12775, Lone Mountain.
21-25 Side, ventral, dorsal, anterior, and posterior views × 1.5, USNM 157050, loc. 10762, northern Simpson Park Range.
26 Interior of posterior portions of articulated pedicle and brachial valves × 3, USNM 157049, loc. 10754, Cortez Range.
27 Interior of posterior portions of articulated pedicle and brachial valves × 3, USNM 157048, loc. 10754, Cortez Range.
28 Interior of brachial valve × 5, USNM 157047, loc. 10754, Cortez Range.

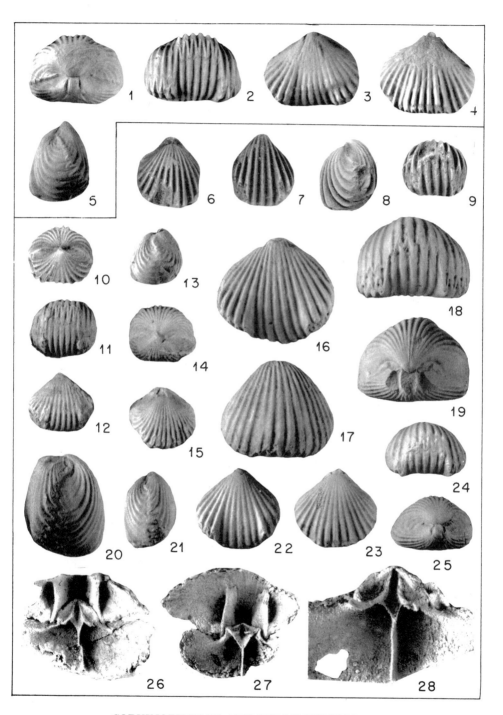

CORVINOPUGNAX AND TRIGONIRHYNCHIA

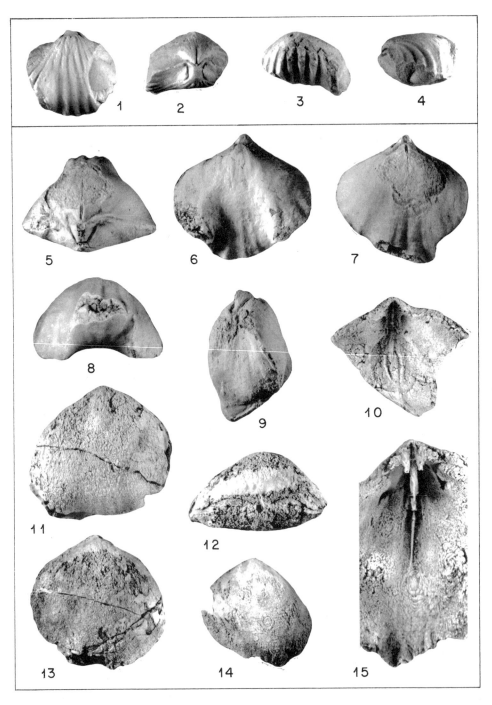

ASTUTORHYNCHA AND "LEIORHYNCHUS"

JOHNSON, PLATE 40
Geological Society of America Memoir 121

PLATE 40. ASTUTORHYNCHA AND "LEIORHYNCHUS"

Figures 1-4 *Astutorhyncha* cf. *prosperpina* (Barrande).
Eurekaspirifer pinyonensis Zone, northern Roberts Mountains.
1-4 Ventral, posterior, anterior, and side views of internal mold × 1, USNM 157054, loc. 10769.

Figures 5-15 *"Leiorhynchus"* sp.
Eurekaspirifer pinyonensis Zone, figs. 5-9, Lone Mountain; figs. 10-15, loc. 10754, Cortez Range.
5-9 Posterior, ventral, dorsal, anterior, and side views of internal mold × 1.5, USNM 157055, loc. 12775. In fig. 5 the thin plate in the groove is the silicified middle layer of the median septum which could not be removed.
10 Interior of pedicle valve × 1.5, USNM 157056.
11-13 Ventral, posterior, and dorsal views × 1, USNM 157057.
14, 15 Exterior × 1, and interior × 2 of brachial valve, USNM 157058. In fig. 15, note rodlike crura.

PLATE 41. ATRYPA

Figures 1, 2 *Atrypa* sp.
    Spinoplasia Zone, loc. 10712, Monitor Range.
    1, 2 Dorsal and ventral views × 4, USNM 157059.

Figures 3, 4 *Atrypa* sp. A.
    Acrospirifer kobehana Zone, northern Roberts Mountains.
    3, 4 Dorsal and ventral views × 1, USNM 157060, loc. 10772.

Figures 5-17 *Atrypa nevadana* Merriam.
    Eurekaspirifer pinyonensis Zone.
    5-9 Ventral, dorsal, side, anterior, and posterior views of the holotype × 1, USNM 93671, northern Roberts Mountains.
    10, 11 Dorsal and ventral views × 2, USNM 157061, loc. 10774, northern Roberts Mountains.
    12-16 Anterior, ventral, posterior, side, and dorsal views of internal mold × 1.5, USNM 157062, loc. 10720, Garden Valley quadrangle.
    17 Dorsal view × 1.5, USNM 157063, locality unknown, Roberts Mountains. Note marginal frill.

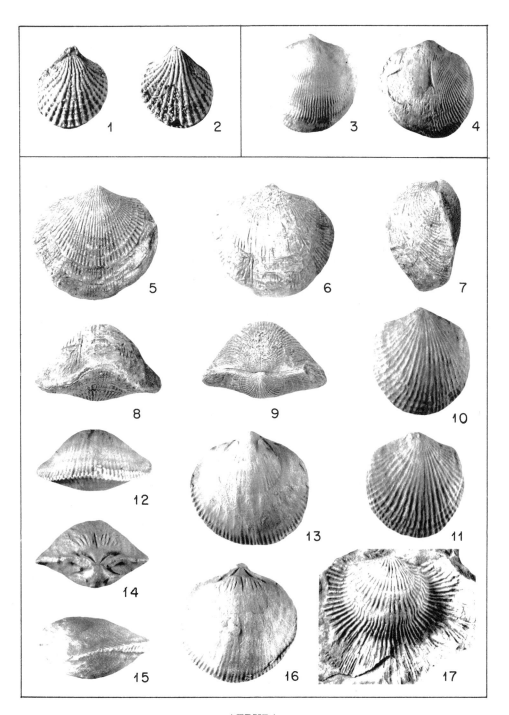

ATRYPA

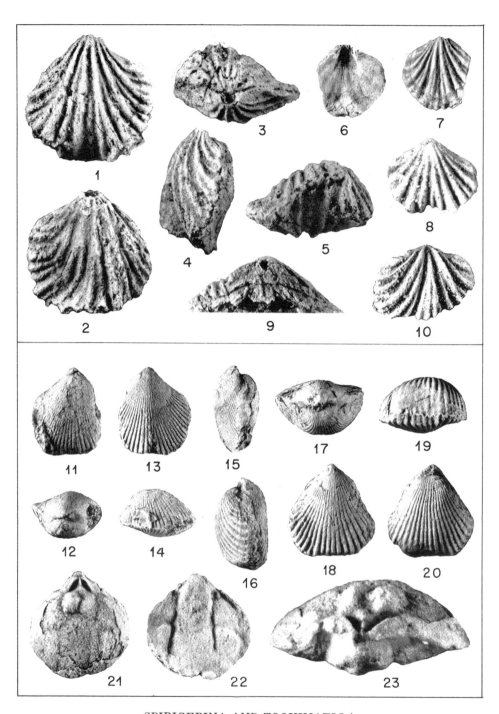

SPIRIGERINA AND TOQUIMAELLA

JOHNSON, PLATE 42
Geological Society of America Memoir 121

PLATE 42. SPIRIGERINA AND TOQUIMAELLA

Figures 1-10 *Spirigerina supramarginalis* (Khalfin).
Quadrithyris Zone, Coal Canyon.
1-5 Ventral, dorsal, posterior, side, and anterior views × 2, USNM 157064, loc. 10758.
6, 7 Interior and exterior of pedicle valve × 1.5, USNM 157065, loc. 10758.
8 Ventral view × 2, USNM 157066, loc. 10757.
9 Dorsal view of posterior of pedicle valve × 4, USNM 157067, loc. 10758.
10 Exterior of brachial valve × 2, USNM 157068, loc. 10758.

Figures 11-23 *Toquimaella kayi* Johnson.
Quadrithyris Zone, loc. 10808, Toquima Range.
11-15 Dorsal, posterior, ventral, anterior, and side views × 2, paratype, Columbia Univ. no. 28791.
16-20 Side, posterior, ventral, anterior, and dorsal views × 2, paratype, Columbia Univ. no. 28790.
21-23 Dorsal and ventral views × 2, and posterior view × 4 of internal mold of the holotype, USNM 155508.

PLATE 43. ATRYPINA AND BIFIDA

Figures 1-4 *Atrypina* cf. *simpsoni* Johnson, n. sp.
Upper Roberts Mountains Formation, northern Roberts Mountains.
1, 2 Ventral and dorsal views × 4, USNM 157069, loc. 12858.
3, 4 Interior and exterior views of pedicle valve × 3, USNM 157070, loc. 12855.

Figures 5-16 *Atrypina simpsoni* Johnson, n. sp.
Quadrithyris Zone, loc. 10758, Coal Canyon.
5-7, 12 Posterior, anterior, side, and ventral views × 5, paratype, USNM 157071.
8 Interior of pedicle valve × 5, paratype, USNM 157072.
9, 10 Dorsal and ventral views × 5, holotype, USNM 157073.
11, 15 Ventral and dorsal views × 5, paratype, USNM 157074.
13, 14 Dorsal and ventral views × 5, paratype, USNM 157075.
16 Dorsal view × 4, paratype, USNM 157076.

Figures 17-27 *Bifida* sp.
Eurekaspirifer pinyonensis Zone, loc. 10754, Cortez Range.
17, 22-25 Side, posterior, anterior, dorsal, and ventral views × 5, USNM 157077.
18, 19 Exterior and interior of pedicle valve × 5, USNM 157078.
20, 21 Exterior and interior of pedicle valve × 5, USNM 157079.
26, 27 Interior and exterior of pedicle valve × 10, USNM 157080.

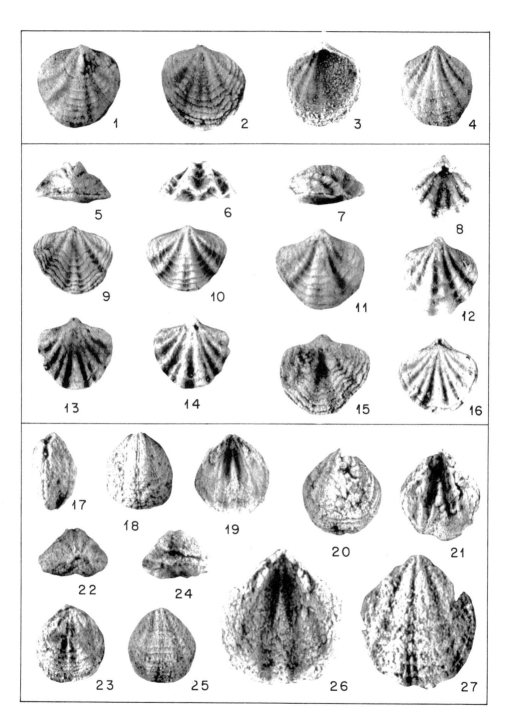

ATRYPINA AND BIFIDA

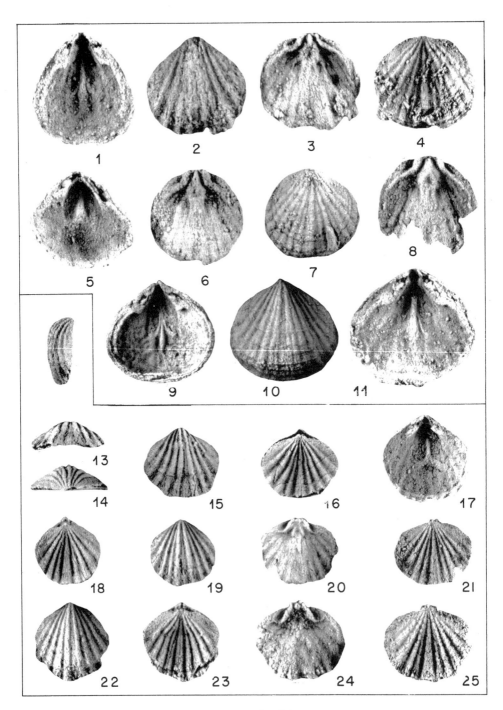

COELOSPIRA

PLATE 44. COELOSPIRA

Figures 1-11 *Coelospira concava* (Hall).
  Spinoplasia Zone, loc. 10713, Rabbit Hill.
  1 Interior of pedicle valve × 5, USNM 157081.
  2 Ventral view × 5, USNM 157082.
  3, 4 Interior and exterior views of brachial valve × 5, USNM 157083.
  5 Interior of pedicle valve × 5, USNM 157084.
  6, 7 Interior and exterior of brachial valve × 5, USNM 157085.
  8 Interior of brachial valve × 5, USNM 157086.
  9-10 Interior and exterior of pedicle valve × 5, USNM 157087.
  11 Interior of pedicle valve × 5, USNM 157088.

Figures 12-25 *Coelospira pseudocamilla* Johnson, n. sp.
  Trematospira Zone, loc. 10726, Sulphur Spring Range.
  12-16 Side, anterior, posterior, ventral, and dorsal views × 4, of the holotype, USNM 157089.
  17 Interior of pedicle valve of a paratype × 4, USNM 157090.
  18, 19 Dorsal and ventral views of a paratype × 5, USNM 157091.
  20, 21 Interior and exterior of brachial valve of a paratype × 4, USNM 157092.
  22, 23 Ventral and dorsal views of a paratype × 5, USNM 157093.
  24, 25 Interior and exterior of brachial valve of a paratype × 5, USNM 157094.

PLATE 45. LEPTOCOELINA AND LEPTOCOELIA

Figures 1-12 *Leptocoelina squamosa* Johnson, n. gen. and n. sp.
Spinoplasia Zone, figs. 1-5, loc. 10760, Coal Canyon; figs. 6-12, loc. 10713, Rabbit Hill.
1-5 Ventral, dorsal, side, anterior, and posterior views × 2, holotype, USNM 157095.
6, 7 Exterior and interior of pedicle valve × 2, USNM 157096.
8, 9 Rubber impression of interior of pedicle valve and interior of pedicle valve × 2, USNM 157097.
10, 11 Interior and exterior of brachial valve × 3, USNM 157098.
12 Exterior of pedicle valve × 3, USNM 157099.

Figures 13-19 *Leptocoelia murphyi* Johnson, n. sp.
13-15 Side, anterior, and dorsal views × 2, USNM 157100, loc. 10713, *Spinoplasia* Zone, Rabbit Hill.
16 Rubber impression of interior of pedicle valve × 1.5, USNM 157101, loc. 10713, *Spinoplasia* Zone, Rabbit Hill.
17 Interior of brachial valve × 2, USNM 157102, loc. 10730, *Trematospira* Zone, Sulphur Spring Range.
18, 19 Dorsal and ventral views × 2, USNM 157103, loc. 10734. *Trematospira* Zone, Sulphur Spring Range. Note bifurcating plication on the fold of the brachial valve and complementary plication orginating by implantation in the sulcus of the pedicle valve.

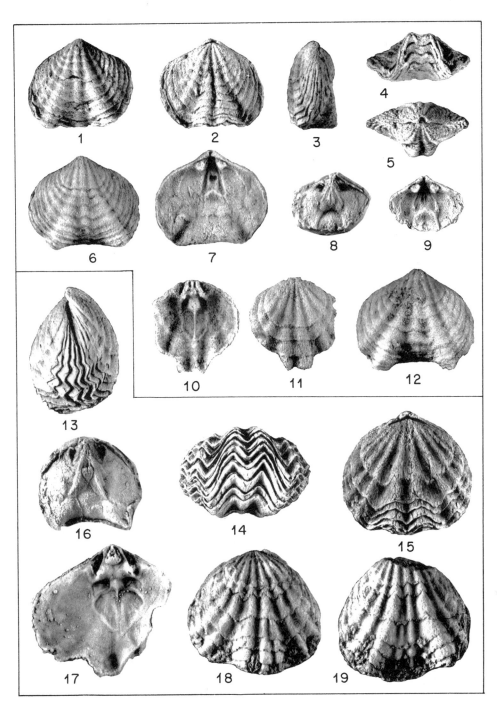

LEPTOCOELINA AND LEPTOCOELIA

JOHNSON, PLATE 45
Geological Society of America Memoir 121

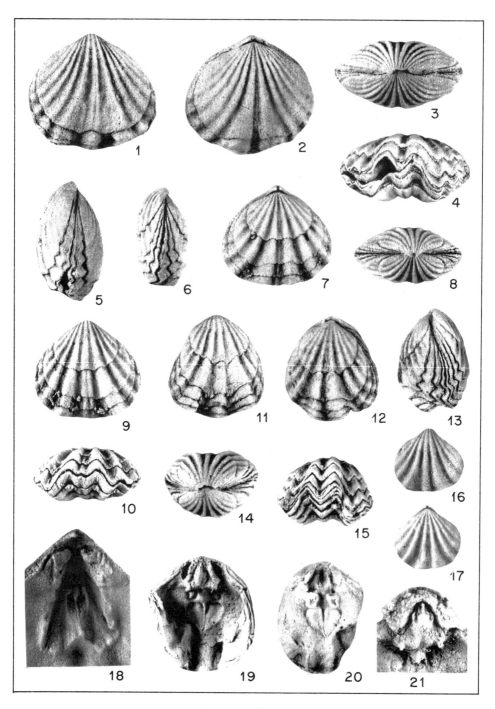

LEPTOCOELIA

PLATE 46. LEPTOCOELIA

Figures 1-21 *Leptocoelia murphyi* Johnson n. sp.
Trematospira Zone.
1-5 Ventral, dorsla, posterior, anterior, and side views × 1.5, paratype, USNM 157104, loc. 10726, Sulphur Spring Range.
6-10 Side, dorsal, posterior, ventral, and anterior views × 1.5, holotype, USNM 157105, loc. 10726, Sulphur Spring Range.
11-15 Ventral, dorsal, side, posterior, and anterior views × 1.5, paratype, USNM 157106, loc. 10726, Sulphur Spring Range.
16, 17 Ventral and dorsal views × 1.5, USNM 157107, loc. 10730, Sulphur Spring Range.
18 Portion of interior of pedicle valve × 3, paratype, USNM 157108, loc. 10726, Sulphur Spring Range. Note grooved lateral edges of the diductor muscle impressions.
19, 20 Interior of brachial valve and rubber impression of interior of brachial valve × 2, USNM 140382, loc. 10715, Roberts Mountains. Note arrangement of sharply impressed posterior and anterior adductor muscle scars.
21 Cardinalia of brachial valve with portion of pedicle valve umbo and beak attached × 3, paratype, USNM 157109, loc. 10726, Sulphur Spring Range.

PLATE 47. LEPTOCOELIA

Figures 1-14 *Leptocoelia infrequens infrequens* (Walcott).
Eurekaspirifer pinyonensis Zone.
1-3 Ventral, dorsal, and posterior views of internal mold × 3, USNM 157110, loc. 10761, northern Simpson Park Mountains.
4-8 Ventral, dorsal, posterior, anterior, and side views of the holotype × 1.5, USNM 13843, Lone Mountain.
9-13 Side, ventral, dorsal, posterior, and anterior views × 1.5, USNM 157111, loc. 10761, northern Simpson Park Mountains.
14 Ventral view × 1.5, USNM 157112, loc. 12771, Roberts Mountains.

Figures 15-27 *Leptocoelia infrequens globosa* Johnson, n. subsp.
Eurekaspirifer pinyonensis Zone, Cortez Range.
15-19 Anterior, dorsal, ventral, posterior, and side views of the holotype × 3, USNM 157113, loc. 10756.
20-24 Ventral, posterior, dorsal, anterior, and side views of a paratype × 3, USNM 157114, loc. 10756.
25 Interior of brachial valve × 2, USNM 157115, loc. 10754.
26, 27 Interior and exterior of pedicle valve × 2, USNM 157116, loc. 10754.

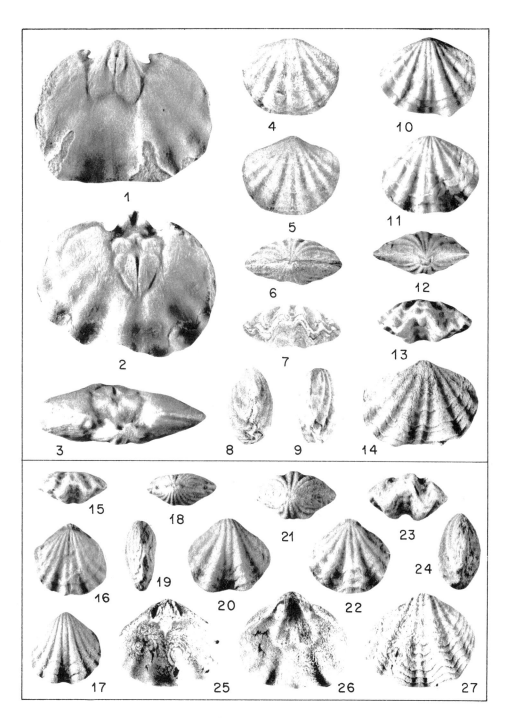

LEPTOCOELIA

JOHNSON, PLATE 47
Geological Society of America Memoir 121

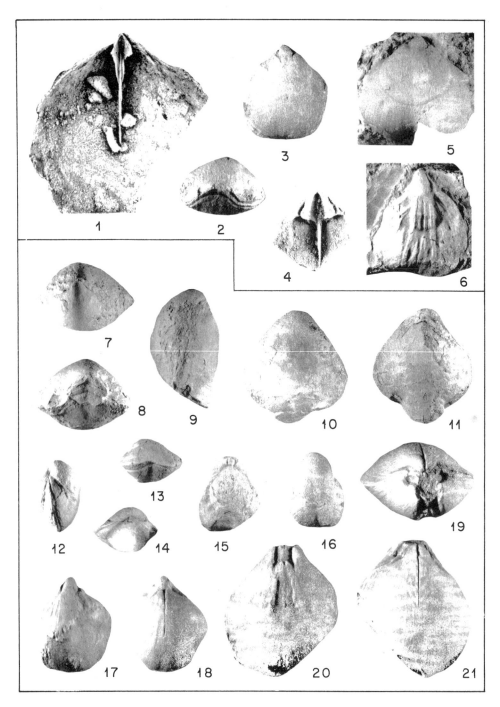

MERISTINA

JOHNSON, PLATE 48
Geological Society of America Memoir 121

PLATE 48. MERISTINA

Figures 1-6 *Meristina* sp. A.
  Quadrithyris Zone.
  1 Interior of brachial valve × 3, USNM 157117, loc. 10757, Coal Canyon.
  2, 3 Anterior and ventral views of internal mold × 1.5, USNM 157118, loc. 10808, Toquima Range.
  4 Interior of brachial valve × 3, USNM 157119, loc. 10758, Coal Canyon.
  5 Internal mold of pedicle valve × 1.5, USNM 157120, loc. 10808, Toquima Range.
  6 Internal mold of pedicle valve × 1.5, USNM 157121, loc. 10808, Toquima Range. Note relatively strongly impressed trapezoidal diductor muscle field.

Figures 7-21 *Meristina* sp. B.
  Eurekaspirifer pinyonensis Zone.
  7-11 Posterior, anterior, side, ventral, and dorsal views × 1 USNM 157122, loc. 10762, Coal Canyon.
  12-16 Side, anterior, posterior, dorsal, and ventral views × 1, USNM 157123, loc. 13267, Sulphur Spring Range.
  17, 18 Ventral and dorsal views of internal mold × 1, USNM 157124, loc. 10783, Roberts Mountains.
  19-21 Posterior, ventral, and dorsal views of internal mold × 2, USNM 157125, loc. 10705, northern Roberts Mountains. Note elongate trapezoidal platform within the ventral muscle field in figure 20.

PLATE 49. MERISTELLA

Figures 1-6 *Meristella robertsensis* Merriam.
    Acrospirifer kobehana Zone.
    1-5 Ventral, anterior, dorsal, posterior, and side views × 1 of the holotype, USNM 96369, southern Roberts Mountains.
    6 Interior of brachial valve × 2, USNM 157126, loc. 10738, Sulphur Spring Range.

Figures 7-25 *Meristella* cf. *robertsensis* Merriam.
    7 Ventral view × 1, USNM 157127, loc. 10735, *Trematospira* Zone, Sulphur Spring Range.
    8-12 Anterior, dorsal, posterior, ventral, and side views × 1, USNM 157128, loc. uncertain, ?*Eurekaspirifer pinyonesis* Zone, northern Roberts Mountains.
    13-17 Side, anterior, posterior, ventral, and dorsal views × 1, USNM 157129, loc. 10744, *Trematospira* Zone, Sulphur Spring Range.
    18-22 Side, anterior, posterior, ventral, and dorsal views × 1, USNM 157130, loc. 10746, *Trematospira* Zone, Sulphur Spring Range.
    23, 25 Interior of brachial valve × 1.5, and × 3, USNM 157137A, loc. 10742, *Trematospira* Zone, Sulphur Spring Range.
    24 Interior of pedicle valve × 1.5, USNM 157137B, loc. 10742. This valve articulates with USNM 157137A.

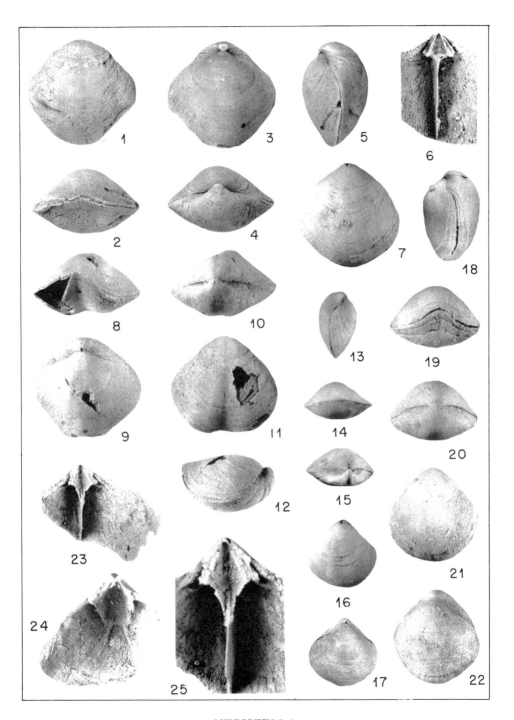

MERISTELLA

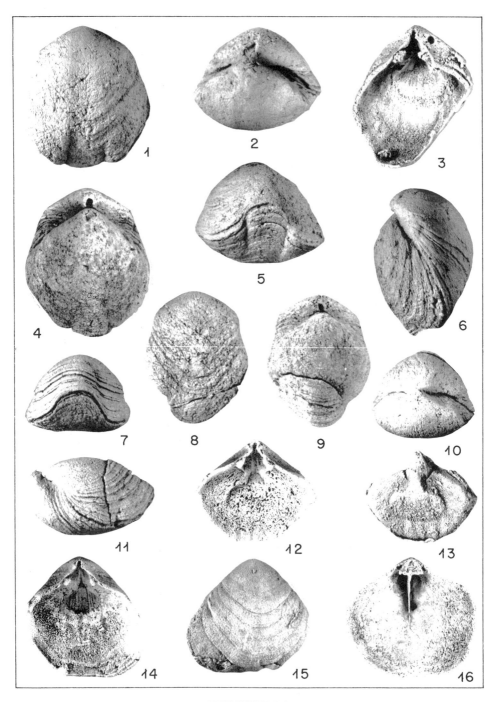

MERISTELLA

PLATE 50. MERISTELLA

Figures 1-16 *Meristella* cf. *robertsensis* Merriam.
1, 2, 4-6 Ventral, posterior, dorsal, anterior, and side views × 1.5, USNM 157131, loc. 10726, *Trematospira* Zone, Sulphur Spring Range.
3 Interior of pedicle valve × 1.5, USNM 157132, loc. 10726, *Trematospira* Zone, Sulphur Spring Range.
7-11 Anterior, ventral, dorsal, posterior, and side views × 1.5, USNM 157133, loc. 10726, *Trematospira* Zone, Sulphur Spring Range.
12, 13 Interior of pedicle valve and rubber impression × 2, USNM 157134, loc. 10713, *Spinoplasia* Zone, Rabbit Hill.
14, 15 Interior and exterior of pedicle valve × 1.5, USNM 157135, loc. 10713, *Spinoplasia* Zone, Rabbit Hill.
16 Interior of brachial valve × 1.5, USNM 157136, loc. 10726, *Trematospira* Zone, Sulphur Spring Range.

PLATE 51. NUCLEOSPIRA

Figures 1-16 *Nucleospira* sp. B.
    1, 2 Exterior and interior of brachial valve × 3, USNM 157138, loc. 10713, *Spinoplasia* Zone, Rabbit Hill.
    3, 4 Interior and exterior of pedicle valve × 2, USNM 157139, loc. 10713, *Spinoplasia* Zone, Rabbit Hill.
    5-9 Side, dorsal, posterior, ventral, and anterior views × 2, USNM 157140, loc. 10777, *Acrospirifer kobehana* Zone, northern Roberts Mountains.
    10, 11 Two views of interior of brachial valve × 3, USNM 157141, loc. 10777, *Acrospirifer kobehana* Zone, northern Roberts Mountains.
    12 Exterior of shell × 5, showing fine spinules, USNM 157142, loc. 10731, *Trematospira* Zone, Sulphur Spring Range.
    13 Interior of pedicle valve × 3, USNM 157143, loc. 10777, *Acrospirifer kobehana* Zone, northern Roberts Mountains.
    14-16 Side view × 2, interior × 3, and dorsal view × 2 of brachial valve and beak of the pedicle valve, USNM 157144, loc. 10714, *Trematospira* Zone, northern Roberts Mountains.

Figures 17-24 *Nucleospira subsphaerica* Johnson, n. sp.
    Eurekaspirifer pinyonensis Zone; figs. 17, 19-24, Sulphur Spring Range; fig. 18, Cortez Range.
    17 Dorsal view × 1.25, USNM 157145, loc. 10711.
    18 Interior of brachial valve × 3, USNM 157146, loc. 10752.
    19 Interior of pedicle valve × 2, USNM 157147, loc. 13267.
    20-24 Ventral, dorsal, side, anterior, and posterior views × 1.5, holotype, USNM 157148, loc. 10711.

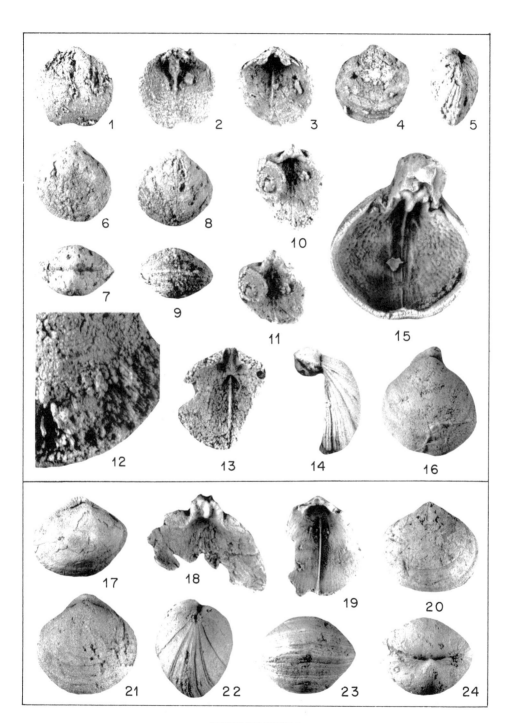

NUCLEOSPIRA

JOHNSON, PLATE 51
Geological Society of America Memoir 121

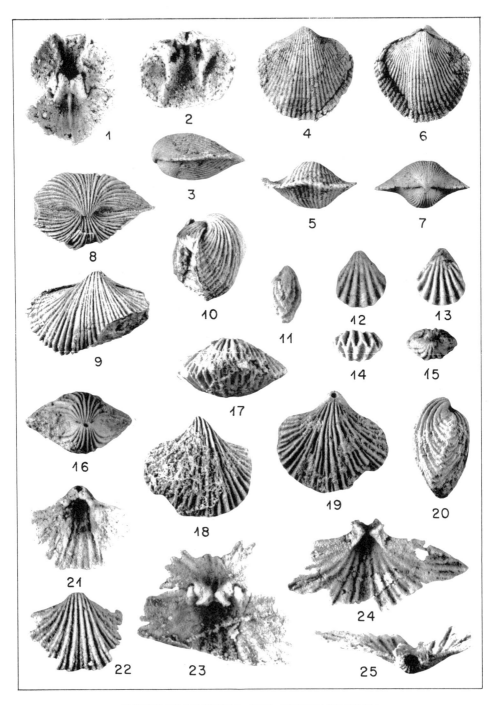

PSEUDOPARAZYGA AND TREMATOSPIRA

PLATE 52. PSEUDOPARAZYGA AND TREMATOSPIRA

Figures 1-7 *Pseudoparazyga cooperi* (Merriam) n. gen.
Trematospira Subzone, McColley Canyon, loc. 12794.
1 Interior of articulated valves × 3, USNM 157155.
2 Interior of pedicle valve × 3, USNM 157156.
3-7 Side, ventral, anterior, dorsal, and posterior views × 2, USNM 157157.

Figures 8-25 *Trematospira perforata* Hall.
Trematospira Zone, figures 8-10, 21-25, McColley Canyon; figures 11-20, Cortez Range.
8-10 Posterior, ventral, and side views × 1.25, USNM 157149, loc. 12793.
11-15 Side, ventral, dorsal, anterior, and posterior views of a small specimen × 4, USNM 157150, loc. 10750.
16-20 Posterior, anterior, ventral, dorsal, and side views × 2, USNM 157151, loc. 10750.
21, 22 Interior and exterior of pedicle valve × 2, USNM 157152, loc. 10731.
23 Interior of posterior portions of both valves × 3, USNM 157153, loc. 10727.
24, 25 Interior and posterior views of brachial valve × 3, USNM 157154, loc. 10727.

PLATE 53. PSEUDOPARAZYGA

Figures 1-18 *Pseudoparazyga cooperi* (Merriam) n. gen.
  1-5 Ventral, dorsal, anterior, posterior, and side views of the holotype, × 1, USNM 96372, northern Roberts Mountains.
  6-8 Ventral, posterior, and dorsal views × 1.5, USNM 157158, loc. 10726, *Trematospira* Subzone, Sulphur Spring Range.
  9 Interior of posterior part of pedicle valve × 4, USNM 157159, loc. 10713, *Spinoplasia* Zone. Note the plate-like pedicle collar.
  10, 11 Interior and posterior of brachial valve × 2, USNM 157160, loc. 10713, *Spinoplasia* Zone, Rabbit Hill.
  12, 13 Interior and posterior views of brachial valve × 3, USNM 157161, loc. 10713, *Spinoplasia* Zone, Rabbit Hill.
  14-16 Anterior, ventral, and dorsal views × 1, USNM 157162, loc. 10742, *Trematospira* Subzone, Sulphur Spring Range.
  17, 18 Rubber impression of interior of pedicle valve and interior of pedicle valve × 1.5, USNM 157163, loc. 10713, *Spinoplasia* Zone, Rabbit Hill.

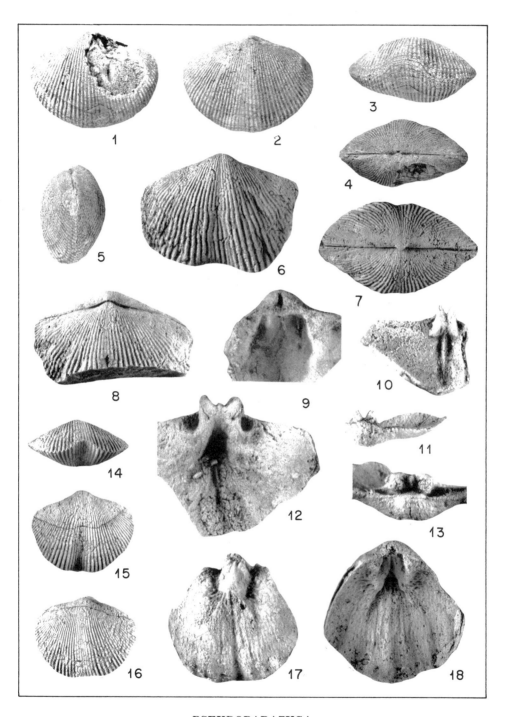

PSEUDOPARAZYGA

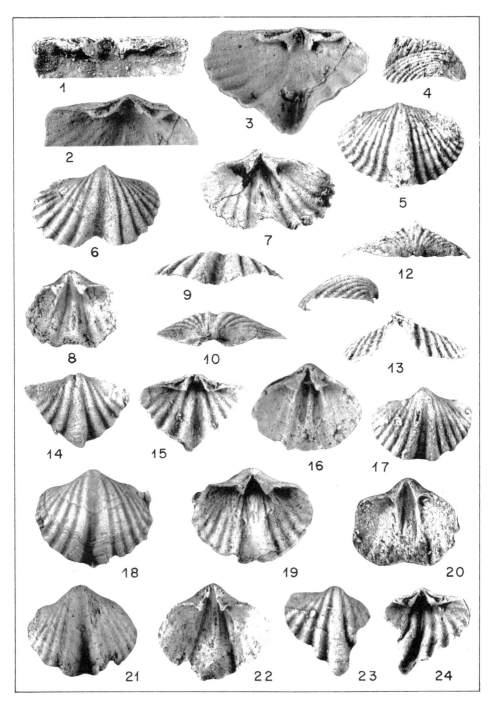

HOWELLELLA

PLATE 54. HOWELLELLA

Figures 1-24 *Howellella cycloptera* (Hall).
  Spinoplasia Zone, Rabbit Hill and vicinity.
  1 Anterior view of dorsal interior × 3, USNM 157164, loc. 12324.
  2, 3 Interior and anterior views of brachial valve × 3, USNM 157165, loc. 12324. Note crural plates in figure 3.
  4, 5, 12, 13 Side, dorsal, posterior, and anterior views of brachial valve × 1.25, USNM 157166, loc. 12324.
  6, 7, 9, 10, 11 Ventral, interior, anterior, posterior, and side views of pedicle valve × 1.5, USNM 157167, loc. 12324.
  8 Interior of pedicle valve × 2, USNM 157168, loc. 10713.
  14, 15 Exterior and interior of brachial valve × 2, USNM 157169, loc. 10713.
  16 Interior of pedicle valve × 3, USNM 157170, loc. 10713.
  17 Ventral view × 2, USNM 157171, loc. 10713.
  18, 19 Exterior and interior of pedicle valve × 1.5, USNM 157172, loc. 10713.
  20 Interior of pedicle valve × 1.5, USNM 157173, loc. 10713.
  21, 22 Exterior and interior of pedicle valve × 1.5, USNM 157174, loc. 10713.
  23, 24 Exterior and interior of brachial valve × 2, USNM 157175, loc. 10713.

PLATE 55. HOWELLELLA

Figures 1-19 *Howellella* cf. *textilis* Talent.
Eurekaspirifer pinyonensis Zone.
1-5 Ventral, dorsal, side, anterior, and posterior views of internal mold × 1.5, USNM 157176, loc. 10769, northern Roberts Mountains.
7, 8, 10-12 Posterior, anterior, ventral, dorsal, and side views × 2, USNM 157177, loc. 10783, Roberts Mountains.
6, 9 Posterior view × 5, and ventral view × 2 of internal mold prepared from specimen figured in figures 7, 8, 10-12.
13, 14, 17 Anterior and dorsal views × 2, and dorsal view of fine ornament × 5, USNM 157178, loc. 10761, *Leptocoelia infrequens* Subzone, northern Simpson Park Range.
15, 16 Ventral and posterior views × 2 of internal mold prepared from a specimen illustrated in figures 13, 14, 17. Note groove for narrow myophragm.
18, 19 Posterior and ventral views of internal mold × 1.5, USNM 157179, loc. 10723, Garden Valley quadrangle.

Figures 20-22 *Howellella aculeata* (Schnur).
Calceola-Schichten, Gerolstein, Germany. Coll. by. B. Dohm.
20, 21 Posterior view of internal mold of brachial valve × 5 and ventral view of internal mold of pedicle valve × 2, UCLA no. 17387A. Note short crural plates present in the brachial valve and median groove of the long myophragm in the pedicle valve.
22 Dorsal view of one flank of brachial valve × 5, UCLA 17387B. Note swollen spine bases on the growth lamellae.

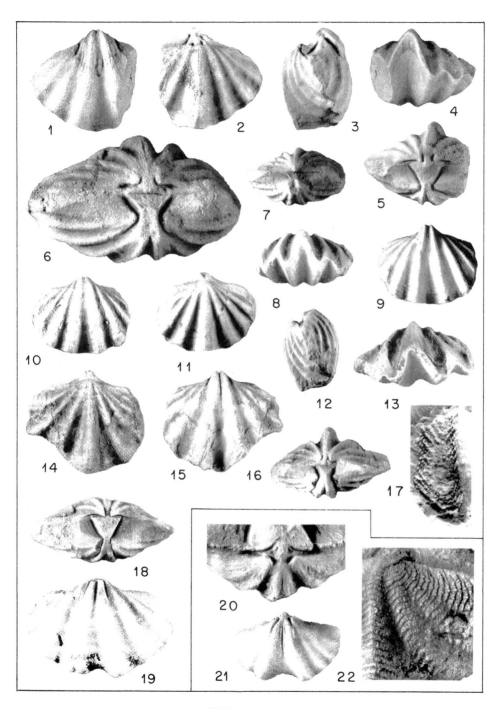

HOWELLELLA

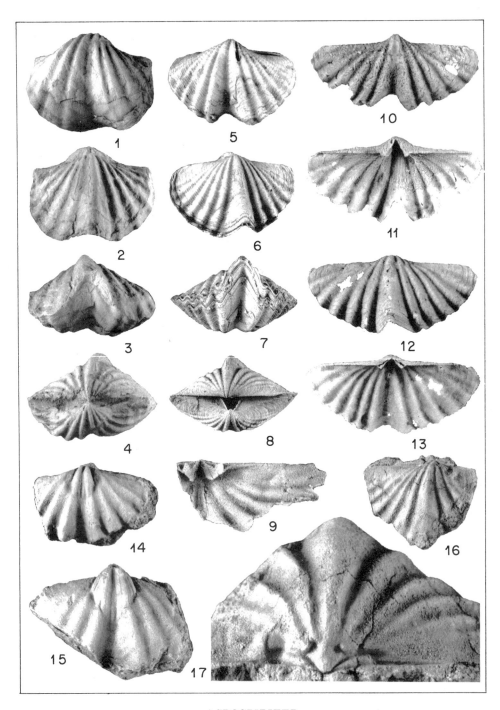

ACROSPIRIFER

PLATE 56. ACROSPIRIFER

Figures 1-4, 14-17 *Acrospirifer* cf. *kobehana* (Merriam).
Zone uncertain, Sulphur Spring Range.
1-4 Ventral, dorsal, anterior, and posterior views × 1, USNM 157184, near Bailey Pass (loc. 10722?), Garden Valley quadrangle.
14 Ventral view of internal mold of pedicle valve × 2, USNM 157185, loc. 10722, Garden Valley quadrangle.
15 Ventral view of internal mold of pedicle valve × 1.5, USNM 157186, loc. 10722, Garden Valley quadrangle.
16, 17 Dorsal view × 1 and posterior view × 3 of internal mold of pedicle valve, USNM 157187, loc. 10722, Garden Valley quadrangle.

Figures 5-13 *Acrospirifer* aff. *murchisoni* (Castelneau).
Trematospira Zone.
5-8 Ventral, dorsal, anterior, and posterior views × 1, USNM 157181, loc. 10745, McColley Canyon.
9 Interior of brachial valve × 3, USNM 157182, loc. 10778, northern Roberts Mountains.
10, 11 Exterior and interior of pedicle valve × 1.5, USNM 157183A, loc. 10778, northern Roberts Mountains.
12, 13 Exterior and interior of brachial valve of specimen illustrated in figures 10 and 11 × 1.5, USNM 157183B, loc. 10778, northern Roberts Mountains.

PLATE 57. ACROSPIRIFER

Figures 1-6 *Acrospirifer* aff. *murchisoni* (Castelneau).
Trematospira Zone, Hanson Creek area, Roberts Mountains (exact locality unknown).
1 Anteroventral view of exterior × 5 showing fimbriate ornament. Same specimen as in figures 2-6, but with anterior growth increments trimmed away.
2-6 Ventral, dorsal, posterior, side, and anterior views × 1, USNM 157180.

Figures 7-11 *Acrospirifer kobehana* (Merriam).
Acrospirifer kobehana Zone, loc. 12778, Sulphur Spring Range.
7-11 Side, dorsal, anterior, ventral, and posterior views × 1.25, USNM 157188. Note the carinate outline of the brachial valve in figure 11.

ACROSPIRIFER

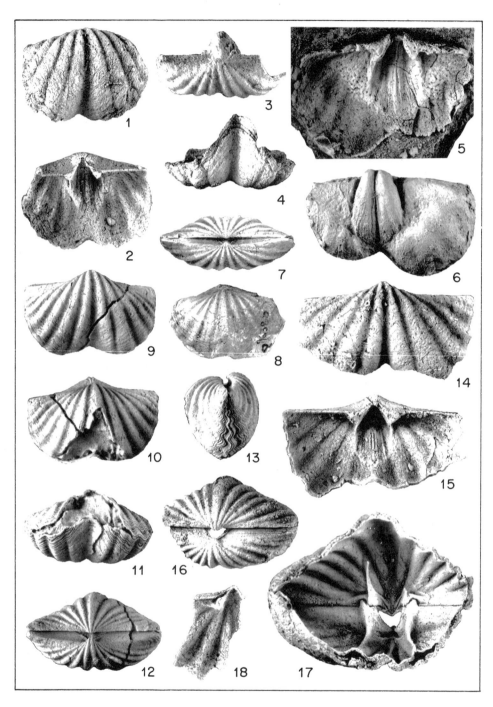

ACROSPIRIFER

PLATE 58. ACROSPIRIFER

Figures 1-18 *Acrospirifer kobehana* (Merriam).
Acrospirifer kobehana Zone.
1-4 Exterior, interior, posterior, and anterior views of pedicle valve × 1, USNM 157189, loc. 10738, Sulphur Spring Range.
5, 6 Interior of pedicle valve and rubber impression × 1.5, USNM 157190, loc. 10780, Roberts Mountains.
7 Posterior view × 1 of the holotype, USNM 96387, southern Roberts Mountains.
8 Dorsal view × 1 of a paratype, USNM 96388A, southern Roberts Mountains.
9-13 Ventral, dorsal, anterior, posterior, and side views × 1, USNM 157191, loc. 10790, southern Roberts Mountains.
14, 15 Exterior and interior views of a pedicle valve × 1.5, USNM 157192, loc. 10777, northern Roberts Mountains.
16, 17 Posterior view × 1, and interior view × 1.5, USNM 157193, loc. 10738, Sulphur Spring Range.
18 Interior of part of brachial valve × 1.5, USNM 157194, loc. 10738, Sulphur Spring Range.

PLATE 59. DYTICOSPIRIFER

Figures 1-16 *Dyticospirifer mccolleyensis* Johnson.
Trematospira Zone, Sulphur Spring Range.

1, 2 Ventral and side views of exterior of the holotype × 1.5, USNM 146512, loc. 10726.
3 Exterior of brachial valve of a paratype × 1.5, USNM 146516, loc. 10726.
4-8 Ventral, anterior, dorsal, posterior, and side views × 1, USNM 146514, loc. 10731.
9 Ventral view of fine ornament in sulcus of a paratype × 8, USNM 146513, loc. 10726.
10-12 Interior showing cardinalia × 2, interior × 1.5, and exterior × 1.5 of brachial valve, USNM 146517, loc. 10731.
13 Interior of pedicle valve × 1.5, USNM 146515, loc. 10731.
14 Exterior of brachial valve of a paratype × 1.5, USNM 146520, loc. 10726.
15 Rubber impression of interior of pedicle valve × 1.5, impression of USNM 146519, loc. 10731.
16 Interior view showing cardinalia of brachial valve × 5, USNM 146518, loc. 10730.

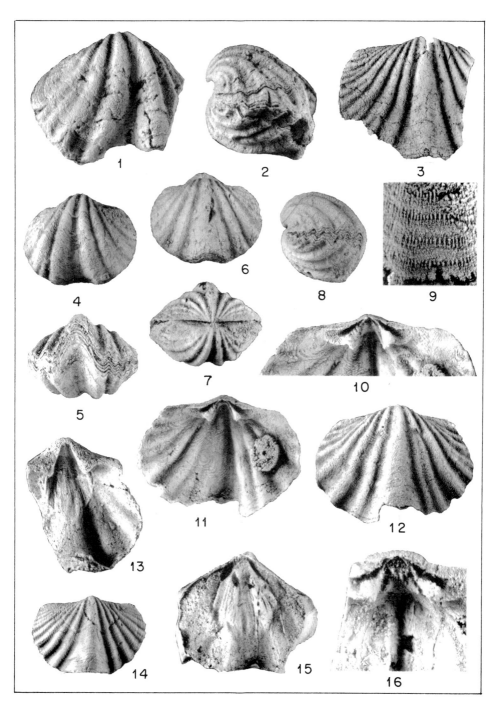

DYTICOSPIRIFER

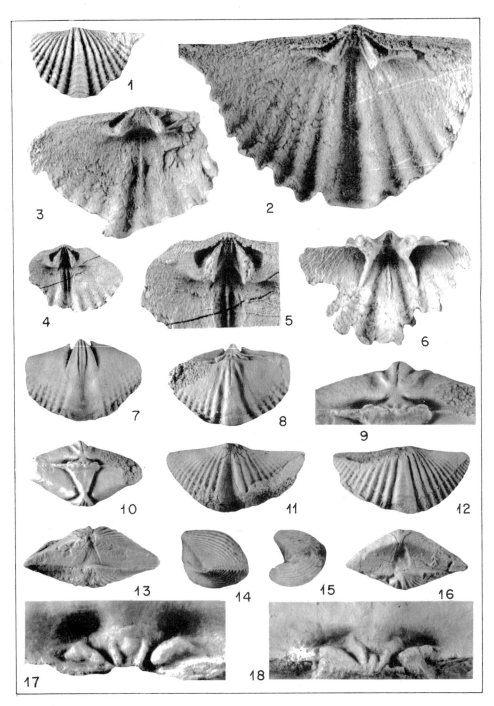

HYSTEROLITES

JOHNSON, PLATE 60
Geological Society of America Memoir 121

PLATE 60. HYSTEROLITES

Figures 1-18 *Hysterolites* sp. A.
Eurekaspirifer pinyonensis Zone.
1, 2 Exterior × 2 and interior × 5 of brachial valve, USNM 157195, loc. 10754, Cortez Range.
3 Interior of brachial valve × 5, USNM 157196, loc. 10754, Cortez Range.
4, 5 Interior of brachial valve × 2 and × 5, USNM 157197, loc. 10754, Cortez Range.
6 Interior of pedicle valve × 5, USNM 157198, loc. 10754, Cortez Range.
7, 8, 10 Ventral, dorsal, and posterior views of internal mold × 1.5, USNM 157199, loc. 10789, Roberts Mountains.
9 Posterior view of internal mold of brachial valve × 3. Same specimen as in figure 10. Note slots for very short crural plates and slight bilobation of the site of diductor attachment.
11-14 Ventral, dorsal, posterior, and side views × 1.5, USNM 157200, loc. 10787, Roberts Mountains.
15, 16 Side and posterior views × 2, USNM 157201, loc. 10793, southern Roberts Mountains.
17 Posterior view of internal mold of brachial valve × 5, USNM 157202, loc. 12775, Lone Mountain.
18 Posterior view of internal mold of brachial valve × 5, USNM 157203, loc. 12775, Lone Mountain.

PLATE 61. HYSTEROLITES

Figures 1-20 *Hysterolites* sp. B.

Eurekaspirifer pinyonensis Zone; figures 1-17, Garden Valley quadrangle; figures 18-20, Roberts Mountains.

1, 2 Posterior view × 3 and dorsal view × 1.5, of internal mold of brachial valve, USNM 157204, loc. 10717. Note relatively large slots of the crural plates.

3-7 Ventral, dorsal, posterior, anterior, and side views × 1.5, USNM 157205, loc. 10717.

8-12 Side, posterior, anterior, ventral, and dorsal views × 1.5, USNM 157206, loc. 10717.

13, 14 Dorsal and ventral views × 1.5, USNM 157207, loc. 10717.

15 Ventral view of internal mold of specimen illustrated in figure 14 × 1.5, loc. 10717.

16 Ventral view × 1.5, USNM 157208, loc. 10717.

17 Posterior view of internal mold of brachial valve × 5, USNM 157209, loc. 10717. Note relatively strong slots of the crural plates.

18-20 Anterior, side, and posterior views × 1.5, USNM 157210, loc. 10787.

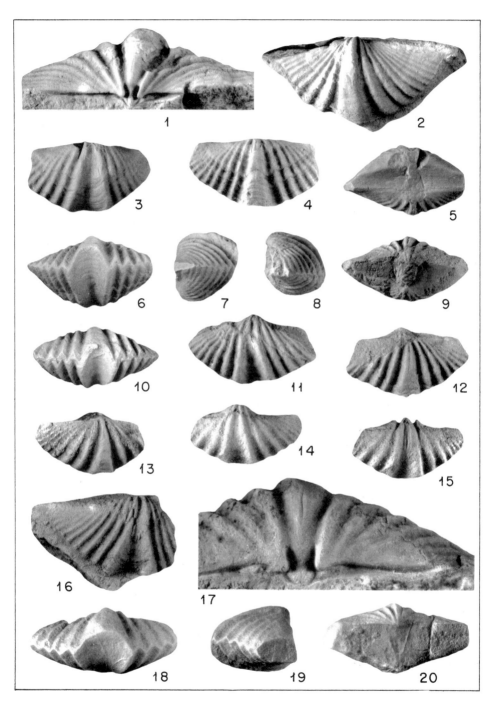

HYSTEROLITES

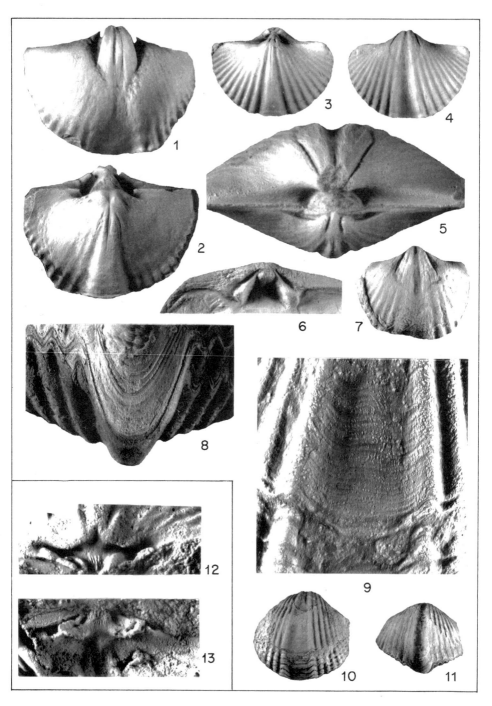

BRACHYSPIRIFER

PLATE 62. BRACHYSPIRIFER

Figures 1-11 *Brachyspirifer pinyonoides* Johnson n. sp.
Eurekaspirifer pinyonensis Zone.
1, 2 Ventral and dorsal views of internal mold × 1.5, USNM 157212, loc. 10788, Roberts Mountains.
3-5 Dorsal and ventral views × 2, and posterior view × 5 of internal mold, USNM 157213, loc. 12775, Lone Mountain. In figure 5 note relatively long, thin, slightly divergent slots for crural plates and note lack of secondary shell material filling the umbonal cavities of the pedicle valve.
6 Interior of posterior portion of brachial valve × 3, USNM 157214, loc. 12775, Lone Mountain.
7 Ventral view of internal mold × 1.5, USNM 157215, loc. 10788, Roberts Mountains. Another view is figure 18 of Plate 63.
8, 11 Anterior view × 5, and ventral view × 1, USNM 157216, loc. 10781, northern Roberts Mountains.
9, 10 Ventral view × 5, and × 1 of exterior of pedicle valve, USNM 157217, loc. 10770, northern Roberts Mountains.

Figures 12, 13 *Brachyspirifer carinatus* (Schnur).
South of Daleiden, on road to Bommert, at northeast end of Hubert, Eifel. Wiltzer Schichten. USNM loc. 10325. Coll. by A. J. Boucot, 1956.
12 Posterior view of internal mold of brachial valve × 5, USNM 157211.
13 Rubber impression of specimen in figure 12 × 5.

PLATE 63. BRACHYSPIRIFER

Figures 1-18 *Brachyspirifer pinyonoides* Johnson, n. sp.
Eurekaspirifer pinyonensis Zone.

1-5 Dorsal, ventral, side, anterior, and posterior views of the holotype × 1.5, USNM 157218, loc. 10762, Coal Canyon.

6, 7 Dorsal and ventral views × 2 of a paratype, USNM 157219, loc. 10762, Coal Canyon.

8, 9, 11 Ventral, dorsal, and posterior views of an internal mold of a paratype × 1.5, USNM 157220, loc. 10762, Coal Canyon.

10 Posterior view of internal mold of brachial valve × 3, of specimen in figure 11. Note absence of slots for crural plates and presence of strongly impressed adductor muscle impressions.

12-14 Ventral and dorsal views × 1.5 and posterior view × 2.5 of internal mold of a paratype, USNM 157221, loc. 10762, Coal Canyon. Note the absence of strongly impressed adductor muscle scars and presence of slots for crural plates in the brachial valve and absence of strongly impressed musculature in the pedicle valve.

15 Posterior view of internal mold of brachial valve × 5, USNM 157222, loc. 10788, Roberts Mountains. Note short slots for crural plates and moderately faceted striate area of diductor attachment.

16 Ventral view of internal mold × 2, USNM 157223, loc. 10788, Roberts Mountains.

17 Ventral view of internal mold × 1.5, USNM 157224, loc. 10788, Roberts Mountains.

18 Posterior view of internal mold of brachial valve × 5, USNM 157215, loc. 10788, Roberts Mountains. Note absence of slots for crural plates. Another view is figure 7 of Plate 62.

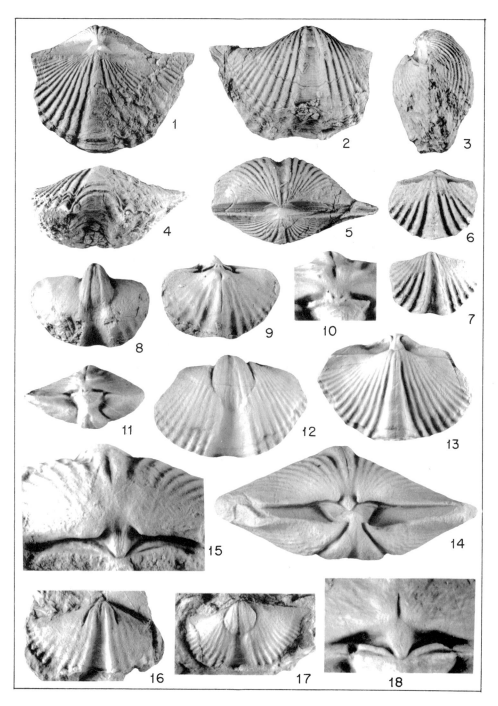

BRACHYSPIRIFER

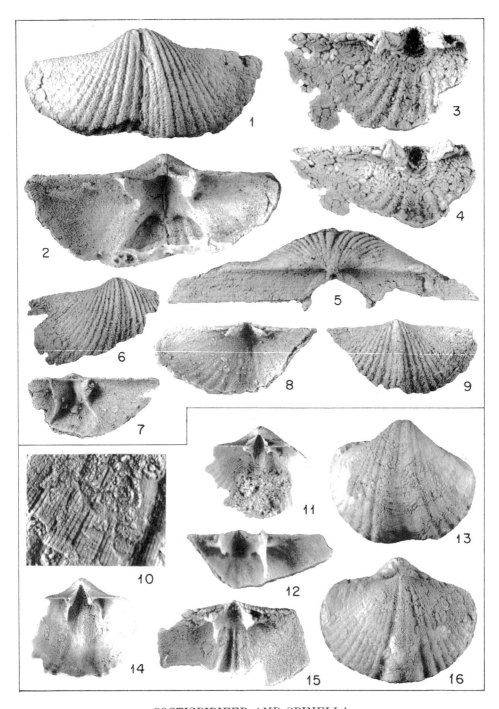

COSTISPIRIFER AND SPINELLA

PLATE 64. COSTISPIRIFER AND SPINELLA

Figures 1-9 *Costispirifer* sp.
Trematospira Zone, *Costispirifer* Subzone.
1, 2 Exterior and interior of pedicle valve × 1.5, USNM 157225, loc. 10773, northern Roberts Mountains.
3, 4 Two views of interior of brachial valve × 2, USNM 157226, loc. 10773, northern Roberts Mountains.
5 Posterior view of pedicle valve × 1.5, USNM 157227, loc. 10741, Sulphur Spring Range. Note trapezoidal interarea.
6-9 Exterior and interior views of pedicle valve and interior and exterior views of brachial valve × 1, USNM 157228A and B, loc. 10773, northern Roberts Mountains.

Figures 10-16 *Spinella talenti* Johnson, n. sp.
Eurekaspirifer pinyonensis Zone, loc. 12775, Lone Lountain.
10 Dorsal view of part of exterior × 5, holotype, USNM 146536.
11, 15 Interior of pedicle valve × 1.5, and interior of brachial valve × 3, USNM 157229A and B.
12, 14 Interior of brachial valve × 3, USNM 146537A, and interior of pedicle valve × 1.5, USNM 146537. In figure 12, note well-developed crural plates attached to the base of the valve.
13 Ventral view × 2, USNM 157230.
16 Dorsal view × 2, USNM 146536. Figure 10 is an enlargement of a portion of left flank of this specimen.

PLATE 65. EUREKASPIRIFER

Figures 1-15 *Eurekaspirifer pinyonensis* (Meek).
    E. pinyonensis Zone, Sulphur Spring Range.
    1, 2 Interior of pedicle valve and rubber impression of interior of pedicle valve × 1.5, USNM 146525, loc. 10728.
    3, 4 Interior of brachial valve and rubber impression of interior of brachial valve × 3, USNM 146521, loc. 10728.
    5 Interior of pedicle valve × 2, USNM 146529, loc. 10729.
    6 Interior of brachial valve × 3, USNM 146522, loc. 10729. Note the long dorsal adminicula.
    7 Interior of pedicle valve × 3, viewed slightly from the anterior, USNM 146528, loc. 10729.
    8, 9 Anterior and oblique views of interior of brachial valve × 2, of USNM 146527, loc. 10728.
    10 Side view of interior of posterior portion of articulated valves × 2, USNM 146526, loc. 10728. Note outline of the dental lamellae and of the crus.
    11, 12 Two views of interior of pedicle valve × 2, USNM 146524, loc. 10728.
    13 Interior of brachial valve × 4, USNM 146523, loc. 10728. Note absence of dorsal adminicula.
    14, 15 Two views of cardinalia of the brachial valve × 5, USNM 146526, loc. 10728. Note the pair of strongly projecting, deeply striate, cardinal process lobes attached to the inner sides of the crural plates flanking the notothyrial cavities.

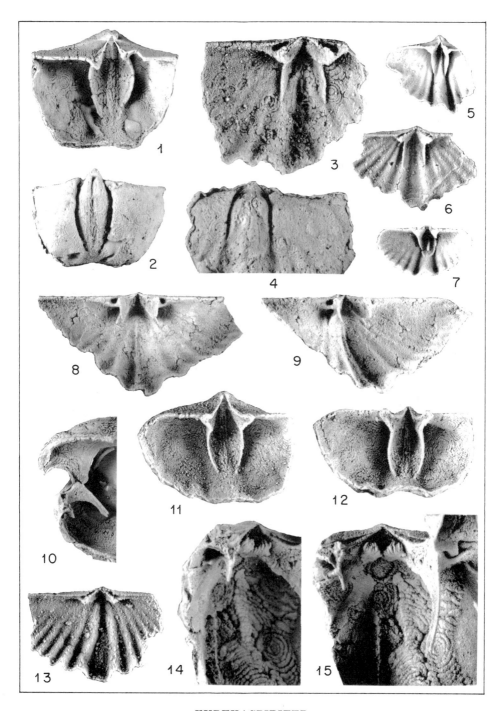

EUREKASPIRIFER

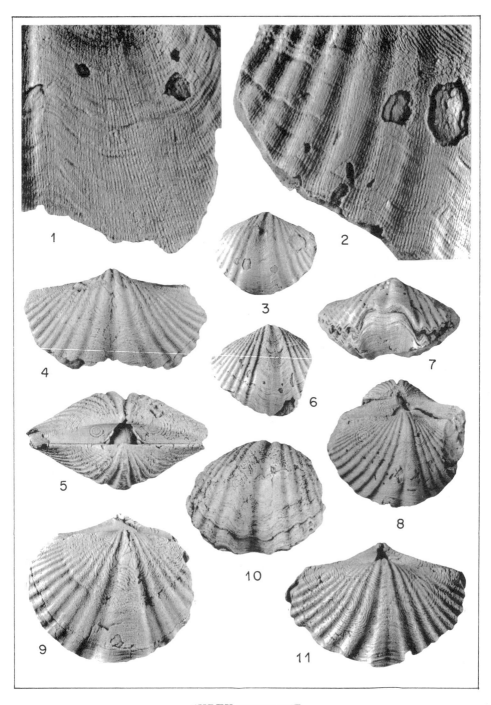

EUREKASPIRIFER

PLATE 66. EUREKASPIRIFER

Figures 1-12 *Eurekaspirifer pinyonensis* (Meek).
E. pinyonensis Zone, Sulphur Spring Range.
1, 6 Ventral view of the fine ornament × 5 and ventral view of exterior × 1, USNM 146530, loc. 10711.
2, 3 Ventral view of fine ornament × 5 and ventral view of exterior × 1, USNM 146531, loc. 10711.
4, 5, 11 Ventral, posterior, and dorsal views × 2, USNM 146535, loc. 10711.
7, 8 Anterior and dorsal views × 1.5, USNM 146533, loc. unknown.
9 Dorsal view × 2, USNM 146532, loc. 10711.
10 Ventral view × 1.5, USNM 146534, loc. 10711.

PLATE 67. QUADRITHYRIS AND RETICULARIOPIS

Figures 1-15 *Quadrithyris* cf. *minuens* (Barrande).
Quadrithyris Zone.
1 Exterior of pedicle valve × 5, USNM 157231, loc. 10758, Coal Canyon. Note fine ornament.
2-6 Anterior, posterior, side, ventral, and interior views of pedicle valve × 2, of specimen in figure 1.
7 Exterior of brachial valve × 2, USNM 157232, loc. 10758, Coal Canyon.
8, 9 Interior and exterior of brachial valve × 3, USNM 157233, loc. 10758, Coal Canyon.
10-12 Exterior of brachial valve and two views of interior of brachial valve × 3, USNM 157234, loc. 10758, Coal Canyon.
13-15 Interior and exterior of pedicle valve × 2, interior of pedicle valve from oblique angle × 3, USNM 157235, loc. 10808, Ikes Canyon, Toquima Range. Note median septum.

Figures 16, 17 *Reticulariopsis?* sp. A.
Quadrithyris Zone, loc. 10757, Coal Canyon.
16 Interior of brachial valve × 5, USNM 157236.
17 Interior of pedicle valve × 3, USNM 157237.

Figures 18-24 *Reticulariopsis* sp. B.
Trematospira Zone, Sulphur Spring Range.
18, 19 Exterior and interior of pedicle valve × 4, USNM 157238, loc. 10736.
20 Exterior of brachial valve showing fine ornament × 8, USNM 157239, loc. 10747.
21-24 Side, dorsal, and two interior views of brachial valve × 3, of specimen in figure 20.

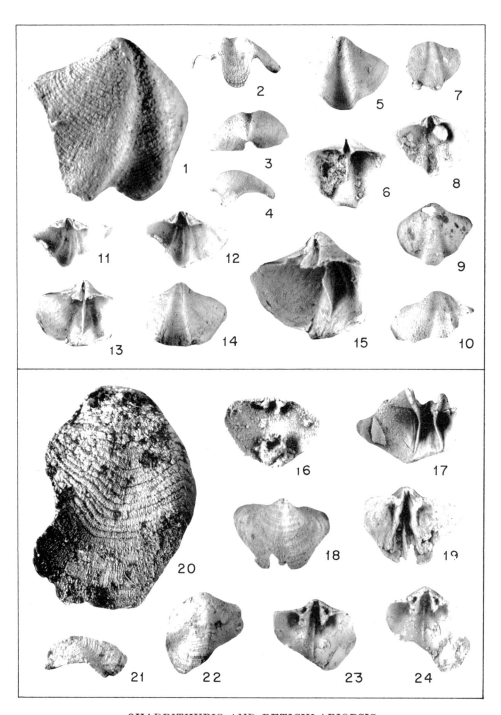

QUADRITHYRIS AND RETICULARIOPSIS

JOHNSON, PLATE 67
Geological Society of America Memoir 121

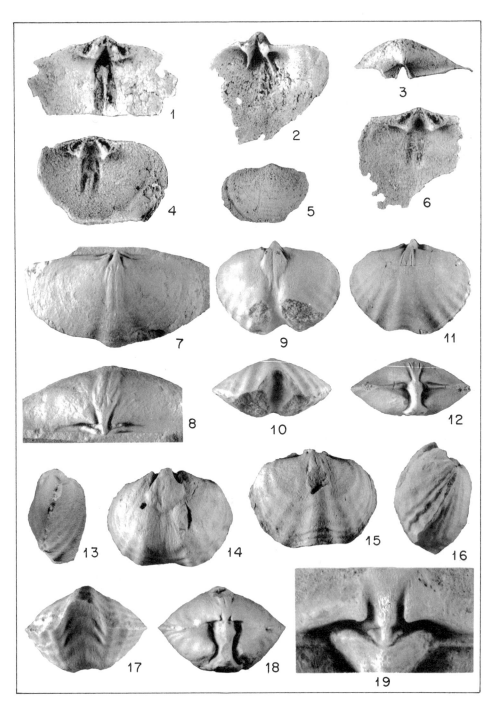

ELYTHYNA

PLATE 68. ELYTHYNA

Figures 1-13 *Elythyna* sp. A.
Eurekaspirifer pinyonensis Zone.
1 Interior of brachial valve × 2, USNM 157240, loc. 10754, Cortez Range.
2, 3 Interior and posterior views of pedicle valve × 2, USNM 157241, loc. 10754, Cortez Range.
4, 5 Interior view × 3 and exterior view × 2 of brachial valve, USNM 157242, loc. 10754, Cortez Range.
6 Interior of brachial valve × 3, USNM 157243, loc. 10754, Cortez Range.
7, 8 Dorsal view × 1.5 and posterior view × 2 of internal mold of brachial valve, USNM 157244, loc. 12393, *Leptocoelia infrequens* Subzone, northern Simpson Park Mountains.
9-13 Ventral, anterior, dorsal, posterior, and side views of internal mold × 1, USNM 157245, loc. 12393, *Leptocoelia infrequens* Subzone, northern Simpson Park Mountains.

Figures 14-19 *Elythyna* sp. B.
Elythyna beds.
14-18 Ventral, dorsal, side, anterior, and posterior views of internal mold × 1.5, USNM 157246, loc. 10767, northern Roberts Mountains.
19 Posterior view of internal mold of brachial valve × 5, USNM 157247, loc. 10792, southern Roberts Mountains. Note long slots of the crural plates.

PLATE 69. SPINOPLASIA AND PLICOPLASIA

Figures 1-13 *Spinoplasia roeni* Johnson, n. sp.
  Spinoplasia Zone, loc. 10713, Rabbit Hill.
  1, 2 Interior and rubber impression of interior of pedicle valve × 5, USNM 157248.
  3, 4 Exterior and interior of pedicle valve × 5, USNM 157249.
  5 Interior of brachial valve × 5, USNM 157250.
  6 Interior of brachial valve × 5, USNM 157251.
  7, 8 Interior and exterior of pedicle valve × 5, holotype, USNM 157252.
  9 Interior of pedicle valve × 5, USNM 157253.
  10 Interior of pedicle valve × 6, USNM 157254.
  11 Interior of pedicle valve × 5, USNM 157255.
  12 Interior of pedicle valve × 6, USNM 157256.
  13 Interior of pedicle valve × 6, USNM 157257.

Figures 14-23 *Plicoplasia cooperi* Boucot.
  Spinoplasia Zone, loc. 10713, Rabbit Hill.
  14, 15 Interior and exterior of brachial valve × 2, USNM 157258.
  16 Posterior part of interior of pedicle valve × 5, USNM 157259.
  17, 21 Exterior and interior of pedicle valve × 4, USNM 157260.
  18 Interior of brachial valve × 5, USNM 157261.
  19, 20 Exterior and interior of brachial valve × 3, USNM 157262.
  22, 23 Interior and exterior of pedicle valve × 4, USNM 157263.

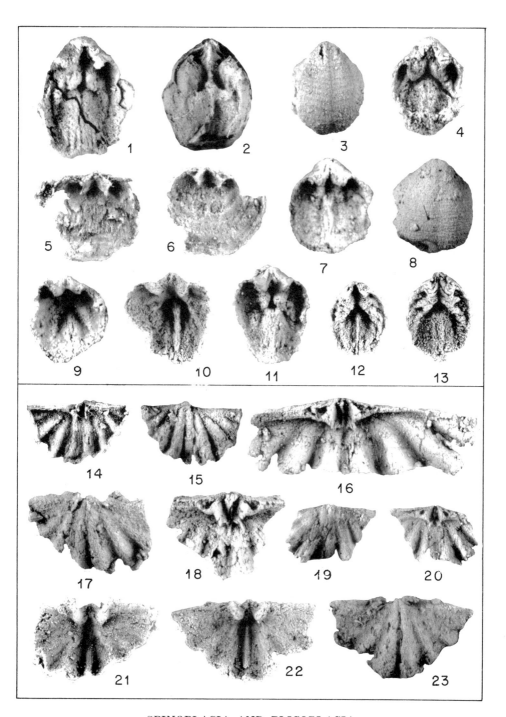

SPINOPLASIA AND PLICOPLASIA

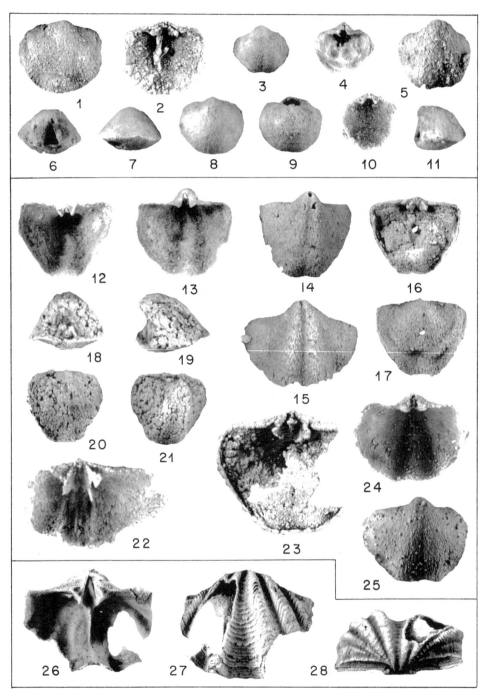

AMBOCOELIA, METAPLASIA, AND MEGAKOZLOWSKIELLA

PLATE 70. AMBOCOELIA, METAPLASIA, AND MEGAKOZLOWSKIELLA

Figures 1-11 *Ambocoelia* sp.
 Quadrithyris Zone, loc. 10758, Coal Canyon.
 1, 2 Exterior and interior of brachial valve × 5, USNM 157264.
 3, 4 Exterior and interior of pedicle valve × 3, USNM 157265.
 5 Exterior of pedicle valve × 5, USNM 157266.
 6-9, 11 Posterior, anterior, two ventral views, and side view × 5, USNM 157267.
 10 Interior of brachial valve × 5, USNM 157268.

Figures 12-25 *Metaplasia* cf. *paucicostata* (Schuchert).
 12-14 Two interior views and exterior view of pedicle valve × 5, USNM 157269, loc. *Trematospira* Zone, Sulphur Spring Range.
 15 Ventral view × 5, USNM 157270, loc. 10726, *Trematospira* Zone, Sulphur Spring Range.
 16, 17 Interior and exterior of brachial valve × 5, USNM 157271, loc. 10726, *Trematospira* Zone, Sulphur Spring Range.
 18-21 Posterior, side, dorsal, and ventral views × 5, USNM 157272, loc. 10743, *Trematospira* Zone, Sulphur Spring Range.
 22 Interior of brachial valve × 8, USNM 157273, loc. 10791, *Acrospirifer kobehana* Zone, southern Roberts Mountains.
 23 Interior of brachial valve × 8, USNM 157274, loc. 10726, *Trematospira* Zone, Sulphur Spring Range.
 24, 25 Interior and exterior of pedicle valve × 6, USNM 157275, loc. 10740, *Acrospirifer kobehana* Zone, Sulphur Spring Range.

Figures 26-28 *Megakozlowskiella* cf. *raricosta* (Conrad).
 Acrospirifer kobehana Zone, Sulphur Spring Range.
 26-28 Interior, ventral, and posterior views of pedicle valve × 1.25, USNM 157276, loc. 10710.

PLATE 71. MEGAKOZLOWSKIELLA

Figures 1-19 *Megakozlowskiella magnapleura* Johnson, n. sp.

1, 2 Anterior view of interior of pedicle valve and posterior portion of brachial valve × 2.5 and posterior view × 1.5, USNM 157277, loc. 10714, *Trematospira* Zone, northern Roberts Mountains. Note divisions of the interarea.

3-7 Side, anterior, posterior, ventral, and dorsal views × 1.5, USNM 157278, loc. 10726, *Trematospira* Zone, Sulphur Spring Range.

8-12 Anterior, side, ventral, dorsal, and posterior views × 1, USNM 157279, loc. 10726, *Trematospira* Zone, Sulphur Spring Range.

13 Ventral view × 1.5, USNM 157280, loc. 10714, *Trematospira* Zone, northern Roberts Mountains.

14 Interior of part of brachial valve × 3, USNM 157281, loc. 10713, *Spinoplasia* Zone, Rabbit Hill.

15, 16 Interior and exterior of pedicle valve × 2, USNM 157282, loc. 10713, *Spinoplasia* Zone, Rabbit Hill. Note extra plications on each lateral flank.

17-19 Posterior, side, and dorsal views × 2, holotype, USNM 157283, loc. 10726, *Trematospira* Zone, Sulphur Spring Range.

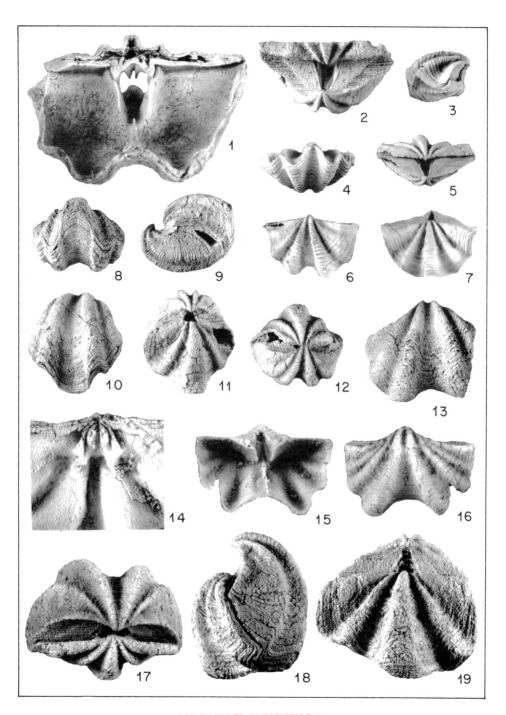

MEGAKOZLOWSKIELLA

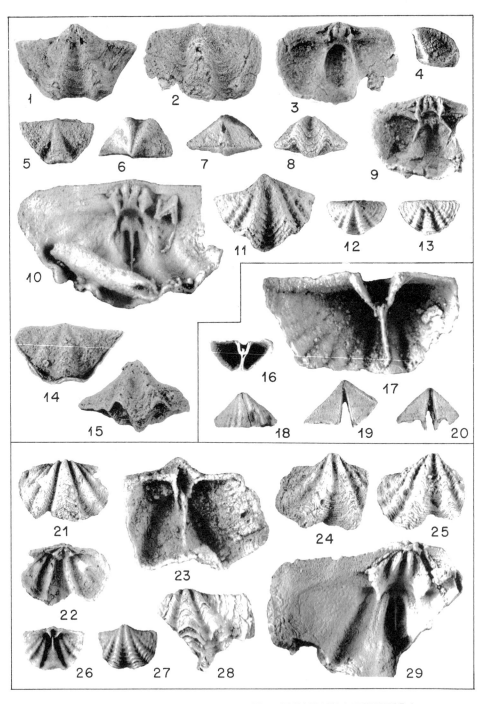

CYRTINAELLA, CYRTINA, AND MEGAKOZLOWSKIELLA

JOHNSON, PLATE 72
Geological Society of America Memoir 121

PLATE 72. CYRTINAELLA, CYRTINA, AND MEGAKOZLOWSKIELLA

Figures 1-9, 14, 15 *Cyrtinaella causa* Johnson, n. sp.
Quadrithyris Zone.
1 Exterior of pedicle valve × 3, USNM 157289, loc. 10808, Toquima Range.
2, 3 Exterior and interior of brachial valve × 3, USNM 157290, loc. 10808, Toquima Range.
4-8 Side, dorsal, ventral, posterior, and anterior views × 1.5, USNM 157291, loc. 10758, Coal Canyon.
9 Interior of brachial valve × 3, USNM 157292, loc. 10808, Toquima Range.
14, 15 Dorsal and anterior views, holotype, × 2, USNM 157293, loc. 10808, Toquima Range.

Figures 10-13 *Cyrtina* sp. A.
Quadrithyris Zone, loc. 10758, Coal Canyon.
10 Interior of brachial valve × 5, USNM 157294.
11 Ventral view × 3, USNM 157295.
12, 13 Ventral and dorsal views × 2, USNM 157296.

Figures 16-20 *Cyrtina* sp. D.
Eurekaspirifer pinyonensis Zone, Cortez Range.
16, 18, 20 Interior, anterior, and posterior views of pedicle valve × 2, USNM 157297, loc. 10754.
17, 19 Interior × 5 and posterior × 2 of pedicle valve, USNM 157298, loc. 10752.

Figures 21-29 *Megakozlowskiella* sp. A.
Quadrithyris Zone, loc. 10758, Coal Canyon.
21, 22 Exterior and interior of brachial valve × 1.5, USNM 157284.
23, 24 Interior × 3 and exterior × 2 of pedicle valve, USNM 157285.
25 Exterior of pedicle valve × 1.5, USNM 157286.
26, 27 Interior and exterior of pedicle valve × 2, USNM 157287.
28, 29, Exterior × 1.5 and interior × 3 of brachial valve, USNM 157288.

PLATE 73. CYRTINA

Figures 1-14 *Cyrtina* cf. *varia* Clarke.
    1, 2 Interior of posterior portion of brachial valve and exterior of brachial valve × 5, USNM 157299, loc. 10713, *Spinoplasia* Zone, Rabbit Hall.
    3 Interior of posterior portion of brachial valve × 4, USNM 157300, loc. 10713, *Spinoplasia* Zone, Rabbit Hill.
    4-8 Dorsal, anterior, side, posterior, and ventral views × 2, USNM 157301, loc. 10778, *Costispirifer* Subzone, northern Roberts Mountains.
    9 Dorsal view × 1.5, USNM 157302, *Trematospira* Zone, Sulphur Spring Range.
    10-14 Ventral, dorsal, side, anterior, and posterior views × 2, USNM 157303, loc. 10732, *Trematospira* Zone, Sulphur Spring Range.

Figures 15-20 *Cyrtina* sp. C.
    Acrospirifer kobehana Zone, Sulphur Spring Range.
    15-19 Ventral, dorsal, anterior, side, and posterior views × 1, USNM 157304, loc. 10739.
    20 Interior of pedicle valve × 1.5, USNM 157305, loc. 10739.

Figures 21-26 *Cyrtina* sp. C?
    Acrospirifer kobehana Zone, northern Roberts Mountains.
    21 Exterior of pedicle valve × 3, USNM 157306, loc. 10777.
    22-26 Ventral, anterior, side, posterior, and dorsal views × 3, USNM 157307, loc. 10777.

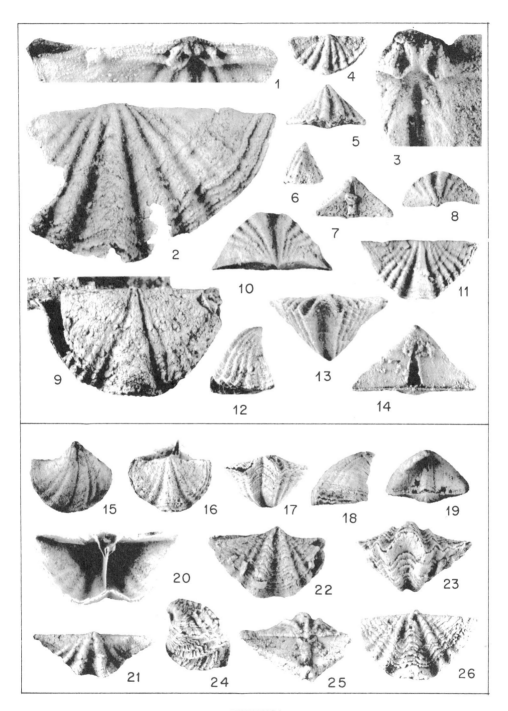

CYRTINA

JOHNSON, PLATE 73
Geological Society of America Memoir 121

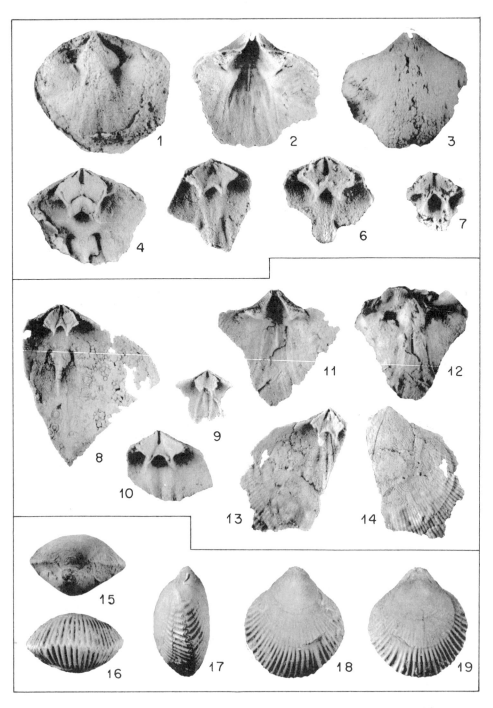

RENSSELAERINA?, RENSSELAERIA, AND MUTATIONELLA?

PLATE 74. RENSSELAERINA?, RENSSELAERIA, AND MUTATIONELLA?

Figures 1-7 *Rensselaerina?* sp.
  Spinoplasia Zone, loc. 10713, Rabbit Hill.
  1-3 Rubber impression, interior, and exterior of pedicle valve × 3, USNM 157309.
  4 Interior of posterior portion of brachial valve × 3, USNM 157310.
  5 Interior of posterior portion of brachial valve × 3, USNM 157311.
  6 Cardinalia of brachial valve × 4, USNM 157312.
  7 Cardinalia of brachial valve × 4, USNM 157313. Note crural plates.

Figures 8-14 *Rensselaeria* sp.
  Trematospira Zone, Sulphur Spring Range.
  8 Interior of brachial valve × 2, USNM 157314, loc. 10745.
  9 Cardinalia of brachial valve × 2, USNM 157315, loc. 10726.
  10 Interior of posterior portion of brachial valve × 3, USNM 157316, loc. 10726.
  11, 12 Interior of pedicle valve and rubber impression of interior × 1.5, USNM 157317, loc. 10726.
  13, 14 Interior and exterior of brachial valve × 2, USNM 157318, loc. 10743.

Figures 15-19 *Mutationella?* sp.
  Eurekaspirifer pinyonesis Zone, Sulphur Spring Range.
  15-19 Posterior, anterior, side, ventral, and dorsal views × 2, USNM 157308, loc. 10704.

PART IV. APPENDIX, REFERENCES CITED AND INDEX

Appendix of Localities

The locality numbers listed below are those of the locality register of the U.S. National Museum, Department of Paleobiology, Washington, D. C. Four-digit combinations pertain to the locality file of the Department of Geology, University of California, Los Angeles. Other designations are informal field numbers.

2567 (USNM loc. 10705)
In Dry Creek, approx. 2 mi N. of "k" in "Creek" and 1.5 mi due E. from BM 6095, sec. 24, T. 24 N., R. 50 E., Roberts Creek Mtn. quad., Eureka Co., Nevada. McColley Canyon Fm.
Collector: B. Troxel, 1951.

3198 (USNM loc. 10704)
Brown to pinkish weathering platy ls, exposed in saddle 0.35 mi W. of SE cor. or sec. 36, T. 27 N., R. 52 E., Sulphur Spring Range, Mineral Hill quad., Eureka Co., Nevada. McColley Canyon Fm.
Collectors: Geo. Cleveland and Geo. Brown, 1952.

3505, I4 (USNM loc. 10706)
On crest of Hill 8149, Sulphur Spring Range, 1 mi S. of Bald Mountain, Mineral Hill quad., Eureka Co., Nevada.
Collectors: Summer Field Class, July, 1951.

3507, I9 (USNM loc. 10707)
In second saddle on crest of Sulphur Spring Range, SE. of Hill 7881, ¼ mi SE. of Bald Mtn., Mineral Hill quad., Eureka Co., Nevada.
Collectors: Summer Field Class, July, 1951.

3508, N53-10 (USNM loc. 10708)
At 7700 ft elevation, ¼ mi directly W. of Hill 7690, 1 mi NE. of Bald Mtn., Mineral Hill quad., Eureka Co., Nevada.
Collector: C. A. Nelson, July, 1953.

3520, E5 (USNM loc. 10709)
At SE. end of long saddle on ridge trending NW. from Hill 7836, on crest of Sulphur Spring Range, W. of Williams Spring, Mineral Hill quad., Eureka Co., Nevada.
Collectors: Summer Field Class, July, 1951.

3522, M4-3 (USNM loc. 10710)
In highest saddle on ridge trending SW. from Hill 7836, Sulphur Spring Range, W. of Williams Spring, Mineral Hill quad., Eureka Co., Nevada.
Collectors: Summer Field Class, July, 1953.

3527, E1-3 (USNM loc. 10711)
In saddle immediately south of "O" in 8081, in SE. cor. of sec. 36, T. 27 N., R. 52 E., Mineral Hill quad., Eureka Co., Nevada.
Collector: C. A. Nelson, July, 1953.

4351, J60-1 (USNM loc. 10712)
North end of double-topped ridge, elev. 7560 ft, 2500 ft N., 1800 ft W. of SE. cor. of sec. 35, T. 16 N., R. 49 E., east slope of Monitor Range, Horse Heaven Mtn. quad., Eureka Co., Nevada. Rabbit Hill Limestone.
Collectors: J. G. Johnson and J. B. Roen, 1959; A. J. Boucot, 1963.

4354, J61-1 (USNM loc. 10713)
South end of Rabbit Hill, north flank of Whiterock Canyon, elev. 7080 ft, 4400 ft S., 900 ft W. of SE. cor. of sec. 35, T. 16 N., R. 49 E., east slope of Monitor Range, Horse Heaven Mtn. quad., Eureka Co., Nevada. Rabbit Hill Limestone.
Collectors: J. G. Johnson and J. B. Roen, 1959; A. J. Boucot, 1963.

4386, J63-1 (USNM loc. 10714)
West flank of Willow Creek, elev. 6680 ft, 2600 ft S., 1400 ft E. of SE. cor. of sec. 22, T. 24 N., R. 50 E., Roberts Creek Mtn. quad., Eureka Co., Nevada. McColley Canyon Fm.
Collector: J. G. Johnson, 1959.

4387, W3-59 (USNM loc. 10715)
East flank of Willow Creek, 25 ft above the base of the McColley Canyon Fm. and below *Costispirifer*. Roberts Creek Mtn. quad., Eureka Co., Nevada.
Collector: E. L. Winterer, 1959.

4770, A1-50 (USNM loc. 10716)
Elev. 7240 ft, 1500 ft N., 4500 ft W. of SW. cor. of sec. 1, T. 24 N., R. 52 E., Garden Valley quad., Eureka Co., Nevada. McColley Canyon Fm.
Collectors: Student group A, 1950.

4771, B2-50 (USNM loc. 10717)
Elev. 7480 ft, near top of hill 7517, 4900 ft N., 6700 ft W. of SW. cor. of sec. 13, T. 24 N., R. 52 E., Garden Valley quad., Eureka Co., Nevada. McColley Canyon Fm.
Collectors: Student group B, 1950.

4772, C4-50 (USNM loc. 10718)
Elev. 6560 ft, 1700 ft N., 4500 ft W. of SW. cor. of sec. 13, T. 24 N., R. 52 E., Garden Valley quad., Eureka Co., Nevada. McColley Canyon Fm.
Collectors: Student group C, 1950.

4773, E4-50 (USNM loc. 10719)
Elev. 7000 ft, 5000 ft N., 4800 ft W. of SW. cor. of sec. 35, T. 24 N., R. 52 E., Garden Valley quad., Eureka Co., Nevada. McColley Canyon Fm.
Collectors: Student group E, 1950.

Appendix of Localities

4774, E5-50 (USNM loc. 10720)
Elev. 6280 ft, 4600 ft N., 5800 ft W. of SW cor. of sec. 12, T. 23 N., R. 52 E., Garden Valley quad., Eureka Co., Nevada. McColley Canyon Fm.
Collectors: Student group E, 1950.

4775, F10-50 (USNM loc. 10721)
Elev. 6280 ft. 4600 ft N., 5800 ft W. of SW cor. of sec. 12, T. 23 N., R. 52 E., Garden Valley quad., Eureka Co., Nevada. McColley Canyon Fm.
Collectors: Student group G, 1950.

4776, F12-50 (USNM loc. 10722)
Elev. 7160 ft, 4300 ft N., 11600 ft W. of SW cor. of sec. 12, T. 23 N., R. 52 E., Garden Valley quad., Eureka Co., Nevada. McColley Canyon Fm.
Collectors: Student group F, 1950.

4777, G4-50 (USNM loc. 10723)
Elev. 7040 ft, 1000 ft S., 12700 ft W. of SW cor. of sec. 12, Garden Valley quad., Eureka Co., Nevada. McColley Canyon Fm.
Collectors: Student group G, 1950.

4778, G6-50 (USNM loc. 10724)
Elev. 7080 ft, 1300 ft S., 11700 ft W. of SW cor. of sec. 12, Garden Valley quad., Eureka Co., Nevada. McColley Canyon Fm.
Collectors: Student group G, 1950.

4779, K9-50 (USNM loc. 10725)
Elev. 6360 ft, 1600 ft S., 11800 ft W. of NW cor. of sec. 1, T. 22 N., R. 52 E., Garden Valley quad., Eureka Co., Nevada. McColley Canyon Fm.
Collectors: Student group K, 1950.

4780, UCR 3485, KC, A73-1 (USNM loc. 10726)
Elev. 7250 ft, 2800 ft N., 1300 ft W. of SE cor. of sec. 36, T. 27 N., R. 52 E., head of McColley Canyon, Sulphur Spring Range, Mineral Hill quad., Eureka Co., Nevada, McColley Canyon Fm.
Collectors: M. A. Murphy, 1956; Murphy and Boucot, 1963.

4781, A73-3 (USNM loc. 10727)
Elev. 7300 ft, 2700 ft N., 1300 ft W. of SE cor. of sec. 36, T. 27 N., R. 52 E., Mineral Hill quad., Eureka Co., Nevada. Fifteen feet above base of McColley Canyon Fm.
Collector: M. A. Murphy, 1963.

4782, B-2 (USNM loc. 10728)
Elev. 7050 ft, 2100 ft due W. of SE cor. of sec. 24, T. 27 N., R. 52 E., on ridge crest, Sulphur Spring Range, Mineral Hill quad., Eureka Co., Nevada. McColley Canyon Fm.
Collector: unknown.

4783, R77, MCPIN (USNM loc. 10729)
Crest of ridge east of McColley Canyon, elev. 7550 ft, 2200 ft N., 600 ft W. of SE cor. of sec. 36, T. 27 N., R. 52 E., Mineral Hill quad., Eureka Co., Nevada. McColley Canyon Fm.
Collector: M. A. Murphy, 1963.

4784, A75-1 (USNM loc. 10730)
Elev. 7400 ft to 7500 ft along line of section az. 160° between 2200 ft N., 1600 ft W., and 1500 ft N., 1400 ft W. of SE cor. of sec. 36, T. 27 N., R. 52 E., Eureka Co., Nevada. 1-5 ft above base of McColley Canyon Fm. Collector: M. A. Murphy, 1963.

4785, A76-5 (USNM loc. 10731)
Elev. 7400 ft to 7500 ft along line of section az. 160° between 2200 ft N., 1600 ft W., and 1500 ft N., 1400 ft W. of SE cor. of sec. 36, T. 27 N., R. 52 E., Eureka Co., Nevada. 15-30 ft above base of McColley Canyon Fm. Collector: M. A. Murphy, 1963.

4786, A77-1+28 (USNM loc. 10768)
Elev. 7400 ft to 7500 ft along line of section az. 160° between 2200 ft N., 1600 ft W., and 1500 ft N., 1400 ft W. of SE cor. of sec. 36, T. 27 N., R. 52 E., Eureka Co., Nevada. 18 ft above base of McColley Canyon Fm. Collector: M. A. Murphy, 1963.

4787, A77-2+28 (USNM loc. 10732)
Elev. 7400 ft to 7500 ft along line of section az. 160° between 2200 ft N., 1600 ft W., and 1500 ft N., 1400 ft W. of SE cor. of sec. 36, T. 27 N., R. 52 E., Eureka Co., Nevada. 34 ft above base of McColley Canyon Fm. Collector: M. A. Murphy, 1963.

4788, A77-2+32 (USNM loc. 10733)
Elev. 7400 ft to 7500 ft along line of section az. 160° between 2200 ft N., 1600 ft W., and 1500 ft N., 1400 ft W. of SE cor. of sec. 36, T. 27 N., R. 52 E., Eureka Co., Nevada. 35 ft above base of McColley Canyon Fm. Collector: M. A. Murphy, 1963.

4789, A77-2+64 (USNM loc. 10734)
Elev. 7400 ft to 7500 ft along line of section az. 160° between 2200 ft N., 1600 ft W., and 1500 ft N., 1400 ft W. of SE cor. of sec. 36, T. 27 N., R. 52 E., Eureka Co., Nevada. 41 ft above base of McColley Canyon Fm. Collector: M. A. Murphy, 1963.

4790, A77-3+10 (USNM loc. 10735)
Elev. 7400 ft to 7500 ft along line of section az. 160° between 2200 ft N., 1600 ft W., and 1500 ft N., 1400 ft W. of SE cor. of sec. 36, T. 27 N., R. 52 E., Eureka Co., Nevada. 52 ft above base of McColley Canyon Fm. Collector: M. A. Murphy, 1963.

4791, A77-3+44 (USNM loc. 10736)
Elev. 7400 ft to 7500 ft along line of section az. 160° between 2200 ft N., 1600 ft W., and 1500 ft N., 1400 ft W. of SE cor. of sec. 36, T. 27 N., R. 52 E., Eureka Co., Nevada. 61 ft above base of McColley Canyon Fm. Collector: M. A. Murphy, 1963.

4792, A77-4+79 (USNM loc. 10737)
Elev. 7400 ft to 7500 ft along line of section az. 160° between 2200 ft N., 1600 ft W., and 1500 ft N., 1400 ft W. of SE cor. of sec. 36, T. 27 N., R. 52 E., Eureka Co., Nevada. 102 ft above base of McColley Canyon Fm. Collector: M. A. Murphy, 1963.

4793, A78-1+0 (USNM loc. 10738)
Elev. 7400 ft to 7500 ft along line of section az. 120° between 1800 ft N., 1150 ft W., and 1600 ft N., 800 ft W. of SE cor. of sec. 36, T. 27 N., R. 52 E., Eureka Co., Nevada. 185 ft above base of McColley Canyon Fm.
Collector: M. A. Murphy, 1963.

4794, A78-1+26 (USNM loc. 10739)
Elev. 7400 ft to 7500 ft along line of section az. 120° between 1800 ft N., 1150 ft W., and 1600 ft N., 800 ft W. of SE cor. of sec. 36, T. 27 N., R. 52 E., Eureka Co., Nevada. 202 ft above base of McColley Canyon Fm.
Collector: M. A. Murphy, 1963.

4795, A78-1+28 (USNM loc. 10740)
Elev. 7400 ft to 7500 ft along line of section az. 120° between 1800 ft N., 1150 ft W., and 1600 ft N., 800 ft W. of SE cor. of sec. 36, T. 27 N., R. 52 E., Eureka Co., Nevada. 203 ft above base of McColley Canyon Fm.
Collector: M. A. Murphy, 1963.

4796, A78-3 (USNM loc. 10741)
Elev. 7400 ft to 7500 ft along line of section az. 120° between 1800 ft N., 1150 ft W., and 1600 ft N., 800 ft W. of SE cor. of sec. 36, T. 27 N., R. 52 E., Eureka Co., Nevada. McColley Canyon Fm.
Collector: M. A. Murphy, 1963.

4797, A79-1+16 (USNM loc. 10742)
Elev. 7450 ft to 7350 ft along line of section az. 80° between 1900 ft N., 1500 ft W., and 2000 ft N., 1100 ft W. of SE cor. of sec. 36, T. 27 N., R. 52 E., Eureka Co., Nevada. 20 ft above base of McColley Canyon Fm.
Collector: M. A. Murphy, 1963.

4798, A79-1+22 (USNM loc. 10743)
Elev. 7450 ft to 7350 ft along line of section az. 80° between 1900 ft N., 1500 ft W., and 2000 ft N., 1100 ft W. of SE cor. of sec. 36, T. 27 N., R. 52 E., Eureka Co., Nevada. 22 ft above base of McColley Canyon Fm.
Collector: M. A. Murphy, 1963.

4799, A79-2 (USNM loc. 10744)
Elev. 7450 ft to 7350 ft along line of section az. 80° between 1900 ft N., 1500 ft W., and 2000 ft N., 1100 ft W. of SE cor. of sec. 36, T. 27 N., R. 52 E., Eureka Co., Nevada. 48 ft above base of McColley Canyon Fm.
Collector: M. A. Murphy, 1963.

4800, A79-2+30 (USNM loc. 10745)
Elev. 7450 ft to 7350 ft along line of section az. 80° between 1900 ft N., 1500 ft W., and 2000 ft N., 1100 ft W. of SE cor. of sec. 36, T. 27 N., R. 52 E., Eureka Co., Nevada. 52 ft above base of McColley Canyon Fm.
Collector: M. A. Murphy, 1963.

4801, A79-2+40 (USNM loc. 10746)
Elev. 7450 ft to 7350 ft along line of section az. 80° between 1900 ft N., 1500 ft W., and 2000 ft N., 1100 ft W. of SE cor. of sec. 36, T. 27 N., R. 52 E., Eureka Co., Nevada. 55 ft above base of McColley Canyon Fm.
Collector: M. A. Murphy, 1963.

4802, A79-2+68 (USNM loc. 10747)
Elev. 7450 ft to 7350 ft along line of section az. 80° between 1900 ft N., 1500 ft W., and 2000 ft N., 1100 ft W. of SE cor. of sec. 36, T. 27 N., R. 52 E., Eureka Co., Nevada. 58 ft above base of McColley Canyon Fm.
Collector: M. A. Murphy, 1963.

4803, A79-3+47 (USNM loc. 10748)
Elev. 7450 ft to 7350 ft along line of section az. 80° between 1900 ft N., 1500 ft W., and 2000 ft N., 1100 ft W. of SE cor. of sec. 36, T. 27 N., R. 52 E., Eureka Co., Nevada. 70 ft above base of McColley Canyon Fm.
Collector: M. A. Murphy, 1963.

4804, A79-3+64 (USNM loc. 10749)
Elev. 7450 ft to 7350 ft along line of section az. 80° between 1900 ft N., 1500 ft W., and 2000 ft N., 1100 ft W. of SE cor. of sec. 36, T. 27 N., R. 52 E., Eureka Co., Nevada. 73 ft above base of McColley Canyon Fm.
Collector: M. A. Murphy, 1963.

4805, 5-1-49, (USNM loc. 10750)
Exact location unknown; probably SW slope of Mt. Tenabo, sec. 4, T. 26 N., R. 48 E., Cortez quad., west slope of Cortez Range, Eureka Co., Nevada. Wenban Limestone.
Collectors: Student group 5, 1949.

4806, 6-14-49 (USNM loc. 10751)
Elev. 8200 ft, 3000 ft N., 600 ft W. of SE cor. of sec. 4, T. 26 N., R. 48 E., Cortez quad., Eureka Co., Nevada. Wenban Limestone.
Collectors: Student group 6, 1949.

4807, HC-1 (USNM loc. 10752)
Elev. 7950 ft, 1000 ft N., 700 ft W. of SE cor. of sec. 4, T. 26 N., R. 48 E., Crotez quad., Eureka Co., Nevada. Wenban Limestone.
Collectors: A. J. Boucot and E. F. Lawrence, 1963; Boucot and H. K. Erben, 1965.

4808, 9-104-49 (USNM loc. 10753)
Elev. 7050 ft, 2500 ft N., 4000 ft E. of SW cor. of sec. 21, T. 26 N., R. 48 E., Cortez quad., Eureka Co., Nevada. Wenban Limestone.
Collectors: Student group 9, 1949.

4809, 9-106-49, FC-34 (USNM loc. 10754)
Elev. 6875 ft, 800 ft N., 2900 ft E. of SW cor. of sec. 21, T. 26 N., R. 48 E., Cortez quad., Eureka Co., Nevada. Wenban Limestone.
Collectors: A. J. Boucot and H. Masursky, 1963.

4810, 9-107-49 (USNM loc. 10755)
Elev. 6950 ft, 500 ft N., 4200 ft E. of SW cor. of sec. 21, T. 26 N., R. 48 E., Cortez quad., Eureka Co., Nevada. Wenban Limestone.
Collectors: Student group 9, 1949.

4811, 9-108-49 (USNM loc. 10756)
Elev. 6550 ft, 2500 ft E., 1700 ft N. of SE cor. of sec. 20, T. 26 N., R. 48 E.,

Cortez quad., west slope of Cortez Range, Eureka Co., Nevada. Wenban Limestone.
Collectors: Student group 9, 1949.

4812, DWLB, HF-34 (USNM loc. 10757)
West flank, mouth of Coal Canyon, elev. 6320 ft, 1600 ft N., 1800 ft W. of SE cor. of sec. 17, T. 25 N., R. 49 E., Horse Creek Valley quad., Eureka Co., Nevada. Windmill Limestone.
Collectors: M. A. Murphy, 1957; E. L. Winterer, 1957, 1962; N. G. Lane, 1962; H. Masursky, 1960, 1961; A. J. Boucot, 1963.

4813, WUB, A37-1, J52-2 (USNM loc. 10758)
Big ledge, east flank and near mouth of Coal Canyon, elev. 6320 ft, 1500 ft N., 1100 ft W. of SE cor. of sec. 17, T. 25 N., R. 49 E., Horse Creek Valley quad., northern Simpson Park Mts., Eureka Co., Nevada. Windmill Limestone.
Collectors: J. G. Johnson, 1957, 1958; M. A. Murphy, 1957; E. L. Winterer, 1957, 1962; H. Masursky, 1960; N. G. Lane, 1962; A. J. Boucot, 1963.

4814, J52-5 (USNM loc. 10759)
Elev. 6600 ft, 700 ft W., 1200 ft N. of SE cor. of sec. 17, T. 25 N., R. 49 E., Horse Creek Valley quad., northern Simpson Park Range, Eureka Co., Nevada. Lower Rabbit Hill Ls.
Collector: J. G. Johnson, 1958.

4815, A36-2, J52-6 (USNM loc. 10760)
Elev. 6760 ft, 450 ft W., 1150 ft N. of SE cor. of sec. 17, T. 25 N., R. 49 E., Horse Creek Valley quad., northern Simpson Park Range, Eureka Co., Nevada, Upper Rabbit Hill Ls.
Collectors: J. G. Johnson and M. A. Murphy, 1958.

4816, J37-4 (USNM loc. 10761)
Elev. 6560 ft, 2300 ft W., 1500 ft N. of SE cor. of sec. 16, T. 25 N., R. 49 E., Horse Creek Valley quad., northern Simpson Park Range, Eureka Co., Nevada. McColley Canyon Fm., *infrequens* Subzone.
Collector: J. G. Johnson, 1958.

4817, J16-2 (USNM loc. 10762)
Elev. 6800 ft, 2450 ft W., 650 ft N. of SE cor. of sec. 16, T. 25 N., R. 49 E., Horse Creek Valley quad., northern Simpson Park Range, Eureka Co., Nevada. McColley Canyon Fm., *pinyonensis* Zone.
Collector: J. G. Johnson, 1957.

4820, Mca (USNM loc. 10808)
North flank of Ikes Canyon, elev. 8320 ft, 5000 ft N., 7000 ft W. of SE cor. of sec. 18, T. 14 N., R. 47 E., Dianas Punch Bowl quad., Nye Co., Nevada. McMonnigal Limestone, lower beds.
Collectors: Marshall Kay; A. J. Boucot, 1963.

4821, Mcb (USNM loc. 10765)
Ridge south of Ikes Canyon, elev. 8240 ft, 200 ft N., 7400 ft W. of SE cor. of sec. 18, T. 14 N., R. 47 E., Dianas Punch Bowl quad., Nye Co.,

Nevada. McMonnigal Limestone, lower beds.
Collector: Marshall Kay.

4822, Mcc (USNM loc. 10766)
Ridge south of Ikes Canyon, elev. 8280 ft, 200 ft N., 8700 ft W. of SE cor. of sec. 18, T. 14 N., R. 47 E., Dianas Punch Bowl quad., Nye Co., Nevada. McMonnigal Ls.
Collector: Marshall Kay.

4823, A2-50 (USNM loc. 10767)
Elev. 6280 ft, 800 ft S., 3200 ft E. of SW cor. of sec. 13, T. 24 N., R. 50 E., Roberts Creek Mtn. quad., Eureka Co., Nevada. McColley Canyon Fm.
Collectors: Student group A, 1950.

4824, B3-50 (USNM loc. 10769)
Elev. 6320 ft, 1900 ft S., 3000 ft E. of SW cor. of sec. 13, T. 24 N., R. 50 E., Roberts Creek Mtn. quad., Eureka Co., Nevada. McColley Canyon Fm.
Collectors: Student group B, 1950.

4825, C16-50 (USNM loc. 10770)
Elev. 6400 ft, 2700 ft S., 3200 ft E. of SW cor. of sec. 13, T. 24 N., R. 50 E., Roberts Creek Mtn. quad., Eureka Co., Nevada. McColley Canyon Fm.
Collectors: Student group C, 1950.

4826, D16-50, E17-50, 9-2-58 (USNM loc. 10771)
In Dry Creek, elev. 6760 ft, 5800 ft N., 3800 ft W. of SE cor. of sec. 36 (unsurveyed), Roberts Creek Mtn. quad., McColley Canyon Fm.
Collectors: Student groups D and E, 1950; group 9, 1958.

4827, 9-6-58 (USNM loc. 10772)
West flank of Dry Creek, elev. 6320 ft, 3200 ft E., 1300 ft S. of SE cor. of sec. 14, T. 24 N., R. 50 E., Roberts Creek Mtn. quad., Eureka Co., Nevada. McColley Canyon Fm.
Collectors: Student group 9, 1958.

4828, 11-4-58, J63-5 (USNM loc. 10773)
Elev. 7420 ft, on west slope of ridge crest between Willow and Birch Creeks, 1700 ft S., 1200 ft W. of SE cor. of sec. 22, T. 24 N., R. 50 E., Roberts Creek Mtn. quad., northern Roberts Mts., Eureka Co., Nevada. Fault block of basal McColley Canyon Fm.
Collectors: Student group 11, 1958, and J. G. Johnson, 1959.

4829, 11-7-58 (USNM loc. 10774)
Crest of ridge connecting Cooper Peak and Western Peak, elev. 8580 ft, 9300 ft on a radius approx. ESE of SE cor. of sec. 31, T. 24 N., R. 50 E., Roberts Creek Mtn. quad., Eureka Co., Nevada. McColley Canyon Fm.
Collectors: Student group 11, 1958.

4830, W7-59 (USNM loc. 10775)
Elev. 6800 ft, 2100 ft E., 1700 ft S. of SE cor. of sec. 22, T. 24 N., R. 50 E., Roberts Creek Mtn. quad., Eureka Co., Nevada. McColley Canyon Fm.
Collector: E. L. Winterer, 1959.

4831, W8-59 (USNM loc. 10776)
Elev. 6880 ft, 2300 ft E., 1700 ft S. of SE cor. of sec. 22, T. 24 N., R. 50 E.,

Roberts Creek Mtn. quad., Eureka Co., Nevada. McColley Canyon Fm.
Collector: E. L. Winterer, 1959.

4832, J58-2 (USNM loc. 10777)
Elev. 6840 ft, 2300 ft E., 3200 ft S. of SE cor. of sec. 15, T. 24 N., R. 50 E., Roberts Mts., Eureka Co., Nevada. McColley Canyon Fm., near base.
Collector: J. G. Johnson, 1959.

4833, J59-1 (USNM loc. 10778)
East flank of Willow Creek, 50-100 ft east of stream bed, elev. 6600 ft, 1800 ft S., 1500 ft E. of SE cor. of sec. 22, T. 24 N., R. 50 E., Roberts Creek Mtn. quad., northern Roberts Mts., Eureka Co., Nevada. McColley Canyon Fm.
Collector: J. G. Johnson, 1959.

4834, J62-1 (USNM loc. 10779)
Elev. 6680 ft, 2100 ft E., 2100 ft S. of SE cor. of sec. 22, T. 24 N., R. 50 E., Roberts Creek Mtn. quad., Eureka Co., Nevada. McColley Canyon Fm.
Collector: J. G. Johnson, 1959.

4835, J64-2 (USNM loc. 10780)
Elev. 7120 ft on east flank of Birch Creek near stream divide at head of creek, 8000 ft S., approx. 2500 ft W. of SE cor. of sec. 22, T. 24 N., R. 50 E., Roberts Creek Mtn. quad., northern Roberts Mts., Eureka Co., Nevada. Approx. 20 ft above base of McColley Canyon Fm.
Collector: J. G. Johnson, 1959.

4836, J64-3 (USNM loc. 10781)
Elev. 7340 ft on east flank of Birch Creek near stream divide at head of creek, 7800 ft S., approx. 1900 ft W. of SE cor. of sec. 22, T. 24 N., R. 50 E., Roberts Creek Mtn. quad., northern Roberts Mts., Eureka Co., Nevada. Near middle of *pinyonensis* Zone, McColley Canyon Fm.
Collector: J. G. Johnson, 1959.

4837, R76 (USNM loc. 10782)
Elev. 8860 ft, NE slope of spur of Cooper Peak, descending to Willow Creek, T. 23 N., R. 50 E., Roberts Creek Mountain quad., Eureka Co., Nevada. McColley Canyon Fm.
Collector: M. A. Murphy, 1957.

4838, 12-1-58 (USNM loc. 10783)
Elev. 8320 ft, crest of N-S ridge between Birch and Kelly Cks., 5800 ft ESE of Western Peak, T. 23 N., R. 50 E., Roberts Creek Mtn, quad., Eureka Co., Nevada. McColley Canyon Fm.
Collectors: Student group 12, 1958.

4839, 12-4-58 (USNM loc. 10784)
Elev. 8800 ft, 4300 ft WNW of BM 8095 which is approx. 1 mile E. of Cooper Peak, T. 23 N., R. 50 E., Roberts Creek Mtn. quad., Eureka Co., Nevada. McColley Canyon Fm.
Collectors: Student group 12, 1958.

4840, 12-5-58 (USNM loc. 10785)
Elev. 8120 ft, 6700 ft NW of BM 8095, 6800 ft WSW of BM 7437 on spur N.

of Cooper Peak, T. 23 N., R. 50 E., Roberts Creek Mtn. quad., Eureka Co., Nevada. McColley Canyon Fm.
Collectors: Student group 12, 1958.

4841, 14-3-58 (USNM loc. 10786)
Elev. 8160 ft, 6900 ft NW of BM 8095, 6750 ft WSW of BM 7437, on spur N. of Cooper Peak, T. 23 N., R. 50 E., Roberts Creek Mtn. quad., Eureka Co., Nevada. McColley Canyon Fm.
Collectors: Student group 14, 1958.

4842, W19-59 (USNM loc. 10787)
One mile SW of peak of Roberts Creek Mtn. and 5500 ft ESE of hill 8600, elev. 8200 ft, T. 23 N., R. 50 E., Roberts Creek Mtn. quad., Eureka Co., Nevada. McColley Canyon Fm.
Collector: E. L. Winterer, 1959.

4843, W20-59 (USNM loc. 10788)
Approx. 3200 ft NW of peak of Roberts Creek Mtn., elev. 9280 ft on crest of E-W spur, T. 23 N., R. 50 E., Roberts Creek Mtn. quad., Eureka Co., Nevada. McColley Canyon Fm.
Collector: E. L. Winterer, 1959.

4844, W27-59 (USNM loc. 10789)
Elev. 7160 ft on crest of small spur approx. ¾ mi. W. of Roberts Creek and 5400 ft WSW of hill 7554, T. 22 N., R. 50 E., Roberts Creek Mtn. quad., Eureka Co., Nevada. McColley Canyon Fm.
Collector: E. L. Winterer, 1959.

4845, W31-59 (USNM loc. 10790)
Elev. 6700 ft, 1800 ft N., 1000 ft E. of NW cor. of sec. 19, T. 22 N., R. 51 E., Roberts Creek Mtn. quad., Eureka Co., Nevada. McColley Canyon Fm.
Collector: E. L. Winterer, 1959.

4846, W34-59 (USNM loc. 10791)
Elev. 6700 ft, 1900 ft N., 1100 ft E. of NW cor. of sec. 19, T. 22 N., R. 51 E., Roberts Creek Mtn. quad., Eureka Co., Nevada. McColley Canyon Fm.
Collector: E. L. Winterer, 1959.

4847, S4-59 (USNM loc. 10792)
Elev. 8480 ft on crest of spur, approx. 9000 ft W. of Roberts Creek, 2900 ft NE of hill 8788, T. 22 N., R. 50 E., Roberts Creek Mtn. quad., Eureka Co., Nevada. McColley Canyon Fm.
Collector: F. Siders, 1959.

4848, S8-59 (USNM loc. 10793)
Elev. 8320 ft, approx. 11200 ft W. of hill 7554, 2300 ft WNW of hill 8788, T. 22 N., R. 50 E., Roberts Creek Mtn. quad., Eureka Co., Nevada. McColley Canyon Fm.
Collector: F. Siders, 1959.

4852, 1-3-58 (USNM loc. 10796)
Elev. 6960 ft, 4200 ft N., 3100 ft W. of SE cor. of sec. 4, T. 24 N., R. 49 E., Roberts Creek Mtn. quad., Eureka Co., Nevada. Formation unknown.
Collectors: Student group 1, 1958.

Appendix of Localities

A91-1 (USNM loc. 12328)
Elev. 6440 ft, 900 ft due N. of SE cor. of sec. 22, T. 24 N., R. 50 E., Willow Creek, Roberts Creek Mtn. quad., Eureka Co., Nevada. Roberts Mountains Fm.
Collectors: A. J. Boucot and M. A. Murphy, summer, 1964.

HF-53 (USNM loc. 12393)
East of Coal Canyon, elev. 6760 ft, 800 ft N., 2400 ft W. of SE cor. of sec. 16, T. 25 N., R. 49 E., N. Simpson Park Range, Horse Creek Valley quad., Eureka Co., Nevada. McColley Canyon Fm.
Collector: H. Masursky, 1959.

J52-10 (USNM loc. 12344)
East of Coal Canyon, elev. 6500 ft, 1000 ft W., 200 ft N. of SE cor. of sec. 17, T. 25 N., R. 49 E., N. Simpson Park Range, Eureka Co., Nevada. Upper shaly member of Windmill Ls.
Collector: J. G. Johnson, June, 1958.

HFC-211 (USNM loc. 12726)
Elev. 6400 ft, 1000 ft W., 3900 ft N. of SE cor. of sec. 20, T. 25 N., R. 49 E., Windmill Window, N. Simpson Park Range, Horse Creek Valley quad., Eureka Co., Nevada. McColley Canyon Fm.
Collector: H. Masursky, July, 1962.

A79- (whole number lost) (USNM loc. 12769)
Elev. 7450 ft to 7350 ft along line of section az. 80°, between 1900 ft N., 1500 ft W. and 2000 ft N., 1100 ft W. of SE cor. of sec. 36, T. 27 N., R. 52 E., Eureka Co., Nevada. McColley Canyon Fm.
Collector: M. A. Murphy, 1963.

A79-4+19 (USNM loc. 12770)
Elev. 7450 ft to 7350 ft along line of section, az. 80°, between 1900 ft N., 1500 ft W. and 2000 ft N., 1100 ft W. of SE cor. of sec. 36, T. 27 N., R. 52 E., Eureka Co., Nevada. 84 ft above base of McColley Canyon Fm.
Collector: M. A. Murphy, 1963.

1-3-57 (USNM loc. 12771)
East flank of north Simpson Park Range, probably in sec. 4, T. 24 N., R. 49 E., Eureka Co., Nevada. McColley Canyon Fm.
Collectors: Student group 1, 1957.

5-16-58 (USNM loc. 12772)
West flank of Roberts Mts., T. 23 N., R. 49 or 50 E., Eureka Co., Nevada. McColley Canyon Fm.
Collectors: Student group 5, 1958.

11-8-58 (USNM loc. 12773)
North flank of Roberts Mts., crest of ridge approx. 780 ft N., 1900 ft E. of SE cor. of sec. 22, T. 24 N., R. 50 E., Eureka Co., Nevada. McColley Canyon Fm.
Collectors: Student group 11, 1958.

W2-59 (USNM loc. 12774)
Elev. 7200 ft, 1100 ft N., 2400 ft E. of SE cor. of sec. 22, T. 24 N., R. 50 E.,

Roberts Creek Mtn. quad., Eureka Co., Nevada. *Pinyonensis* Zone, McColley Canyon Fm.
Collector: E. L. Winterer, 1959.

LM-7 (USNM loc. 12775)
West slope of Lone Mountain, T. 20 N., R. 51 E., Bartine Ranch quad., Eureka Co., Nevada. McColley Canyon Fm.
Collectors: UCLA Staff, 1950?

A-8899, Berkeley (USNM loc. 12776)
Vaughn Gulch Ls., 95 ft interval of ls. and interbedded calcareous shale between 1065 and 1160 ft above base (total 1500 ft), NE¼ of sec. 8, T. 13 S., R. 36 E., Independence quad., California.
Collector: R. H. Waite, 1951-53.

MCW (USNM loc. 12777)
West flank of McColley Canyon, 50-100 ft W. of creek, elev. 7200 ft, 3200 ft N., 1400 ft W. of SE cor. of sec. 36, T. 27 N., R. 52 E., Sulphur Spring Range, Mineral Hill quad., Eureka Co., Nevada. *Costispirifer* Subzone, McColley Canyon Fm.
Collector: A. J. Boucot, 1963.

MCE (USNM loc. 12778)
East flank of McColley Canyon, 200-300 ft E. of creek, elev. 7250 ft, 3100 ft N., 1000 ft W. of SE cor. of sec. 36, T. 27 N., R. 52 E., Sulphur Spring Range, Mineral Hill quad., Eureka Co., Nevada. *Kobehana* Zone, McColley Canyon Fm.
Collector: A. J. Boucot, 1963.

Basal McColley, loc. A (USNM loc. 12792)
Head of McColley Canyon, E-central part of sec. 36, T. 27 N., R. 52 E., Mineral Hill quad., Sulphur Spring Range, Eureka Co., Nevada. *Trematospira* Subzone, McColley Canyon Fm.
Collector: A. J. Boucot, 1963.

Basal McColley, loc. B (USNM loc. 12793)
Head of McColley Canyon (west of Canyon), E-central part of sec. 36, T. 27 N., R. 52 E., Mineral Hill quad., Sulphur Spring Range, Eureka Co., Nevada. *Trematospira* Subzone, McColley Canyon Fm.
Collector: A. J. Boucot, 1963.

UCR 3477 (USNM loc. 13267)
East flank of McColley Canyon, float, sec. 36, T. 27 N., R. 52 E., Mineral Hill quad., Eureka Co., Nevada. At least two levels in *pinyonensis* Zone, McColley Canyon Fm.
Collector: M. A. Murphy, ca. 1953.

WCE (USNM loc. 13268)
East flank of Willow Creek, float from more than 25 ft above base of McColley Canyon Fm., elev. 6600 ft, 1800 ft S., 1500 ft E. of SE cor. of sec. 22, T. 24 N., R. 50 E., Roberts Creek Mtn. quad., northern Roberts Mts., Eureka Co., Nevada.
Collector: A. J. Boucot, 1963.

APPENDIX OF LOCALITIES

FF-10 (USNM loc. 13269)
Elev. 7500 ft, 485-535 ft upsection to NE from base of McColley Canyon Fm. 1600 ft S., 1900 ft W. of VABM 7936, Lone Mtn., Bartine Ranch quad., Nevada.
Collector: E. Gronberg, 1966.

FF-9 (USNM loc. 13270)
Elev. 7480 ft, 435-485 ft upsection to NE from a point at the base of McColley Canyon Fm., 1600 ft S., 1900 ft W. of VABM 7936, Lone Mtn., Bartine Ranch quad., Nevada.
Collector: E. Gronberg, 1966.

FF-4 (USNM loc. 13271)
Elev. 7420 ft, 185-235 ft upsection to NE from a point at the base of McColley Canyon Fm. 1600 ft S., 1900 ft W. of VABM 7936, Lone Mtn., Bartine Ranch quad., Nevada.
Collector: E. Gronberg, 1966.

SPI-1617 (USNM loc. 13659)
Lower Windmill Ls., 1617 ft above the base of the Roberts Mts. Fm., elev. 6580 ft, 50 ft S., 2350 ft W. of SE cor. of sec. 17, T. 25 N., R. 49 E., N. Simpson Park Range, Horse Creek Valley quad., Eureka Co., Nevada.
Collectors: M. A. Murphy and G. Klapper, 1967.

Ubrh (USNM loc. 13660)
Float from Rabbit Hill Ls., low on east slope of Coal Canyon, SE¼ of sec. 17, T. 25 N., R. 50 E., northern Simpson Park Range, Eureka Co., Nevada.
Collector: A. J. Boucot, 1967.

SPI-1716 (USNM loc. 13670)
Lower Windmill Ls., 1716 ft above base of Roberts Mts. Fm., elev. 6540 ft, 50 ft S., 2250 ft W. of SE cor. of sec. 17, T. 25 N., R. 49 E., west flank of Coal Canyon, northern Simpson Park Range, Eureka Co., Nevada.
Collector: G. Klapper, 1967.

Nw-12 (USNM loc. 17272)
About 15 miles north of Wells, elev. approx. 7000 ft, on NW-trending spur NW of Antelope Peak, 3 mi N., 4 mi W. of SE cor. of T. 40 N., R. 62 E., Wells 1:250,000 sheet, Elko Co., Nevada.
Collector: B. L. Peterson, 1967.

References Cited

Alekseeva, R. E., 1967, Brachiopods and stratigraphy of the Lower Devonian of the northeast USSR: Akad. nauk SSSR, Siberian Div., Inst. Geol. and Geophy., p. 1-162, 16 Pls. Moscow.

Amos, Arturo, and Boucot, A. J., 1963, A revision of the brachiopod family Leptocoeliidae: Palaeontology, v. 6, pt. 3, p. 440-457, Pls. 62-65.

Amsden, T. W., 1949, Stratigraphy and paleontology of the Brownsport Formation (Silurian) of western Tennessee: Peabody Mus. Nat. Hist. Bull. 5, 138 p., 34 Pls.

——— 1951, Brachiopods of the Henryhouse Formation (Silurian) of Oklahoma: Jour. Paleontology, v. 25, no. 1, p. 69-96, Pls. 15-20.

——— 1953, Some notes on the Pentameracea, including a description of one new genus and one new subfamily: Washington Acad. Sci. Jour., v. 43, no. 5, p. 137-147.

——— 1958a, Stratigraphy and paleontology of the Hunton Group in the Arbuckle Mountain Region; Part II. Haragan articulate brachiopods; Part III. Supplement to the Henryhouse brachiopods: Oklahoma Geol. Survey Bull. 78, 157 p., 14 Pls.

——— 1958b, Stratigraphy and paleontology of the Hunton Group in the Arbuckle Mountain region; Part V. Bois d'Arc articulate brachiopods: Oklahoma Geol. Survey Bull. 82, 110 p., 5 Pls.

——— 1960, Hunton stratigraphy: Oklahoma Geol. Survey Bull. 84, 311 p., 17 Pls., 3 panels.

——— 1962, Additional fossils from the Bois d'Arc Formation in the southeastern part of the Arbuckle Mountains region: Oklahoma Geology Notes, v. 22, no. 8, p. 212-216.

——— 1964, Brachial plate structure in the brachiopod family Pentameridae: Palaeontology, v. 7, pt. 2, p. 220-239, Pls. 40-43.

——— 1968, Articulate brachiopods of the St. Clair Limestone (Silurian), Arkansas and the Clarita Formation (Silurian), Oklahoma: Paleont. Soc. Mem. 1, 117 p., 20 Pls.

Amsden, T. W., Boucot, A. J., and Johnson, J. G., 1967 *Conchidium* and its separation from the subfamily Pentamerinae: Jour. Paleontology, v. 41, no. 4, p. 861-867, Pls. 106-108.

Amsden, T. W., and Ventress, W.P.S., 1963, Early Devonian Brachiopods of Oklahoma: Oklahoma Geol. Survey Bull. 94, 239 p., 21 Pls.

Andronov, S. M., 1961, On some representatives of the family Pentameridae from the Devonian deposits in the region of the north Urals: Akad. nauk SSSR, Trudy Geol. Inst., v. 55, 136 p., 32 Pls.

Archiac, d', and De Verneuil, M. E., 1842, On the fossils of the older deposits in the Rhenish Provinces; preceded by a general survey of the fauna of the Paleozoic rocks and followed by a tabular list of the organic remains of the Devonian System in Europe: Geol. Soc. London. Trans., 2nd ser., v. 6, pt. 2, p. 303-410, Pls. 25-38.

Asselberghs, Etienne, 1946, L'Éodévonien de l'Ardenne et des Régions voisines: Univ. Louvain, Inst. Géol. Mém., v. 14, 598 p., 10 Pls.

Bancroft, B. B., 1945, The brachiopod zonal indices of the stages Costonian to Onnian in Britain: Jour. Paleontology, v. 19, no. 3, p. 181-252, Pls. 22-38.

Barrande, Joachim, 1847, Ueber die brachiopoden der silurischen Schichten von Böhmen: Wein, W. Haidinger, Naturwissenschaft-liche Abhandlungen, v. 1, p. 357-475, Pls. 14-22.

——1848, Uber die Brachiopoden der silurischen Schichten von Boehmen: ibid., v. 2, no. 2, p. 153-256, Pls. 15-23.

—— 1879, Système Silurien du centre de la Bohême, v. 5, Brachiopodes: Paris, Prague, 226 p., 153 Pls.

Barrois, Charles, Pruvost, P., and Dubois. G., 1922, Description de la faune Siluro-Dévonienne de Liévin: Soc. Géol. du Nord Mém., v. 6, fasc. 2, p. 65-234, Pls. 10-17.

Bayle, C. E., 1878, Explication de la carte géologique de la france; Paris, v. 4, pt. 1, 176 Pls.

Beecher, C. E., 1890, On *Leptaenisca,* a new genus of brachiopod from the Lower Helderberg Group: Am. Jour. Sci., 3rd. ser., v. 40, p. 238-246. Pl. 9.

Berdan, J. M., 1949, Brachiopoda and ostracoda of the Manlius and Cobleskill Limestones of New York: Unpub. PhD thesis, Yale Univ., New Haven, Connecticut (available on microfilm or by interlibrary loan).

Berdan, J. M., Berry, W.B.N., Boucot, A. J., Cooper, G. A., Jackson, D. E., Johnson, J. G., Klapper, Gilbert, Lenz, A. C., Martinsson, Anders, Oliver, W. A., Jr., Rickard, L. V., and Thorsteinsson, R., 1969, Siluro-Devonian boundary in North America: Proc. Third International Symposium on the Silurian-Devonian Boundary, and the stratigraphy of the Lower and Middle Devonian, Leningrad, U.S.S.R. 1968. (in Russian); also Geol. Soc. Amer. Bull., v. 80, no. 11. p. 2165-2174.

Berry, W.B.N., 1967, *Monograptus hercynicus nevadensis* n. subsp., from the Devonian in Nevada: U.S. Geol. Survey Prof. Paper 575-B, p. B26-B31.

—— 1968, Siluro-Devonian graptolite sequence in the Great Basin (abs.): Geol. Soc. America Program, 64th Ann. Meeting, Cordilleran Section, Tucson, p. 35, 36.

Berry, W.B.N., and Roen, J. B., 1963, Early Wenlock graptolites from Roberts Mountains Formation, Tuscarora Mountains, Nevada: Jour. Paleontology, v. 37, no. 5, p. 1123-1126.

Billings, Elkanah, 1860, Description of some new species of fossils from the lower and middle Silurian rocks of Canada: Canadian Nat., v. 5, p. 49-69.

References Cited

Bischoff, Günther, and Sannemann, Dietrich, 1958, Unterdevonische Conodonten aus dem Frankenwald: Notizbl. hess. L-Amt. Bodenforsch., v. 86, p. 87-110, Pls. 12-15.

Bouček, Bedrich, 1966, Eine neue und bisher jungste Graptolithen-Fauna aus dem böhmischen Devon: Neues Jahrb. Geologie u. Paläontologie Monatasch. v. 3, p. 161-168, 2 Pls.

Boucot, A. J., 1957, Revision of some Silurian and Early Devonian spiriferid genera and erection of Kozlowskiellinae, new subfamily: Senckenbergiana Lethaea, v. 38, no. 5/6, p. 311-334, 3 Pls.

——— 1958, *Kozlowskiellina*, new name for *Kozlowskiella* Boucot, 1957: Jour. Paleontology, v. 32, no. 5, p. 1030.

——— 1959, Early Devonian Ambocoeliinae (Brachiopoda): Jour. Paleontology, v. 33, no. 1, p. 16-24, Pls. 1, 2.

——— 1960, Lower Gedinnian brachiopods of Belgium: Univ. Louvain, Inst. Géol. Mém., v. 21, p. 283-324, 3 Tables, Pls. 9-18.

Boucot, A. J., and Gauri, K. L., 1966, *Quadrikentron* (Brachiopoda) a new subgenus of *Strophochonetes*: Jour. Paleontology, v. 40, no. 5, p. 1023-1026, Pl. 119.

Boucot, A. J., and Gill, E. D., 1956 *Australocoelia*, a new Lower Devonian brachiopod from South Africa, South America, and Australia: Jour. Paleontology, v. 30, no. 5, p. 1173-1178, Pl. 126.

Boucot, A. J., and Harper, C. W., 1968, Silurian to Lower Middle Devonian Chonetacea: Jour. Paleontology, v. 42, no. 1, p. 143-176, Pls. 27-30.

Boucot, A. J., Hazzard, J. C., & Johnson, J. G., 1969, Reworked Lower Devonian fossils, Nopah Range, Inyo County, California: Am. Assoc. Petroleum Geologists Bull., v. 53, no. 1, p. 163-167.

Boucot, A. J., and Johnson, J. G., 1964, Devonian brachiopods from the Mina Plomosas-Placer de Guadalupe area, Chihuahua, Mexico, and their paleogeographic significance: West Texas Geol. Soc., 1964 Field Trip Guidebook, Pub. no. 64-50, p. 104-108.

——— 1967a, Species and distribution of *Coelospira* (Brachiopoda): Jour. Paleontology, v. 41, no. 5, p. 1226-1241, Pls. 163-166.

——— 1967b, Paleogeography and correlation of Appalachian Province Lower Devonian sedimentary rocks: Tulsa Geol. Soc. Digest, v. 35, p. 35-87, 2 Pls.

——— 1968, Brachiopods of the Bois Blanc Formation in New York: U.S. Geol. Survey Prof. Paper 584-B, p. 1-27, Pls. 1-8.

Boucot, A. J., Johnson, J. G., Harper, C. W., and Walmsley, V. G., 1966, Silurian brachiopods and gastropods of southern New Brunswick: Canada Geol. Survey Bull. 140, 45 p., 18 Pls.

Boucot, A. J., Johnson, J. G., and Staton, R. D., 1964, On some atrypoid, retbioid, and athyridoid Brachiopoda: Jour. Paleontology, v. 38, no. 5, p. 805-822, Pls. 125-128.

Boucot, A. J., Johnson, J. G., and Talent, J. A., 1967, Lower and Middle Devonian faunal provinces based on Brachiopoda: Internat. Symposium on the Devonian System, Calgary, Canada, v. 2, p. 1239-1254.

Boucot, A. J., Johnson, J. G., and Walmsley, V. G., 1965, Revision of the Rhipidomellidae (Brachiopoda) and the affinities of *Mendacella* and *Dalejina*: Jour. Paleontology, v. 39, no. 3, p. 331-340, Pls. 45, 46.

Boucot, A. J., Martinsson, Anders, Thorsteinsson, R., Walliser, O. H., Whittington, H. B., and Yochelson, Ellis, 1960, A Late Silurian fauna from the Sutherland River Formation, Devon Island, Canadian Arctic Archipelago: Canada Geol. Survey Bull. 65, 51 p., 10 Pls.

Bowen, Z. P., 1967, Brachiopoda of the Keyser Limestone (Silurian-Devonian) of Maryland and adjacent areas: Geol. Soc. America Mem. 102, 103 p., 8 Pls.

Branson, E. B., and Mehl, M. G., 1931, Fishes of the Jefferson Formation of Utah: Jour. Geology, v. 39, p. 509-531, Pls. 1-3.

Bryant, W. L., 1932, Lower Devonian fishes of Beartooth Butte, Wyoming: Am. Philos. Soc. Proc., v. 71, p. 225-254, Pls. 1-10.

—— 1933, The fish fauna of Beartooth Butte, Wyoming, Part I, The Heterostraci and Osteostraci: Am. Philos. Soc. Proc., v. 72, p. 285-314, Pls. 1-21.

—— 1934, The fish fauna of Beartooth Butte, Wyoming, Parts II and III: Am. Philos. Soc. Proc., v. 73, p. 127-162, Pls. 1-26.

—— 1935, *Cryptaspis* and other Lower Devonian fossil fishes from Beartooth Butte, Wyoming: Am. Philos. Soc. Proc., v. 75, p. 111-128, Pls. 1-18.

Bublitschenko, N. L., 1956, Nekotoryie novye predstavitely brachiopod Devona i Karbona rudnogo Altaya i Sary-arka: Akad. Nauk Kazakhshay S.S.R., Alma-Ata Izvest. Ser. Geol., v. 23, p. 93-103, 1 Pl.

Carlisle, Donald, Murphy, M. A., Nelson, C. A., and Winterer, E. L., 1957, Devonian stratigraphy of Sulphur Springs and Pinyon Ranges, Nevada: Am. Assoc. Petroleum Geologists Bull., v. 41, no. 10, p. 2175-2191.

Castelnau, Francis de, 1843, Essai Système Silurien l'Amérique Septentrionale: Paris, p. 36-44, 4 Pls.

Caster, K. E., 1939, A Devonian fauna from Colombia: Bulls. Am. Paleontology, v. 24, (no. 83) 218 p., 14 Pls.

—— 1945, New names for two homonyms: Jour. Paleontology, v. 19, no. 3, p. 319.

Clark, D. L., and Ethington, R. L., 1966, Conodonts and biostratigraphy of the Lower and Middle Devonian of Nevada and Utah: Jour. Paleontology, v. 40, no. 3, p. 659-689, Pls. 82-84.

Clarke, J. M., 1900, The Oriskany Fauna of Becraft Mountain, Columbia County, N.Y.: New York State Mus. Mem., v. 3, no. 3, 128 p., 9 Pls.

—— 1908, Early Devonic history of New York and eastern North America: New York State Mus. Mem. 9, pt. 1, 366 p., 48 Pls.

—— 1909, Early Devonic history of New York and eastern North America: New York State Mus. Mem. 9, pt. 2, 250 p., 34 Pls.

Cloud, P. E., Jr., 1942, Terebratuloid brachiopods of the Silurian and Devonian: Geol. Soc. America Spec. Paper 38, 182 p., 26 Pls.

Conrad, T. A., 1838, Report on the paleontological department of the survey: New York State Geol. Survey 2nd Ann. Rept., p. 107-119.

——1839, Descriptions of new species of organic remains: New York State Geol. Survey 3rd Ann. Rept., p. 57-66.

—— 1841, Fifth annual report on the paleontology of the state of New York: New York State Geol. Survey, 5th Ann. Rept. p. 25-57.

Conrad, T. A., 1842, Observations on the Silurian and Devonian systems of the United States, with descriptions of new organic remains: Acad. Nat. Sci. Philadelphia, v. 8 p. 228-280.

Cooper, G. A., 1942, New genera of North American brachiopods: Washington Acad. Sci. Jour., v. 32, no. 8, p. 228-235.

—— 1944, Phylum Brachiopoda, in Shimer, H. W. and Shrock, R. R., Index Fossils of North America: p. 277-365, Pls. 105-143.

—— 1945, New species of brachiopods from the Devonian of Illinois and Missouri: Jour. Paleontology, v. 19, no. 5, p. 479-489, Pls. 63, 64.

—— 1955, New genera of middle Paleozoic brachiopods: Jour. Paleontology, v. 29, no. 1, p. 45-63, Pls. 11-14.

—— 1956, Chazyan and related brachiopods: Smithsonian Misc. Colln., no. 4253; pt. 1, p. 1-1024; pt. 2, 1025-1245, Pls. 1-269.

—— 1962, Pseudopunctate brachiopods (abs.): Geol. Soc. America Spec. Paper 68, p. 155-156.

Cooper, G. A., and others, 1942, Correlation of the Devonian sedimentary formations of North America: Geol. Soc. America Bull., v. 53, no. 12, pt. 1, p. 1729-1793, 1 Pl.

Copper, Paul, 1967, The shell of Devonian Atrypida (Brachiopoda): Geol. Mag., v. 104, no. 2, p. 123-131, Pls. 5, 6.

Dahmer, Georg, 1951, Die Fauna der nach-Ordovizischen Gleider der Verse-Schichten mit Ausschluss der Trilobiten, Crinoiden, und Anthozoen: Palaeontographica Abt. A, v. 101, 152 p., 12 Pls.

Dalman, J. W., 1828, Uppstallning och Beskrifning af di i Sverige funne Terebratuliter: K. svenska Vetensk. Akad. Handl., (1827), p. 85-155, Pls. 1-6.

Davidson, Thomas, 1858, British Carboniferous Brachiopoda, pt. 5: Palaeont. Soc. London Mon., v. 11, p. 49-80, Pls. 9-16.

—— 1882, Monograph of the British Fossil Brachiopoda, part 5, Silurian and Devonian supplements, part 1: p. 1-134, Pls. 1-7.

Defrance, M.J.L., 1824, Dictionnaire des Sciences Naturelles: v. 32.

Denison, R. H., 1952, Early Devonian fishes from Utah, Part I, Osteostraci: Fieldiana—Geology, v. 11, p. 263-287, Figs. 50-60.

—— 1953, Early Devonian fishes from Utah, Part II, Heterostraci: Fieldiana—Geology, v. 11, p. 291-355, Figs. 61-85.

—— 1956, A Review of the habitat of the earliest Vertebrates: Fieldiana —Geology, v. 11, no. 8, p. 357-457.

—— 1958, Early Devonian fishes from Utah, Part III, Arthrodira: Fieldiana—Geology, v. 11, no. 9, p. 461-551.

Dorf, Erling, 1934, Stratigraphy and paleontology of a new Devonian formation at Beartooth Butte, Wyoming: Jour. Geol., v. 42, p. 720-737, Figs. 1-7.

Drot, Jeannine, 1964, Rhynchonelloidea et spiriferoidea Siluro-Dévoniens du Maroc pré-Saharien: Marocco, Notes et Mém. du Cerv. Géol., no. 178, p. 1-286, Pls. 1-24.

Dunbar, C. O., 1917 Rensselaerina, a new genus of Lower Devonian brachiopods: Am. Jour. Sci., 4th ser., v. 43, p. 466-470, Pl. 2.

Eaton, Amos, 1832, Geological Text-Book, 2nd ed.: Albany, 134 p., map.

Ehlers, G. M., and Wright, J. D., 1955, The type species of Spinocyrtia

Fredericks and a new species of this brachiopod genus from southwestern Ontario: Michigan Univ. Mus. Paleontology Contr., v. 13, no. 1, 32 p., 11 Pls.

Erben, H. K., *Editor*, 1962, Symposiums-Band, 2. Internationale Arbeitstagung über die Silur/Devon-Grenze und die Stratigraphie von Silur und Devon, Bonn-Bruxelles 1960: Stuttgart, E. Schweizerbarti'sche Verlagsbuchhandlung, 315 p.

Fagerstrom, J. A., 1966, Biostratigraphic significance of rhipidomellid brachiopods in the Detroit River Group (Devonian): Jour. Paleontology, v. 40, no. 5, p. 1236-1238.

Fenton, C. L., 1931, Studies of evolution in the genus *Spirifer*: Wagner Free Inst. Sci. Pub., v. 2, 430 p., 50 Pls.

Fischer-de-Waldheim, G., 1830, Oryctographie du gouvernment de Moscou, 1st ed.: Moscow, 26 p., 60 Pls.

Foerste, A. F., 1909, Fossils from the Silurian formations of Tennessee, Indiana and Illinois: Denison Univ. Sci. Lab. Bull., v. 14, art. 6, p. 61-116, Pls. 1-14.

Frederiks, George, 1912, Bemerkungen über einige oberpalaeozoische Fossilien von Krasnoufimsk: Prot. Osch. Est. Kazan, 1911-1912, no. 269, p. 1-9, 1 Pl.

────── 1916, Über einige ober Palaozoic Brachiopoden von Eurasien: Comite Geol. Mem., N. S., v. 156, p. 1-87.

Fuchs, A., 1919, Beitrag zur Kenntnis der Devonfauna der Verse-und Hobräcker Schichten des sauerländischen Faciesgebietes: Jb. Preuss. Geol. Landesanstalt, v. 39, pt. 1, p. 58-95, Pls. 5-9.

George, T. N., 1931, *Ambocoelia* Hall and certain similar British Spiriferidae: Geol. Soc. London Quart. Jour., v. 87, pt. 1, p. 30-61, Pls. 3-5.

Gill, T., 1871, Arrangement of the families of Molluscs prepared for the Smithsonian Institution: Smithsonian Misc. Colln., no. 227, 49 p.

Girty, G. H., 1904, New molluscan genera from the Carboniferous: U.S. Natl. Mus. Proc., v. 27, p. 721-736.

Gray, J. E., 1840, Synopsis of the contents of the British Museum, 44th ed.: London, 308 p.

Gronberg, E. C., 1967, Stratigraphy of the Nevada Group at Lone Mountain and Table Mountain, central Nevada: Ms. thesis, Univ. California, Riverside, California. 83 p., 11 Pls. (available on interlibrary loan).

Haas, Winfried, 1969, Lower Devonian trilobites from central Nevada and northern Mexico: Jour. Paleontology, v. 43, p. 641-659, Pls. 81-84.

Hall, James, 1852, Containing descriptions of the organic remains of the lower middle division of the New York System: New York Geol. Survey, Nat. Hist. New York, Palaeontol., v. 2, 362 p., 85 Pls.

────── 1857, Descriptions of Palaeozoic fossils: New York State Cab. 10th Ann. Rept., pt. C, Appendix, p. 41-186.

────── 1859, 1861, Palaeontology of New York: New York Geol. Survey, v. 3, p. 1-532 (1859), Pls. 1-120 (1861).

────── 1860a, Observations on Brachiopoda: New York State Cab., 13th Ann. Rept., p. 65-71.

────── 1860b, Descriptions of new species of fossils from the Hamilton Group of western New York, with notices of others from the same horizon in Iowa and Indiana: New York State Cab. 13th Ann. Rept., p. 76-94.

——— 1863a, Contributions to palaeontology: New York State Cab. 16th Ann. Rept. p. 3-226, Pls. 1-11.

——— 1863b, Notice of some new species of fossils from a locality of the Niagara group in Indiana; with a list of identified species from the same place: Albany Inst. Trans., v. 4, p. 195-228.

——— 1867a, Notice of Volume IV of the Palaeontology of New York: New York State Cab. Nat. Hist., 20th Ann. Rept., p. 145-168.

——— 1867b, Descriptions and figures of the fossil Brachiopoda of the upper Helderberg, Hamilton, Portage, and Chemung Groups: New York Geol. Survey, Palaeontology, v. 4, 428 p., Pls. 1-63.

——— 1879, The fauna of the Niagara Group: New York State Mus. Nat. Hist. 28th Rept., p. 98-203, Pls. 3-34.

Hall, James, and Clarke, J. M., 1892, An Introduction to the study of the genera of Palaeozoic Brachiopoda, Part I: New York Geol. Survey, Palaeontology of New York, v. 8, 367 p., Pls. 1-20.

——— 1893, 1895 (1894), An introduction to the study of the genera of Palaeozoic Brachiopoda: New York Geol. Survey, Palaeontology of New York, v. 8, pt. 2, p. 1-317 (1893); p. 319-394, Pls. 21-84 (1895).

Harper, C. W., Jr., Johnson, J. G., and Boucot, A. J., 1967, The Pholidostrophiinae (Brachiopoda; Ordovician, Silurian, Devonian): Senckenbergiana Lethaea, v. 48, no. 5, p. 403-461, Pls. 1-10.

Havlíček, Vladimir, 1950, Ramenonožci českého Ordoviku: Ústřed. Ústavu Geol. Rozpravy, v. 13, 135 p., 12 Pls.

——— 1953, O několika nových ramenonožcích českého a moravského stredniho devonu: Ustred. Ustavu Geol. Věstnik, v. 28, p. 4-9. Pls. 1, 2.

——— 1956, The brachiopods of the Braník and Hlubočepy Limestones in the immediate vicinity of Prague: Ústřed. Ústavu Geol. Sborník, v. 22, p. 535-665, 12 Pls.

——— 1959, Spiriferidae v ceskem siluru a devonu: Ústřed. Ústavu Geol. Rozpravy, v. 25, 275 p., 28 Pls.

——— 1960, Bericht über die Ergebnisse der Revision der böhmischen altpaläozoischen Rhynchonelloidea: Ústřed. Ústavu Geol. Věstník, v. 35, p. 241-244.

——— 1961, Rhynchonelloidea des böhmischen älteren Paläozoikums (Brachiopoda): Ústřed. Ústavu Geol. Rozpravy, v. 27, 211 p., 27 Pls.

——— 1963, *Zlichorhynchus hiatus* n. g. et n. sp., neuer Brachiopode, vom Unterdevon Böhmens: Ústřed. Ústavu Geol. Věstník, v. 38, p. 403-404, Pl. 1.

——— 1965, Superfamily Orthotetacea (Brachiopoda) in the Bohemian and Moravian Paleozoic: Ústřed. Ústavu Geol. Věstník, v. 40, no. 4, p. 291-294.

——— 1967, Brachiopoda of the suborder Strophomenidina in Czechoslovakia: Ústřed. Ústavu Geol. Rozpravy, v. 33, 235 p., 52 Pls.

Helmbrecht, W., and Wedekind, R., 1923, Versuch einer biostratigraphischen Gleiderung der Siegener Schichten auf Grund von Rensselaerien und Spiriferen: Gluckauf, v. 59, no. 41, p. 949-953.

Hisinger, W., 1827, Gottland, Geognostiskt beskrifvit: Kongl. Vetenskaps Akad. Handl. (1826), p. 311-336, Pls. 7, 8.

House, M. R., 1962, Observations on the ammonoid succession of the North American Devonian: Jour. Paleontology, v. 36, no. 2, p. 247-284, Pls. 43-48.

—— 1965, Devonian goniatites from Nevada: Neues Jahrb. Geologie u. Paläontologie, Abh., v. 122, no. 3, p. 337-342, Pl. 32.

Imbrie, John, 1959, Brachiopods of the Traverse Group (Devonian) of Michigan. Part 1, Dalmanellacea, Pentameracea, Strophomenacea, Orthotetacea, Chonetacea, and Productacea: Am. Mus. Nat. History Bull., v. 116, p. 345-410, Pls. 48-67.

Ivanova, E. A., 1962, Ecology and development of the Silurian and Devonian Brachiopods of the Kuznetzk, Minussinsk, and Tuva basins: Izdat. Akad. Nauk, SSSR, Trudy Palaeont. Inst. v. 88, 152 p., 20 Pls.

Jackson, D. E., and Lenz, A. C., 1963, A new species of *Monograptus* from the Road River Formation, Yukon: Palaeontology, v. 6, pt. 4, p. 751-753.

Jaeger, Hermann, 1965, Review of Symposiums-Band der 2. Internationalen Arbeitstagung über die Silur/Devon-Grenze and die Stratigraphie von Silur und Devon, Bonn-Bruxelles 1960: Geologie, v. 14, no. 3, p. 348-364, 1 chart.

Johnson, J. G., 1962a, Brachiopod faunas of the Nevada Formation (Devonian) in central Nevada: Jour. Paleontology, v. 36, no. 1, p. 165-169.

——1962b, Lower Devonian-Middle Devonian boundary in central Nevada: Am. Assoc. Petroleum Geologists Bull., v. 46, no. 4, p. 542-546.

——1965, Lower Devonian stratigraphy and correlation, northern Simpson Park Range, Nevada: Canadian Petroleum Geologists Bull., v. 13, no. 3, p. 365-381.

——1966a, Middle Devonian brachiopods from the Roberts Mountains, central Nevada: Palaeontology, v. 9, pt. 1, p. 152-181, Pls. 23-27.

——1966b, Two new spiriferid brachiopod genera from the Lower Devonian of Nevada: Jour. Paleontology, v. 40, no. 5, p. 1043-1050, Pls. 127-129.

——1966c, *Parachonetes*, a new Lower and Middle Devonian brachiopod genus: Palaeontology, v. 9, pt. 3, p. 365-370, Pls. 62, 63.

——1967, *Toquimaella*, a new genus of karpinskiinid brachiopod: Jour. Paleontology, v. 41, no. 4, p. 874-880, Pl. 111.

——1968, A new species of *Vagrania* (Devonian, Brachiopoda) from Nevada: Jour. Paleontology, v. 42, no. 5, p. 1200-1204, Pl. 159.

Johnson, J. G., and Boucot, A. J., 1967, *Gracianella*, a new Late Silurian genus of atrypoid brachiopods: Jour. Paleontology, v. 41, no. 4, p. 868-873, Pls. 109, 110.

—— 1970, Brachiopods and age of the Tor Limestone of central Nevada: Jour. Paleontology, v. 44, no. 2.

Johnson, J. G., Boucot, A. J., and Gronberg, E. C., 1968, A new genus of stringocephaloid brachiopod from the Middle Devonian of Nevada: Jour. Paleontology, v. 42, no. 2, p. 406-414, Pls. 59, 60.

Johnson, J. G., Boucot, A. J., and Murphy, M. A., 1967 (1968) Lower Devonian faunal succession in central Nevada: Internat. Symposium on the Devonian System, v. 2, Calgary, Canada, 1967, p. 679-691.

Johnson, J. G., and Murphy, M. A., 1969, Age and position of Lower Devonian graptolite zones relative to the Appalachian standard succession: Geol. Soc. America Bull., v. 80, no. 7, p. 1275-1282.

Johnson, J. G., and Reso, Anthony, 1964, Probable Ludlovian brachiopods from the Sevy Dolomite of Nevada: Jour. Paleontology, v. 38, no. 1, p. 74-84, Pls. 19-20.

Johnson, J. G., and Talent, J. A., 1967a, *Muriferella*, a new genus of Lower Devonian septate dalmanellid: Royal Soc. Victoria Proc., v. 80, pt. 1, p. 43-50, Pls. 9, 10.

——1967b, Cortezorthinae, a new subfamily of Siluro-Devonian dalmanellid brachiopods: Palaeontology, v. 10, pt. 1, p. 142-170, Pls. 19-22.

Johnson, M. S., and Hibbard, D. E., 1957, Geology of the Atomic Energy Commission Nevada Proving Grounds Area, Nevada: U.S. Geol. Survey Bull. 1021-K, p. 333-384, Pls. 32, 33.

Johnston, Joan, 1941, Studies in Silurian Brachiopoda, I. Description of a new genus and species: Linnean Soc. New South Wales Proc., v. 66, pts. 3-4, p. 160-168, Pl. 7.

Kaplun, L. I., 1961, Lower Devonian brachiopods of the northern Pribalkhasch, *in* Materials for the geology and guide fossils of Kazakhstan, pt. 1 (26), Stratigraphy and paleontology: Kazakh SSR, Minist. Geol. Okran. Nedr., p. 64-114, Pls. 7-19.

Kay, Marshall, 1960, Paleozoic continental margin in central Nevada, western United States: 21st Internat. Geol. Congress, Norden, pt. 12, Regional Paleogeography, p. 94-103.

Kay, Marshall, and Crawford, J. P., 1964, Paleozoic facies from the miogeosynclinal to the eugeosynclinal belt in thrust slices, central Nevada: Geol. Soc. America Bull., v. 75, no. 5, p. 425-454, 6 Pls.

Kayser, Emanuel, 1878, Die fauna der ältesten Devon-Ablagerungen des Harzes: Abh. geol. Specialkt. von Preuss., p. 1-296, 36 Pls.

——1889, Die Fauna des Hauptquarzits und der Zorger Schiefer des Unterharzes: Abh. König. Preuss. Geol. L. A., n. f., v. 1, 140 p., 24 Pls.

Khalfin, L. L., 1948, Fauna i Stratigrafiya Devonskikh Otlozhenii Gornogo Altaia: Izvestiya Tomskogo Ordena Trudovogo Krasnogo Znameni Politek. Inst. imeni S. M. Kirova, v. 65, no. 1, 464 p., 36 Pls.

Khodalevich, A. N., 1951, Lower Devonian and Eifelian brachiopods of the Ivdel and Serov districts of the Sverdlovsk region: Sverdlovsk Mining Inst. Trudy, v. 18, 169 p., 30 Pls.

Kindle, E. M., 1912, The Onondaga fauna of the Allegheny Region: U.S. Geol. Survey Bull. 508, 114 p., 13 Pls.

——1938, The correlation of certain Devonian faunas of eastern and western Gaspé: Bulls. Am. Paleontology, v. 24 (no. 82), p. 35-86. Pls. 3, 4.

King, William, 1846, Remarks on certain genera belonging to the class Palliobranchiata: Nat. Hist. London Ann. Mag., v. 18, p. 26-42, 83-94.

——1850, A monograph of the Permian fossils of England: Palaeont. Soc. London Mon., v. 3, pt. 1, xxxviii + 258 p., 28 Pls.

Kirk, Edwin, 1927, New American occurrence of *Stringocephalus*: Am. Jour. Sci., 5th ser., v. 13, p. 219-222.

Klapper, Gilbert, 1968, Lower Devonian conodont succession in central Nevada (abs.): Geol. Soc. America Program, 64th Ann. Meeting, Cordilleran Section, Tucson, p. 72, 73.

Klapper, Gilbert, and Ormiston, A. R., 1969, Lower Devonian conodont sequence, Royal Creek, Yukon Territory and Devon Island, Canada; with a section on Devon Island stratigraphy: Jour. Paleontology, v. 43, no. 1, p. 1-27, Pls. 1-6.

Klapper, Gilbert, and Ziegler, Willi, 1967, Evolutionary development of the *Icriodus latericrescens* group (Conodonta) in the Devonian of Europe and North America: Palaeontographica, pt. A, v. 127, no. 1-3, p. 68-83, Pls. 8-11.

Kozlowski, Roman, 1929, Les Brachiopodes Gothlandiens de la Podolie Polonaise: Palaeontologia Polonica, v. 1, 254 p., 12 Pls.

——1946, *Howellella*, a new name for *Crispella* Kozlowski, 1929: Jour. Paleontology, v. 20, no. 3, p. 295.

Kulkov, N. P., 1963, Brachiopodi Solovikhinskikh Sloev Nizhnego Devona gornogo Altaia: Akad. nauk SSSR, Moscow, 132 p., 9 Pls.

Lenz, A. C., 1966, Upper Silurian and Lower Devonian paleontology and correlations, Royal Creek, Yukon Territory; a preliminary report: Canadian Petroleum Geologists Bull., v. 14, no. 4, p. 604-612.

——1967a, *Thliborhynchia*, a new Lower Devonian rhynchonellid from Royal Creek, Yukon, Canada: Jour. Paleontology, v. 41, no. 5, p. 1188-1192, Pl. 161.

——1967b (1968), Upper Silurian and Lower Devonian biostratigraphy, Royal Creek, Yukon Territory, Canada: Internat. Symposium on the Devonian System, v. 2, Calgary, Canada, 1967, p. 587-599.

Léveillé, Charles, 1835, Aperçu géologique de quelques localitiés tres riches en coquilles sur les frontières de France et de Belgique: Soc. Géol, France Mém., v. 2, p. 29-40, Pl. 2.

Lindström, G., 1861, Bidrag till kännedomen on Gotlands Brachiopoder: Kongl. Vetenskaps-Akad. Förhandlinger, v. 17 (1860), p. 337-382, Pls. 12, 13.

Linnaeus, Caroli, 1758, Systema Naturae, 10th ed.: Holmiae, v. 1, 823 p.

McAllister, J. F., 1952, Rocks and structures of the Quartz Spring Area, northern Panamint Range, California: California Div. Mines Spec. Rept. 25, 38 p.

M'Coy, Frederick, 1844, A synopsis of the characters of the Carboniferous limestone fossils of Ireland: Dublin, 207 p., 29 Pls.

McEwan, E. D., 1939, Convexity of articulate brachiopods as an aid to identification: Jour. Paleontology, v. 13, p. 617-620.

McLaren, D. J., 1962, Middle and early Upper Devonian rhynchonelloid brachiopods from western Canada: Canada Geol. Survey Bull. 86, p. 1-122, 18 Pls.

McLearn, F. H., 1924, Paleontology of the Silurian rocks of Arisaig, Nova Scotia: Canada Geol. Surv. Mem. 137, 179 p., 30 Pls.

Mansuy, H., 1908, Contribution à la Carte Géologique de l'Indochine, Paléontologie: Indochina Serv. Mines, (Hanoi-Haiphong), 73 p., 18 Pls.

——1916, Faunes Paléozoiques du Tonkin Septentrional: Indochina Serv. Géol. Mém., v. 5, pt. 4, p. 1-71, Pls. 1-8.

Martin, William, 1809, Petrificata Derbiensia: 52 Pls., Wigan and London.

Meek, F. B., 1870, Descriptions of fossils collected by the U.S.G.S. under the charge of Clarence King: Acad. Nat. Sci. Philadelphia Proc., no. 1, p. 56-64.

——1877, Palaeontology: U.S. Army Engr. Dept. Prof. Papers, no. 18— Rept. Geol. Expl. 40th Parallel, v. 4, pt. 1, p. 1-197, Pls. 1-17.

Merriam, C. W., 1940, Devonian stratigraphy and paleontology of the Roberts Mountains region, Nevada: Geol. Soc. America Spec. Paper 25, 114 p., 16 Pls.

——1954, Review of Silurian-Devonian boundary relations in the Great Basin (abs.): Geol. Soc. America Bull., v. 65, no. 12, pt. 2, p. 1284, 1285.

——1963, Paleozoic rocks of Antelope Valley, Eureka and Nye Counties, Nevada: U.S. Geol. Survey Prof. Paper 423, 67 p., 2 Pls.

Merriam, C. W., and Anderson, C. A., 1942, Reconnaissance survey of the Roberts Mountains, Nevada: Geol. Soc. America Bull., v. 53, no. 12, pt. 1, p. 1675-1728.

Meyer, Oskar-Erich, 1913, Die Devonischen Brachiopoden von Ellesmereland: Videnskaps - Selskabet i Kristiania, Rept. 2nd Norwegian Arctic Exped. in the "Fram" 1898-1902, v. 29, p. 1-43, Pls. 1-8.

Miller, S. A., 1889, North American geology and Paleontology: Cincinnati, 664 p.

Muir-Wood, H. M., 1925, Notes on the Silurian brachiopod genera *Delthyris*, *Uncinulina*, and *Meristina*: Nat. Hist. London Ann. Mag. v. 9, p. 83-95, 11 Figs.

——1962, On the morphology and classification of the brachiopod suborder Chonetoidea: British Museum (Nat. Hist.), 132 p., 16 Pls.

——1965, Chonetidina, *in* Moore, R. C., *Editor*, Treatise on invertebrate paleontology, Part H, Brachiopoda: Lawrence, Kansas, Geol. Soc. America and Univ. Kansas Press, p. H412-H439, Figs. 272-296.

Muir-Wood, H. M., and Cooper, G. A., 1960, Morphology, classification, and life habits of the Productoidea (Brachiopoda): Geol. Soc. America Mem. 81, p. 1-447, 135 Pls.

Murphy, M. A., and Gronberg, E. C., 1970, Stratigraphy of the lower Nevada Group (Devonian) north and west of Eureka, Nevada: Geol. Soc. America Bull., v. 81, p. 127-136.

Nalivkin, D. V., 1937, Brachiopoda of the Upper and Middle Devonian and Lower Carboniferous of northeastern Kazakhstan: Central Geol. Prosp. Inst., Trudy, v. 99, 200 p., 39 Pls.

Nikiforova, O. I., 1937, Upper Silurian Brachiopoda of the central Asiatic part of the USSR: Central Geol. Prospect. Inst., Mon. Pal. USSR, v. 35, 93 p., 14 Pls.

——1954, Stratigraphy and brachiopods of the Silurian deposits of Podolia: All-Union Scientific-Research Geological Institute (VSEGEI) Trudy, 218 p., pt. 1, 19 Pls; pt. 2, 18 Pls.

Nolan, T. B., Merriam, C. W., and Williams, J. S., 1956, The stratigraphic section in the vicinity of Eureka, Nevada: U.S. Geol. Survey Prof. Paper 276, 77 p., 2 Pls.

Oehlert, D. P., 1887, Brachiopodes, *in* Fischer, P. H., Manuel de Conchyliologue et de Paléontologie Conchyliologique au histoirie naturelle des Mollusques vivants et fossiles: Paris, F. Savy, p. 1189-1334. Pl. 15.

———1890, Note sur différents groupes établis dans le genre Orthis et en particulier sur *Rhipidomella* Oehlert (=*Rhipidomys* Oehlert, *olim*): Jour. Conchyliologie, ser. 3, v. 30, p. 366-374.

———1901, Fossiles Dévoniens de Santa-Lucia (Province de Léon, Espagne): Soc. Geol. France Bull., 4th Ser., v. 1, p. 233-250, Pl. 6.

Oliver, W. A., Jr., 1956, Stratigraphy of the Onondaga Limestone in eastern New York: Geol. Soc. America Bull., v. 67, no. 11, p. 1441-1474.

——— 1964, New occurrences of the rugose coral *Rhizophyllum* in North America: U.S. Geol. Survey Prof. Paper 475-D, p. 149-158.

Öpik, A. A., 1934, Über Klitamboniten: Tartu. Univ. Geol. Inst. Pub., no. 39, p. 1-239, 48 Pls.

Orbigny, Alcide d', 1849 Prodrôme de paléontologie stratigraphique universelle: Paris, v. 1, 394 p.

Paeckelmann, Werner, 1931, Versuch einer zusammenfassenden Systematik der Spiriferidae King: Neues Jahrb. Mineralogie, Geologie u. Paläontologie, v. 67, pt. B, p. 1-64.

Patte, Etienne, 1926, Études Paléontologiques Relatives à la Géologie de L'est du Tonkin (Paléozoique et Trias): Indochina Serv. Géol. Bull., v. 15, pt. 1, p. 1-240, Pls. 1-12.

Peetz, H. Von, 1901, Beitrage zur Kenntniss der fauna aus den devonischen Schichten am Rande des Steinkohlenbassins von Kuznetz: Travaux Sect. geol. cabinet de la majeste St. Petersburg, 394 p., 6 Pls.

Philip, G. M., 1962, The palaeontology and stratigraphy of the Siluro-Devonian sediments of the Tyers area, Gippsland, Vistoria: Royal Soc. Victoria, Proc., v. 75, pt. 2, p. 123-246, Pls. 11-36.

———1965, Lower Devonian conodonts from the Tyers area, Gippsland, Victoria; Royal Soc. Victoria Proc., n. s., v. 79, pt. 1, p. 95-118, Pls. 8-10.

———1966, Lower Devonian conodonts from the Buchan Group, eastern Victoria: Micropaleontology, v. 12, no. 4, p. 441-460, 4 Pls.

Philip, G. M., and Jackson, J. H., 1967, Lower Devonian subspecies of the conodont *Polygnathus linguiformis* Hinde from southeastern Australia: Jour. Paleontology, v. 41, no. 5, p. 1262-1266.

Philip, G. M., and Pedder, A.E.H., 1964, A reassessment of the age of the Middle Devonian of south-eastern Australia: Nature, v. 202, no. 4939, p. 1323, 1324.

Phillips, J., 1841, Figures and descriptions of the Palaeozoic fossils of Cornwall, Devon, and West Somerset: London, 231 p., 60 Pls.

Pitrat, C. W., 1965, Spiriferidina, *in* Moore, R. C. *Editor*, Treatise on Invertebrate Paleontology; Part H., Brachiopoda: Lawrence, Kansas, Geol. Soc. America and Univ. Kansas Press, H667-H728.

Poulsen, Chr., 1943, The fauna of the Offley Island formation, pt. 2, Brachiopoda: Medd. om Grønland, v. 72, no. 3, 60 p., 6 Pls.

Raymond, P. E., 1923, New fossils from the Chapman sandstone: Boston Soc. Nat. Hist. Proc., v. 36, no. 7, p. 467-472, Pl. 6.

Roberts, R. J., Hotz, P. E., Gilluly, James, and Ferguson, H. G., 1958,

Paleozoic rocks of north-central Nevada: Am. Assoc. Petroleum Geologists Bull., v. 42, no. 12, p. 2813-2857.

Roemer, C. F., 1844, Das rheinische Uebergangsgebirge: Hannover, 96 p., 6 Pls.

Rzhonsnitskaya, M. A., 1952, Spiriferidae of the Devonian deposits at the margins of the Kuznetzk Bassin: VSEGEI Trudy (Moscow) 232 p., 25 Pls.

―――1956 (1958), On rhynchonellid systematics: Internat. Geol. Congress, 20th session, Mexico, section 7, p. 107-121.

―――1960a, Die Korrelation der karbonatischen Sedimente des Unterund Mittel Devons in der Sowjetunion und in Westeuropa: Prager, Arbeitstagung über die Stratigraphie des Silurs und des Devons, Praha, p. 123-135.

―――1960b, Order Atrypida *in* Sarycheva, T. G., Osnovi Paleontologii, Mshanki, Brachiopodi: Izdat. Akad. Nauk SSSR, p. 257-264, Pls. 53-56.

―――1962, Devonian intervals of the principal sections of Siberia and their correlation with the Devonian of Europe: Sovetskaya Geologiya, no. 10, p. 16-27 (translation, Internat. Geol. Review, v. 6, no. 6, p. 1085-1098).

―――1964, Stratigraphy and Brachiopoda of the Devonian of the margins of the Kuznetsk Basin: VSEGEI (Leningrad) 48 p.

Sandberg, C. A., 1961, Widespread Beartooth Butte Formation of Early Devonian age in Montana and Wyoming and its paleogeographic significance: Am. Assoc. Petroleum Geologists Bull., v. 45, no. 8, p. 1301-1309.

Sapelnikov, V. P., 1960, A new lower Wenlock genus *Jolvia* (Pentameracea) from the central Urals: Paleont. Zhurn., no. 4 (1960), p. 54-62, Pls. 5, 6.

Sartenaer, Paul, 1961a, Redescription of *Leiorhynchus quadracostatus* (Vanuxem), type species of *Leiorhynchus* Hall, 1860 (Rhynchonellacea): Jour. Paleontology, v. 35, no. 5, p. 963-976, Pls. 111, 112.

―――1961b, Etue nouvelle en deux parties du genre *Camarotoechia* Hall et Clarke, 1893. Premiere Partie: *Atrypa congregata* Conrad, espece-type (1): Inst. Roy. Sci. Nat. Belgique, v. 37, no. 22, 11 p., 1 Pl.

Schlotheim, E. F. von, 1820, 1822, Die Petrefactenkunde auf ihrem jetzigen Standpunkte, 1820; Nachtrage zur Petrefactenkunde, Mit Atlas, 1822: Gotha, 437 p., 29 Pls. (1820); 100 p., 21 Pls. (1822).

Schmidt, Herta, 1964, Neue Gattungen pälazoischer Rhynchonellacea (Brachiopoda): Senckenbergiana Lethaea, v. 45, no. 6, p. 505, 506.

―――1965, Neue Befunde an paläozoischen Rhynchonellacea (Brachiopoda): Senckenbergiana Lethaea, v. 46, no. 1, p. 1-25, 1 Pl.

Schmidt, Herta, and McLaren, D. J., 1965, Paleozoic Rhynchonellacea, *in* Moore, R. C., *Editor*, Treatise on Invertebrate Paleontology, Part H, Brachiopoda: Lawrence, Kansas, Geol. Soc. Amer. and Univ. Kansas Press, p. H552-H597, Figs. 419-477A.

Schmidt, Wolfgang, 1939, Die Grenzschichten Siluro-Devon in Thüringen mit besonderer Berücksichtigung des Downton-Problems: Abh. Preuss. Geol. Landesanstalt, n. f., v. 195, 99 p., 4 Pls.

Schmidt, Wolfgang, 1954, Die ersten Vertebraten-Faunen in deutschen Gedinne: Palaeontographica, pt. A, v. 105, p. 1-47, Pls. 1-6.

Schnur, J., 1853, Zusammenstellung und Beschreibung sammtlicher im Uebergangsgebirge der Eifel vorkommenden Brachiopoden: Paleontographica, v. 3, p. 169-247, Pls. 22-45.

Schuchert, Charles, 1894, A revised classification of the spire-bearing Brachiopoda: Am. Geologist, p. 102-107.

——1913, Brachiopoda *in* Eastman, C. R., *Editor*, Zittel's Text-book of Paleontology, 2nd ed.: London, Macmillan and Co., v. 1, p. 355-420.

Schuchert, Charles, and Cooper, G. A., 1931, Synopsis of the brachiopod genera of the suborders Orthoidea and Pentameroidea with notes on the Telotremata: Am. Jour. Sci., 5th Ser., v. 22, no. 129, p. 241-251.

——1932, Brachiopod genera of the suborders Orthoidea and Pentameroidea: Peabody Mus. Mem., v. 4, pt. 1, 270 p., 29 Pls.

Schuchert, Charles, and LeVene, C. M., 1929, Brachiopoda Foss. Cat., Animalia, pt. 42, 140 p.

Schuchert, Charles, Swartz, C. K., Maynard, T. P., Prosser, C. S., and Rowe, R. B., 1913, Lower Devonian: Maryland Geol. Survey, 560 p., 98 Pls.

Scupin, Hans, 1900, Die Spiriferen Deutschlands: Palaeont. Abh., n. f., v. 4, pt. 3, p. 207-344, Pls. 24-33.

Shaler, N. S., 1865, List of the Brachiopoda from the island of Anticosti sent by the Museum of Comparative Zoology to different institutions in exchange for other specimens, with annotations: Harvard Univ. Mus. Comp. Zoology Bull., v. 1, p. 61-70.

Shirley, Jack, 1938, The fauna of the Baton River beds (Devonian), New Zealand: Geol. Soc. London Quart. Jour., v. 94, p. 384-422. Pls. 40-44.

Siemiradzki, J., 1906, Monografja warstw paleozoicznych Podola: Sprawozdanie Kom. Fizyograficznej A. U., v. 39, p. 87-196.

Solle, Gerhard, 1942, Die Kondelgruppe (Oberkoblenz) im südlichen rheinischen Schiefergebirge. I-III: Senckenberg naturf. Ges. Abh., no. 461, 92 p., 1 Pl.

——1963, *Hysterolites hystericus* (Schlotheim) (Brachiopoda; Unterdevon), die Einstufung der oberen Graptolithen-Schiefer in Thüringen und die stratigraphische Stellung der Zone des *Monograptus hercynicus*: Geol. Jahrb., v. 81, p. 171-220, Pls. 7-9.

Sowerby, J., 1812-1815, The mineral conchology of Great Britain: London, v. 1, 234 p., 102 Pls.

Spriestersbach, Julius, 1942, Lenneschiefer (Stratigraphie, Fazies und Fauna): Abh. Reichsamts. Bodenforsch., n. f., v. 203, 219 p., 11 Pls.

Stainbrook, M. A., 1943, Strophomenacea of the Cedar Valley Limestone of Iowa: Jour. Paleontology, v. 17, no. 1, p. 39-59, Pls. 6, 7.

——1945, Brachiopoda of the Independence Shale of Iowa: Geol. Soc. America Mem. 14, 74 p., 6 Pls.

Steininger, J., 1853, Geognostische Beschreibung der Eifel: Trier, 143 p., 10 Pls.

Stewart, G. A., 1923, The fauna of the Little Saline Limestone in Ste. Genevieve County, *in* The Devonian of Missouri: Missouri Bur. Geology and Mines, 2nd ser., v. 17, p. 213-269, Pls. 57-71.

Struve, Wolfgang, 1964, Über *Alatiformia*-Arten und andere, ausserlich

ahnliche Spiriferacea: Senckenbergiana Lethaea, v. 45, no. 1-4, p. 324-346, Pl. 31.
Swartz, F. M., 1929, The Helderberg group of parts of West Virginia and Virginia: U.S. Geol. Survey Prof. Paper 158, p. 27-75, Pls. 6-9.
Talent, J. A., 1956, Devonian brachiopods and pelecypods of the Buchan Caves Limestone, Victoria: Royal Soc. Victoria Proc., n. s., v. 68, p. 1-56, Pls. 1-5.
―――1963, The Devonian of the Mitchell and Wentworth Rivers: Victoria Geol. Survey Mem. 24, 118 p., 78 Pls.
―――1964 (1965) The Silurian and Early Devonian faunas of the Heathcote District, Victoria. Victoria Geol. Survey Mem. 26, 55 p., 27 Pls.
Tansey, V. O., 1923, The fauna and the correlation of the Bailey Limestone in the Little Saline Creek area of Ste. Genevieve County, Missouri, *in* the Devonian of Missouri: Missouri Bur. Geology and Mines, 2nd ser., v. 17, p. 166-212, Pls. 40-56.
Termier, Henri, and Termier, Geneviéve, 1949, Essai sur l'évolution des Spiriferidés: Marocco, Serv. Géol. Notes et Mém., v. 2, no. 74, p. 85-112.
Vandercammen, Antoine, 1957, Revision du Genre *Gurichella* W. Paeckelmann, 1913: Bruxelles Inst. Roy. Sci. Nat. de Belgique Mem. 138, 50 p., 2 Pls.
―――1963, Spiriferidae du Devonien de la Belgique: Bruxelles Inst. Roy. Sci. Nat. de Belgique Mem. 150, 181 p., 13 Pls.
Vanuxem, Lardner, 1842, Geology of New York, Part III, comprising the survey of the third geological district: Albany, 306 p.
Waagen, William, 1882-1885, Salt Range fossils; Productus limestone fossils, Brachiopoda: Palaeont. Indica, ser. 13, v. 1, p. 329-770, Pls. 25-86.
Waines, R. H., 1962, Devonian sequence in the north portion of the Arrow Canyon Range, Clark County, Nevada (abs.): Geol. Soc. America Spec. Paper 68, p. 62.
Waite, R. H., 1956, Upper Silurian Brachiopoda from the Great Basin: Jour. Paleontology, v. 30, no. 1, p. 15-18, Pl. 4.
Walcott, C. D., 1884, Paleontology of the Eureka District: U.S. Geol. Survey Mon. 8, 298 p., 24 Pls.
Walmsley, V. G., 1965, *Isorthis* and *Salopina* (Brachiopoda) in the Ludlovian of the Welsh Borderland: Palaeontology, v. 8, pt. 3, p. 454-477, Pls. 61-65.
Wang, Y., 1956, Some new brachiopods from the Yükiang Formation of southern Kwangsi Province: Scientia Sinica, v. 5, no. 2, p. 373-388, 3 Pls.
Wedekind, Rudolf, 1926, Die devonische Formation, *in* Solomon, W., Grundzüge der Geologie II: Stuttgart E. Schweizerbart, p. 194-226.
Weller, Stuart, 1903, The Paleozoic faunas: New Jersey Geol. Survey Rept. on Paleontology, v. 3, 462 p., 53 Pls.
White, C. A., 1862, Description of new species of fossils from the Devonian and Carboniferous rocks of the Mississippi Valley: Boston Soc. Nat. Hist. Proc., 9, p. 8-33.
Willard, Bradford, Swartz, F. M., and Cleaves, A. B., 1939, The Devonian of

Pennsylvania: Pennsylvania Geol. Survey, 4th ser., Bull. G 19, 481 p., 32 Pls.

Williams, Alwyn, 1950, New stropheodontid brachiopods: Washington Acad. Sci. Jour., v. 40, no. 9, p. 277-282.

———1953a, The classification of the strophomenoid brachiopods: Washington Acad. Sci. Jour., v. 43, no. 1, p. 1-13.

———1953b, North American and European stropheodontids, their morphology and systematics: Geol. Soc. America Mem. 56, 67 p., 13 Pls.

———1965, Suborder Strophomenidina Öpik, 1934, *in* Moore, R. C., *Editor*, Treatise on Invertebrate Paleontology, Part H, Brachiopoda; Lawrence, Kansas, Geol. Soc. America and Univ. Kansas Press, p. H362-H412, Figs. 231-271.

Williams, Alwyn, and Wright, A. D., 1963, The classification of the "*Orthis testudinaria* Dalman" group of brachiopods: Jour. Paleontology, v. 37, no. 1, p. 1-32, Pls. 1, 2.

Williams, H. S., and Breger, C. L., 1916, The fauna of the Chapman Sandstone of Maine: U.S. Geol. Survey Prof. Paper 89, 347 p., 27 Pls.

Williams, J. Stewart, 1948, Geology of the Paleozoic rocks, Logan quadrangle, Utah: Geol. Soc. America Bull., v. 59, p. 1121-1164.

Winterer, E. L., and Murphy, M. A., 1960, Silurian reef complex and associated facies, central Nevada: Jour. Geology, v. 68, no. 2, p. 117-139, 7 Pls.

Woodward, H. P., 1943, Devonian System of West Virginia: West Virginia Geol. Survey, v. 15, 655 p., 69 Pls.

Woodward, S. P., 1854, A manual of the Mollusca; or, Rudimentary treatise of Recent and fossil shells: London, John Weale, part 2, p. i-xvi, 159-486.

Wright, A. D., 1965, Superfamily Enteletacea Waagen, 1884, *in* Moore, R. C., *Editor*, Treatise on Invertebrate Paleontology, Pt. H, Brachiopoda: Lawrence, Kansas, Geol. Soc. America and Univ. Kansas Press, p. H328-346.

Ziegler, Willi, 1956, Unterdevonische Conodonten, insbesondere aus dem Schönauer und dem Zorgensis-Kalk: Notizbl. Hess. L. Amt. Bodenforsch., v. 84, p. 93-106, Pls. 6, 7.

Index of Genera and Species

Acanthoscapha, sp., 45
Acrospirifer, 51, 54, 57, 183, 184, 188, 189
  sp., 20, 25
  angularis, 186
  cf. *fallax,* 57
  ilsae, 57
  intermedius, 186
  kobehana, 5, 15, 20, 21, 22, 23, 24, 25, 27, 55, 56, 57, 60, 62, 138, 183, 189, **190**, 192; Pls. 57, 58
  cf. *kobehana,* 25, 27; Pl. 56
  murchisoni, 54, 57, 186
  aff. *murchisoni,* 16, 19, 22, 35, 49, 53, 54, 55, 64, **189**, 190, 192; Pls. 56, 57
  pinyonensis, 207
  speciosus, 55
Actinoptera boydi, 61
aculeata, Howellella, 62, 185, 188
acuminatum, Syringaxon, 49
acus, Monograptus, 28
acuticuspidata, Leptaena, 19, 20, 22, 30, 31, 35, 48, 53, 59, 64, 103, 104, 106
acutiplicata, Leptocoelina, 171
Adolfia, 207
aesopeum, Aesopomum, 111, 112
Aesopomum, 46, 101, 107, 111, 112
  sp., 39
  aesopeum, 111
  varistriatus, 2, **111**; Pl. 17
  cf. *varistriatus,* 34, 44, 63; Pl. 17
alatus, Euryspirifer pseudocheehiel, 188
allani, "Schizophoria," 74
altaica, "Levenea," 74
althi, Machaeraria, 142
altiforme, Modiomorpha, 61
Alveolites, sp., 60
Ambocoelia, 46, 183, 212, 213
  sp., 24, 31, 33, 34, 44, 60, 63, **212**; Pl. 70

americana, Scoliostoma, 61
Amphigenia, 223
Amphipora? sp., 49
Amplexus sp., 56
  invaginatus, 61
amplificata, Isorthis, 74
amygdalaeformis, Anodontopsis, 61
Anastrophia, 92, 93
  sp., 35, 38
  magnifica, 46, 93
  cf. *magnifica,* 31, 34, 44, 46, 63, 91, **92**, 93, 94; Pl. 11
  verneuili, 51, 54
ancillans, Ancillotoechia, 144
Ancillotoechia, 141, 144, 146
  aptata, 2, 19, 22, 35, 53, 54, 60, 141, 144, 145, 146; Pl. 38
  cumberlandica, 146
Ancylostrophia, 130
Anetoceras desideratus, 61
angularis, Acrospirifer, 186
angustus, "Brachyspirifer", 199
Anodontopsis amygdalaeformis, 61
Anomia reticularis, 154
Anoplia, 55, 59, 132, 137, 223
  sp. A, 19, 30, 35, 49, 53, 64, **137**, 138; Pl. 34
  elongata, 2, 20, 24, 25, 55, 56, 57, 60, **137**; Pl. 34
  helderbergiae, 137
  nucleata, 137, 138
  pygmaea, 137
Anopliopsis, 137, 138
Anoplotheca acutiplicata, 171
anticlastica, Pleiopleurina, 2, 19, 30, 49, 52, 53, 54, 64, 141, 143, 144
antiqua, Productella, 139
Antirhynchonella, 99
aptata, Ancillotoechia, 2, 19, 22, 35, 53, 54, 60, 141, 144, 145, 146

Apatobolbina? sp., 45
arachne, Radiastrea, 61
arata, Mesodouvillina varistriata, 101, 118
arctimuscula, Leptostrophia, 127
arcuaria, Isorthis, 74
arenosus, Costispirifer, 51
Areostrophia, 107
Astutorhyncha, 148
  proserpina, 148
  cf. prosperpina, 60, 148; Pl. 40
Astylospongia sp., 61
atlantica, Trematospira perforata, 179
atrypa, 153, 154
  sp., 20, 25, 27, 30, 33, 38, 49, 64, **154**, 155; Pl. 41
  sp. A, 23, 56, 60, **155**, 157; Pl. 41
  biconvexa, 171, 173
  elongata, 225
  flabellites, 165
  laevis, 175
  nevadana, 20, 21, 23, 24, 25, 26, 27, 28, 36, 38, 56, 60, 140, 153, 155, **156**, 157, 208; Pl. 41
  cf. nieczlawiensis, 155
  pleiopleura, 143
  reticularis, 31, 34, 44, 63, 153, 156
  sublepida, 160
  supramarginalis, 157
Atrypina, 153, 159, 160
  barrandei, 159
  clintoni, 160
  disparilis, 159
  hami, 159
  imbricata, 159
  simpsoni, 2, 33, 34, 44, 63, **159**, 160; Pl. 43
  cf. simpsoni, Pl. 43
attenuata, Morinorhynchus, 107
audaculus, Mediospirifer, 199
audax, Felinotoechia?, 151
Aulacella, 81
australis, Prionothyris, 139
Australocoelia, 171

Bairdia sp., 45
Bairdiocypris sp., 45
barrandei, Atrypina, 159
bathurstensis, Cortezorthis, 80
Beachia suessana, 51
beckii, Leptostrophia, 19, 53, 60, 125, 127, 128
bella, Meristella, 51
Bellerophon combsi, 61
  perplexa, 61

bellula, Pholidops, 60
Beyrichia (Primitia) occidentalis, 61
biconvexa, Leptocoelia, 10, 168
biconvexa, Leptocoelina?, 171, 173
bicostata, Gypidula?, 96
bicostata, Spirigerina, 158
bicurvatus, Neoprioniodus, 60
Bifida, 154, 160, 162
  sp., 28, 38, 60, 120, **162**; Pl. 43
Billingsastraea nevadensis, 61
biloba, Gypidula?, 96
biplicata, Cyrtinaella, 217, 218
bisinuata, Schizophoria, 87
bohemica, Strophonella, 31, 34, 44, 63, 112, 113
boydi, Actinoptera, 61
Brachyprion, 115
  mertoni, 118
"Brachyprion", sp. 31
  mirabilis, 2, 34, 44, 63, 101, **115**, 116; Pl. 22
Brachyspirifer, 59, 183, 198, 200
  sp., 26, 27
  angustus, 199
  carinatus, 198, 200; Pl. 62
  carinatus crassicosta, 200
  carinatus ignorata, 200
  carinatus latissima, 200
  macronotus, 199
  palmerae, 199
  pinyonoides, 2, 23, 24, 26, 27, 28, 36, 60, 140, **199**, 200; Pls. 62, 63
  ventroplicatus, 199
Branikia, 193
Breviphrentis? sp., 61
buchanensis, Spinella, 205, 206
burgeniana, Chonetes, 135

Calceole heteroclite, 218
Callonema occidentalis, 61
calvini, "Strophodonta", 131
Calymene sp., 44
Camarotoechia, 149
  congregata, 149
"Camarotoechia" cf. modica, 34, 44, 63
Camenidea? sp., 61
"Camerella" sp., 30, 34, 44, 48, 63, 91, **93**, 94; Pls. 11, 13
canalicula, Isorthis, 74
carinatus, Brachyspirifer, 198, 200
carolina, Machaeraria, 141
causa, Cyrtinaella, 2, 31, 34, 44, 63, 217, 218
cayuga, Rensselaeria, 226
Chilidiopsis, 107

Chironiptrum sp., 45
Chonetes, 132, 133, 135
    sp., 20, 24, 25, 27, 28, 33, 56, 60, **135**, 140; Pl. 30
    burgeniana, 135
    cingulata, 133
    filistriata, 133
    cf. *filistriata*, 133
    macrostriatus, 135, 136
    sarcinulatus, 135
    subextensa, 135
cingulatus, *Strophochonetes*, 10, 133
Cladopora sp., 49
claviger, *Phacops* (*Phacops*), 49, 50
Clintiella sp., 45
clintoni, *Atrypina*, 160
clivosa, *Isorthis*, 74
Coelospira, 154, 162, 163
    sp., 20, 25, 34, 44, 56, 60, 63
    concava, 10, 30, 35, 49, 51, 64, 154, **162**, 163, 164; Pl. 44
    dichotoma, 164
    pseudocamilla, 2, 19, 20, 22, 53, 54, 60, 154, **163**, 164; Pl. 44
    virginia, 164
    cf. *virginia*, 33, 34
coeymanensis, *Gypidula*, 16, 97, 98, 100
combensis, *Sanguinolites*?, 61
combsi, *Bellerophon*, 61
comis, *Gypidula*, 98
concava, *Coelospira*, 10, 30, 35, 49, 51, 64, 162, 163
concava, *Leptaenisca*, 50, 101, 154
concava, *Megastrophia*, 120
Conchyliolithus Anomites resuptinatus, 86
concinnus, "*Spirifer*", 51, 194
Condracypris? sp., 45
congregata, *Camarotoechia*, 149
Conocardium sp., 27, 56
    nevadensis, 61
conradi, *Platyceras*, 61
consobrina, *Stropheodonta*, 124
Conularia sp., 49, 61
Coolinia, 107
cooperi, *Plicoplasia*, 30, 49, 64, 215, 216
cooperi, *Pseudoparazyga*, 15, 16, 19, 30, 31, 49, 51, 52, 53, 54, 64, 181, 182
cortezensis, *Cortezorthis*, 26, 27, 36, 59, 60, 73, 79, 80
Cortezorthis, 47, 73, 79, 102
    aff. *bathurstensis*, 80; Pl. 5
    cortezensis, 26, 27, 36, 59, 60, 73, **79**, 80; Pl. 5
    maclareni, 73, 79

corvina, *Corvinopugnax*, 148
Corvinopugnax, 148
    sp., 60, 62, **149**; Pl. 39
Costellirostra peculiaris, 51
    singularis, 51
Costispirifer, 16, 19, 153, 154, 165, 183, 193, 194, 195
    sp., 15, 19, 20, 22, 38, 52, 53, 54, 60, 154, **194**, 195; Pl. 64
    arenosus, 51
    arenosus var. *plaincostatus*, 194
crassicosta, *Brachyspirifer carinatus*, 200
crassiformis, *Salopina*, 33
crispa, "*Terebratula*", 184
Crispella, 184
cumberlandiae, "*Spirifer*", 193, 194
cumberlandica, *Ancillotoechia*, 146
Cyathophyllum sp., 56, 61
cycloptera, *Howellella*, 10, 30, 35, 40, 49, 50, 51, 64, 183, 184, 186, 193
Cymostrophia, 101, 120
    sp., 59, 60, **120**; Pl. 35
    patersoni, 101
    stephani, 120
Cyrtina, 39, 46, 183, 216, 217, 218, 221
    sp., 24, 33, 35, 219
    sp. A, 31, 34, 44, 63, **218**, 219; Pl. 72
    sp. C, 20, 23, 56, 60, **220**, 221; Pl. 73
    sp. D, 38, 60, **221**; Pl. 72
    biplicata, 217
    rostrata, 51, 183
    varia, 64, 219
    cf. *varia*, 19, 20, 22, 30, 40, 49, 53, 183, **219**, 220; Pl. 73
Cyrtinaella, 46, 183, 217
    biplicata, 218
    causa, 2, 31, 34, 44, 63, **217**, 218; Pl. 72
Cyrtoceras nevadense, 61
Cyrtograptus symmetricus, 38
Cystiphyllum sp., 60

Dalejina, 59, 73, 80, 81, 84, 86
    sp., 24, 25, 26, 27, 33, 38, 39
    sp. A, 31, 34, 44, 63, **81**, 82, 83; Pl. 8
    sp. B, 19, 20, 22, 30, 31, 34, 48, 53, 64, 80, 82, **83**; Pl. 7
    sp. C, 20, 23, 27, 36, 38, 56, 58, 59, 60, 80, 82, **83**, 84, 86; Pl. 6
    hanusi, 80, 86
    musculosa, 80
    oblata, 50, 73, 80, 84
"*Dalmanella*" *inostransewi*, 74
Dalmanites sp., 26, 49
    meeki, 61

cf. *meeki*, 53
dartae, Orthostrophella, 71
Dayia, 214
Decoroproetus sp., 50
Delthyris, 188, 198
   elegans, 184
   (*Quadrithyris*) *minuens*, 209
   raricosta, 204
demissa, Stropheodonta, 123
denckmanni, Ozarkodina, 28, 45, 56, 60, 63
Dendrograptus sp., 38
depressa, Levenea, 78
dereimsi, Reticulariopsis, 208
derjawinin, Schizophoria?, 87
desideratus, Anetoceras, 61
devonica, Posidonomya, 61
Devonalosia, 139
Devonochonetes, 133
   filistriata, 133
deweyi, "Parazyga", 181
Dicaelosia cf. *varica*, 33, 34
dichotoma, Coelospira, 164
Discina sp., 60
Discomyorthis, 2, 80, 84
   musculosa, 19, 53, 54, 60, 73, **85**; Pl. 7
disparilis, Atrypina, 159
Dolerorthis, 43, 69, 70, 71, 72
   persculpta, 69
"*Dolerorthis*", 43, 69
   sp., 34, 44, 63, **69**, 71; Pl. 1
Dolichoscapha sp., 45
Dubaria, sp., 31, 44, 63
dubia, Mytilarca, 61
Dyticospirifer, 183, 193
   mccolleyensis, 19, 53, 54, 60, **193**; Pl. 59

Eatonia medialis, 51
Eccentricosta, 133, 136; Pl. 33
elegans, Delthyris, 184
elegantulum, Papiliophyllum, 56
elongata, Anoplia, 2, 20, 24, 25, 55, 56, 57, 60, 137, 138
elongata, Rensselaeria, 225
Elytha sp., 19, 20, 22, 53, 60, 183
Elythyna, 63, 153, 154, 183, 210, 212
   sp., 23, 38
   sp. A, 36, 60, 183, **210**, 212; Pl. 68
   sp. B, **211**, 212; Pl. 68
   salairica, 210
Emanuella sp., 211
Endophyllum sp., 44
enorme, Rhizophyllum, 44
Enteletes, 86

Enterolasma? sp., 44
Eodevonaria, 139
Eognathodus secus, 47
   aff. *secus*, 45
   cf. *sulcatus*, 45
"*Eomartiniopsis*" *rhomboidalis*, 209
equidentata, Hindeodella, 45
equistriata, Pseudoparazyga, 181, 182
erbeni, Erbenoceras, 39
Erbenoceras erbeni, 39
Etymothyris, 57
Eudoxina, 193
Eukloedenella sp., 49
Euomphalus eurekensis, 61
Eurekaspirifer, 183, 207
   pinyonensis, 5, 21, 60, **207**, 208; Pls. 65, 66
eurekensis, Euomphalus, 61
Euryspirifer paradoxus, 55
   pseudocheehiel alatus, 188
excavata, Trichonodella, 45
exiguus, Spathognathodus, 28, 63
expansus, Icriodus, 60
extensus, Plectospathodus, 45

fagerholmi, Levenea, 2, 36, 59, 60, 75, 77, 78
fallaciosa, Trigonirhynchia, 147
fallax, Acrospirifer, 57
Falsipollex? sp., 49
Fardenia, 107
favonia, Leonaspis, tuberculata, 35, 44, 49, 50
Favosites sp., 25, 26, 27, 44, 53, 56, 60
Felinotoechia? *audax*, 151
ferganensis, Schizophoria, 87
festiva, Isorthis, 74
filicosta, Stropheodonta, 2, 19, 20, 22, 30, 48, 53, 56, 64, 101, 123
filistriata, "*Strophochonetes*", 5, 21, 23, 24, 25, 26, 60, 133
flabellites, Leptocoelia, 51, 165, 167
flemingii, Monograptus, 29
formosa, Machaeraria, 141, 142
fornicatimcurvata, Protocortezorthis, 45, 78
foveolatus, Polygnathus, 28, 52, 59, 60, 63
fragilis, Schizophoria, 43, 87
frankenwaldensis, Spathognathodus, 60
furcoides, Thlipsura, 49

gaspensis, Spirifer, 184
gaspensis, Spinoplasia, 213
geinitzianus, Retiolites, 38

Gennaeocrinus maxwelli, 61
gigantia, Nervostrophia?, 130
globosa, Leptocoelia infrequens, 38, 39, 165, 168, 170, 171
Goniophora perangulata, 61
Gracianella, 160
gracilis, Sanguinolites?, 61
Grypophyllum sp., 44
Gurichella, 207
gwenensis, Striatopora, 49
Gypidula, 91, 96, 99
  sp., 28, 33, 34
  bicostata, 96
  biloba, 96
  coeymanensis, 98, 100
  cf. *coeymanensis,* 16, 97
  loweryi, 23, 24, 26, 27, 28, 36, 38, 55, 59, 60, 91, 98, 99, 100, 140; Pls. 13, 14
  multicostata, 98
  pelagica, 6, 97; Pl. 15
  praeloweryi, 2, 15, 21, 23, 56, 60, 91, 97, 98, 155; Pl. 15
  pseudogaleata, 51, 97
  cf. *pseudogaleata,* 31, 34, 44, 46, 63, 96; Pl. 15
  typicalis, 96

hami, Atrypina, 159
hanusi, Dalejina, 80, 86
Hedeina macropleura, 50
helderbergiae, Anoplia, 137, 138
hercynicus, Monograptus, 41, 46, 160
heteroclite, Cyrtina, 218
Heterophrentis cf. *nevadensis,* 61
Hindeodella equidentata, 45
  cf. *priscilla,* 60
Hirnantia, 89
  senecta, 89
Hollinella sp., 45, 49, 61
Howellella, 55, 59, 62, 183, 184, 188, 198
  sp., 27, 31, 34, 38, 44, 63
  aculeata, 185, 188; Pl. 55
  cycloptera, 10, 30, 35, 49, 50, 51, 64, 183, 184, 185, 186, 193; Pl. 54
  cf. *cycloptera,* 40
  koneprusensis, 62
  laeviplicata, 183
  modesta, 215
  pauciplicata, 208
  textilis, 185, 188
  cf. *textilis,* 27, 36, 60, 186, 187, 193; Pl. 55
  tribulis, 10, 186

huddlei, Icriodus latericrescens, 28, 49, 53, 56, 63
Hypotetragona sp., 45
Hypothyridina, 148
hystericus, Hysterolites, 195
Hysterolites, 59, 183, 195, 198
  sp., 21, 26, 27, 28
  sp. A, 24, 38, 60, **195,** 196, 197; Pl. 60
  sp. B, 60, **197,** 198; Pl. 61
  (*Acrospirifer*) *murchisoni,* 189
  hystericus, 195

Icriodus sp., 28, 45, 60
  n. sp., 45, 47
  expansus, 60
  latericrescens, 45, 46, 49, 52, 60
  latericrescens huddlei, 28, 49, 53, 56, 60, 63
  pesavis, 31, 45, 47, 52, 160
  symmetricus, 60
  woschmidti, 47
iddingsi, Megastrophia, 23, 24, 26, 27, 28, 36, 60, 101, 122, 140
ignorata, Brachyspirifer carinatus, 200
ilsae, Acrospirifer, 57
imbricata, Atrypina, 159
impressa, "Orthis", 88
incertus, Spirifer, 206
inclinatus, Spathognathodus, 45
inequicostella, Leptostrophia, 2, 30, 35, 48, 63, 125, 126, 127
infrequens, Leptocoelia, 5, 24, 27, 58, 60, 78, 120, 154, 165, 168, 169, 170, 171
inostranzewi, Isorthis, 74
insolita, Koneprusia, 50
intermedius, Acrospirifer?, 186
interplicata, Dolerorthis, 69
invaginatus, Amplexus, 61
invasor, Mclearnites, 2, 20, 23, 55, 56, 60, 118, 119, 120
Iridistrophia, 107
Isorthis, 74
  sp., 33
  amplificata, 74
  arcuaria, 74
  canalicula, 74
  clivosa, 74
  festiva, 74
  mackenziei, 74
  perelegans, 50, 74
  scutiformis, 74
  szajnochai, 74
  tetragona, 74
  uskensis, 74
Ivanothyris, 188

johnsoni, Spathognathodus, 44, 47
Jolvia, 100

kakvensis, Sieberella, 95
Katunia? sp., 20, 56, 60
kayi, Toquimaella, 31, 34, 38, 44, 46, 47, 57, 63, 153, 160
kayseri, Gypidula, 96
kobehana, Acrospirifer, 5, 15, 16, 20, 21, 22, 23, 24, 25, 27, 55, 56, 60 138, 183, 189, 190, 192
koneprusensis, Howellella, 62
Koneprusia insolita, 50
Kozlowskiella, 201, 217

laeviplicata, "Howellella", 183, 184
laevis, Meristella, 51, 173, 175
laevis, Posidonomya, 61
latericrescens, Icriodus, 45, 46, 49, 52, 60
latior, Phragmostrophia, 130
latissima, Brachyspirifer carinatus, 200
Latonotoechia, 142
Lazutkinia, 193
leda, Brachyprion, 115
Leiorhynchus, 149, 150
   quadracostatus, 149
"Leiorhynchus" sp., 24, 28, 38, 60, 62, **150**, 151; Pl. 40
lens, Schuchertella, 106, 107
lenticularis, Levenea, 78
lenticularis, "Mendacella", 90
lenzi, Polygnathus, 60
Leonaspis sp., 39, 61
   cf. *tuberculatus*, 49
   tuberculata favonia, 35, 44, 49, 50
lepida, Bifida, 160
Leptaena, 101, 102, 106
   sp., 20, 21, 22, 24, 25, 26, 27, 35, 36, 38, 40
   sp. A, 31, 34, 38 44, 63, **102**, 103; Pl. 16
   sp. C, 20, 21, 23, 36, 56, 59, 60, 103, **105**, 106; Pl. 16
   acuticuspidata, 103, 104
   bohemica, 112
   cf. *acuticuspidata*, 19, 20, 22, 30, 31, 35, 48, 53, 64, 103, **104**, 105, 106; Pl. 16
   concava, 102
   nucleata, 137
   rugosa, 102
   ventricosa, 105
Leptaenisca, 101, 102
   sp., 31, 34, 35, 40, 44, 48, 63, **102**; Pl. 17

concava, 50
Leptocoelia, 51, 59, 153, 154, 165, 168, 169, 171, 223
   acutiplicata, 59
   biconvexa, 10, 168
   concava, 162
   flabellites, 51, 165, 167
   imbricata, 159
   infrequens, 5, 27, 58, 60, 78, 120, 154, 165, 168, **169**, 170, 171
   infrequens globosa, 38, 39, 165, 168, **170**, 171; Pl. 47
   infrequens infrequens, 24, 36, 168, **169**, 170; Pl. 47
   murphyi, 2, 19, 20, 22, 30, 35, 49, 52, 53, 64, 154, 164, **165**, 168; Pls. 45, 46
   aff. *murphyi*, 20, 22, 23, 56, 60, 154, 165, 167
   nunezi, 168, 169, 171
   nunezi texana, 171
Leptocoelina, 2, 59, 154, 165, 170, 171, 173
   acutiplicata, 171
   squamosa, 2, 30, 35, 38, 49, 50, 64, 170, **171**, 173; Pl. 45
Leptostrophia, 101, 125, 127, 129
   sp., 20, 22, 125, 127
   sp. A, 34, 44, 63, **125**; Pl. 23
   sp. D, 20, 56, 59, 60, 125, **128**, 129; Pls. 28, 29
   arctimuscula, 127
   beckii, 127
   cf. *beckii*, 19, 53, 60, 125, **127**, 128; Pl. 29
   inequicostella, 2, 30, 35, 48, 63, 125, **126**, 127; Pl. 29
Levenea, 46, 73, 74, 75, 77, 78
   sp., 33, 74, 75
   sp. A, 31, 34, 44, 63, **74**, 77; Pl. 2
   altaica, 74
   depressa, 78
   fagerholmi, 2, 36, 59, 60, 75, **77**, 78; Pl. 2
   lenticularis, 78
   navicula, 2, 19, 21, 22, 30, 31, 35, 39, 48, 53, 56, 64, 73, **75**, 77; Pls. 2, 3
   subcarinata, 50, 73, 74, 78
   taeniolata, 74
Libumella? sp., 45, 49
lineatum, Platystoma, 61
linguiformis, Polygnathus, 60
Lingula sp., 35, 49
   lonensis, 60
Linograptus posthumus, 33, 34, 41

Lissatrypa? sp., 34, 44, 63
lonensis, Lingula, 60
Longispina macrostriatus, 136
loweryi, Gypidula, 23, 24, 26, 27, 28, 36, 38, 55, 59, 60, 91, 97, 98, 99, 100, 140
Loxonema sp., 25, 26, 35, 44, 49, 56
  nobile, 61
  subattenuatum, 61
lubimovi, Sieberella, 95

Machaeraria, 141, 142
  sp., 34, 38, 44, 63, 141, 142, 143; Pl. 36
  formosa, 142
mackenziei, Isorthis, 74
maclareni, Cortezorthis, 73, 79
macronotus, "Brachyspirifer", 199
macropleura, Hedeina, 50
macrostriatus, Microdon, 61
macrostriatus, Parachonetes, 5, 21, 23, 24, 25, 26, 27, 36, 55, 56, 57, 60, 136, 137, 140, 206
magnacosta, Stropheodonta, 2, 19, 30, 35, 49, 53, 56, 64, 101, 124, 125
magnapleura, Megakozlowskiella, 2, 19, 20, 22, 30, 35, 40, 49, 53, 64, 183, 202, 203, 204
magnifica, Anastrophia, 31, 34, 44, 46, 63, 91, 92, 93, 94
magnifica, Leptostrophia, 125
marginalis, Spirigerina, 157
marginaliformis, Spirigerina, 158
maria, Meristella, 173
Martina undifera, 212
masurskyi, Muriferella, 38, 39, 59, 60, 120
maxwelli, Gennaeocrinus, 61
mccolleyensis, Dyticospirifer, 19, 53, 54, 60, 193
Mclearnites, 55, 57, 59, 101, 118, 119, 120
  sp., 25
  sp. B, 38, 39, 59, 60, **119**, 120; Pl. 24
  invasor, 2, 20, 23, 55, 56, 60, **118**, 119, 120; Pl. 24
medialis, Eatonia, 51
medialis, "Spirifer", 199
medioplicata, Rensselaerina, 224
meeki, Dalmanites, 53, 61
Mediospirifer, 199
Megakozlowskiella, 39, 46, 183, 201, 202, 217, 223
  sp. A, 31, 34, 44, 63, **201**, 202, 203; Pl. 72
  magnapleura, 2, 19, 20, 22, 30, 35, 40, 49, 53, 64, 183, **202**, 203, 204; Pl. 71

perlamellosa, 50
cf. *raricosta,* 20, 21, 23, 56, 60, **202**, 204
Megambonia occidualis, 61
Meganteris australis, 139
Megastrophia, 101, 120, 121
  sp., 25, 35
  iddingsi, 23, 24, 26, 27, 28, 36, 60, 101, **122**, 140; Pl. 26
  transitans, 2, 19, 20, 22, 27, 30, 35, 49, 53, 56, 60, 101, **120**, 122, 124; Pl. 25
"Mendacella" lenticularis, 90
Meristella, 175
  sp., 22, 25, 27, 31, 35, 40
  bella, 51
  laevis, 51, 173
  maria, 173
  princeps, 51
  robertsensis, 20, 21, 23, 25, 56, 64, **175**; Pl. 49
  cf. *robertsensis,* 19, 20, 22, 30, 49, 53, **176**; Pls. 49, 50
  (*Whitfieldella*) *nasuta,* 175
Meristina, 173, 174
  sp., 24, 27
  sp. A, 31, 34, 44, 63, **174**, 175; Pl. 48
  sp. B, 36, 60, **174**, 175; Pl. 48
merriami, Phragmostrophia, 20, 21, 23, 24, 26, 27, 28, 36, 38, 55, 60, 120, 130, 131, 132, 208
mertoni, Mclearnites, 118
Mesodouvillina, 116, 117, 120
  sp., 31, 33
  subinterstrialis, 117
  cf. *varistriata,* 34, 44, 63, 101, **117**, 118; Pl. 23
  cf. *varistriata arata,* 101, 118; Pl. 23
Metaplasia, 183, 214
  sp., 20
  nitidula, 215
  paucicostata, 214
  cf. *paucicostata,* 19, 20, 22, 53, 56, 60, 183, **214**, 215; Pl. 70
  pyxidata, 215
Michelinia sp., 49
michelini, Rhipidomella, 85
Microdon macrostriatus, 61
microdon, Monograptus, 6, 33, 34, 41
micropleura, Stropheodonta, 124
minuens, Quadrithyris, 31, 34, 44, 46, 63, 209, 210
mirabilis, "Brachyprion", 2, 34, 44, 63, 101, 115, 116
modesta "Howellella", 184, 213, 215
modica "Camarotoechia", 34, 44, 63

Modiomorpha altiforme, 61
  obtusa, 61
monitorensis, Orthostrophella, 2, 30, 31, 39, 48, 63, 69
Monograptus sp., 29, 33
  acus, 28
  flemingii, 29
  hercynicus, 41, 46, 160
  hercynicus nevadensis, 1, 29, 33, 34, 41, 44, 46, 52
  microdon, 6, 33, 34, 41
  pandus, 28
  parapriodon, 38
  praehercynicus, 6, 33, 41
  cf. *praehercynicus,* 32, 33, 41
  aff. *praehercynicus,* 32
  cf. *praedubius,* 29, 38
  priodon, 38
  aff. *triangulatus,* 38
  ultimus, 29
  vomerinus, 38
  yukonensis, 46, 47, 48, 52, 79, 160
  cf. *yukonensis,* 48
Morinorhynchus, 107
  attenuata, 107
Mucophyllum sp., 44
multicostata, Gypidula, 98
multistriata, Schizophoria, 51, 87
multistriata, Trematospira, 10, 179, 180, 187
murchisoni, Acrospirifer, 5, 16, 19, 22, 35, 49, 53, 54, 55, 57, 64, 186, 189, 190, 192
Muriferella, 59, 74, 90
  masurskyi, 38, 39, 59, 60, **90**, 120; Pl. 10
murphyi, Leptocoelia, 2, 19, 20, 22, 23, 30, 35, 49, 52, 53, 56, 60, 64, 154, 164, 165, 167, 168
musculosa, Discomyorthis, 19, 53, 54, 80, 84, 85
Mutationella, 59, 223, 224
Mutationella? sp., 21, 60, **224**; Pl. 74
Myomphalus? sp., 45
Mytilarca dubia, 61
  (*Plethomytilus*) *oviformis,* 61

nacrea, Pholidostrophia, 129
Nadiastrophia, 102, 130, 131
  superba, 131
nasuta, "Meristella", 175
navicula, Levenea, 2, 19, 21, 22, 30, 31, 35, 39, 48, 53, 56, 64, 75, 77
Neoprioniodus bicurvatus, 60
nerei, Brachyspirifer, 200

Nervostrophia? gigantia, 130
nevadae, Proetus, 56, 61
nevadana, Atrypa, 20, 21, 23, 24, 25, 26, 27, 28, 36, 38, 56, 60, 140, 153, 155, 156, 157, 208
nevadaensis, Schizophoria, 23, 24, 26, 27, 28, 36, 38, 59, 60, 87, 88, 89
nevadaensis "Schuchertella", 24, 59, 60, 108, 109, 110
nevadense, Cyrtoceras, 61
nevadense, Teicherticeras, 61
nevadensis, Billingsastraea, 61
nevadensis, Conocardium, 61
nevadensis, Heterophrentis, 61
nevadensis, Monograptus hercynicus, 1, 29, 33, 34, 41, 43, 44, 46, 52
Nezamyslia sp., 45
nieczlawiensis, Atrypa, 155
nitidula, Metaplasia, 215
nobile, "Loxonema", 61
nodosum, Platyceras, 61
nucleata, Anoplia, 137, 138
Nucleospira, 177, 178, 179
  sp., 19, 20, 22, 23, 24, 27, 31, 33, 38, 40
  sp. A, 34, 44, 63, **177**, 178
  sp. B, 22, 30, 49, 53, 64, **177**, 178, 179; Pl. 51
  subsphaerica, 2, 60, **177**, **178**, **179**; Pl. 51
  cf. *subsphaerica,* 20, 56
nunezi, Leptocoelia, 168, 169, 171

oblata, Dalejina, 50, 80, 84
obtusa, Modiomorpha, 61
occidens, Trigonirhynchia, 5, 21, 24, 26, 28, 36, 38, 60, 140, 141, 147
occidentalis, Callonema, 61
occidentalis, Paracyclas, 61
occidentalis, Welleriopsis?, 45, 49, 61
occidualis, Megambonia, 61
Odontochile sp., 49
orbicularis, Schizodus (Cytherodon), 61
oriskania, Schizophoria, 87
Orthis
  fornicatimcurvata, 78
  impressa, 88
  interplicata, 69
  musculosa, 84, 85
  quadracostatus, 150
  umbonata, 212
Orthoceras sp., 25, 26, 27, 49, 56
Orthostrophella, 51, 69, 71, 72
  sp., 72

monitorensis, 2, 30, 31, 39, 48, 63, 69, 71, 72; Pl. 1
Orthostrophia, 69, 72
   dartae, 71
Otarion (Otarion) sp., 61
   periergum, 49, 50
ovalus, Priononthyris, 51
oviformis, Mytilarca (Plethomytilus), 61
ovoides, Rensselaeria, 225
ovulum, Rensselaeria, 226
Ozarkodina denckmanni, 28, 56, 60, 63
   typica denckmanni, 45, 60

Palaeomanon roemeri, 61
palmerae, "Brachyspirifer", 199
Panderodus simplex, 60
   unicostatus, 60
pandus, Monograptus, 28
Papiliophyllum elegantulum, 56
Parachonetes, 57, 59, 132, 133, 135, 136
   macrostriatus, 5, 21, 23, 24, 25, 26, 27, 36, 55, 56, 57, 60, **136**, 137, 140, 206; Pls. 32, 33
   verneuili, 137; Pl. 33
Paracyclas occidentalis, 61
paradoxus, Euryspirifer, 55
parafragilis, Schizophoria, 2, 31, 34, 43, 44, 63, 86, 87, 89
parapriodon, Monograptus, 38
Parazyga, 181
   deweyi, 181
parryana, "Spirifer", 205, 206
patersoni, Cymostrophia, 101, 120
pauciplicata, "Howellella", 208
pauicostata, Metaplasia, 19, 20, 22, 53, 56, 60, 183, 214, 215
peculiaris, Costellirostra, 51
Pegmarhynchia, 51, 64, 141
   sp., 35, 49
pelagica, Gypidula, 6, 97
Pentamerella, 91
Pentamerus comis, 98
   kayseri, 96
   problematicus, 96
   pseudogaleatus, 96
   sieberi, 94
   verneuili, 92
perangulata, Goniophora, 61
perelegans, Tyersella, 50, 74
perforata, Trematospira, 10, 19, 20, 22, 30, 31, 38, 40, 49, 50, 51, 53, 54, 64, 179, 180
periergum, Otarion (Otarion), 49, 50

perlamellosa, Megakozlowskiella, 50, 201
perplexa, Bellerophon, 61
persculpta, Dolerorthis, 69
pesavis, Icriodus, 31, 45, 47, 52, 160
Phacops sp., 25, 26, 27, 28, 49, 56
   cf. *rana*, 53, 56, 61
   (Phacops) claviger, 49, 50
Phacops (Reedops) sp., 50
Pholidops bellula, 60
Pholidops quadrangularis, 60
Pholidostrophia, 101, 129
   sp., 20, 30, 49, 53, 55, 56, 60, **129**; Pl. 28
Phragmostrophia, 73, 102, 130, 151
   latior, 130
   merriami, 21, 23, 24, 26, 27, 28, 36, 38, 60, 120, 130, **131**, 132, 208; Pls. 26, 27
   cf. *merriami*, 20, 55
pinyonensis, Eurekaspirifer, 5, 21, 60, 140, 207, 208
pinyonoides, Brachyspirifer, 2, 23, 24, 26, 27, 28, 36, 60, 199, 200
planicostatus, Costispirifer, 194
planoconvexa, Platyorthis, 50
Platyceras sp., 27, 49, 56
   conradi, 61
   nodosum, 61
Platyorthis planoconvexa, 50
Platystoma lineatum, 61
Plectatrypa cf. *sibirica*, 46
Plectospathodus extensus, 45
pleiopleura, Pleiopleurina, 144
Pleiopleurina, 51, 54, 143
   anticlastica, 2, 19, 30, 49, 52, 53, 54, 64, 141, **143**, 144; Pls. 36, 37
   pleiopleura, 143, 144
Plethorhyncha, 143
Pleurodictyum sp., 39, 53, 60
   cf. *trifoliatum*, 49
Plicoplasia, 51, 215
   cooperi, 30, 49, 64, **215**, 216; Pl. 69
podolica, Mutationella, 224
Poloniella sp., 45
Polygnathus, 57
   sp., 56
   foveolatus, 28, 52, 59, 60, 63
   lenzi, 60
   aff. *linguiformis*, 60
Posidonomya devonica, 61
   laevis, 61
posthumus, Linograptus, 33, 34, 41
praedubius, Monograptus, 29, 38

praehercynicus, Monograptus, 6, 32, 33, 41
praeloweryi, Gypidula, 2, 15, 21, 23, 56, 60, 91, 97, 98, 155
primaeva, Acrospirifer, 188
primus, Spathognathodus, 60
princeps, Meristella, 51
priodon, Monograptus, 38
Prionothyris, 139
  ovalus, 51
priscilla, Hindeodella, 60
problematica, Sieberella, 34, 38, 44, 63, 96
Productella, 138, 139, 140
  antiqua, 138
  subaculeata, 138
Productus spinulicostae, 138
Proetus sp., 26
  cf. *nevadae,* 56, 61
profunda, Protomegastrophia, 115
proserpina, Astutorhyncha, 60, 148
Protochonetes, 133, 135, 136
Protocortezorthis, 47, 73, 74
  sp., 33, 79; Pl. 4
  fornicatimcurvata, 45
  windmillensis, 34, 44, 45, 63, 73, 78, 79; Pl. 4
Protoleptostrophia, 101
Protomegastrophia, 115
  profunda, 115
Protophragmapora, 79
pseudocamilla, Coelospira, 2, 19, 20, 22, 53, 54, 60, 154, 163, 164
pseudogaleata, Gypidula, 31, 34, 44, 46, 51, 63, 96, 97
Pseudoparazyga, 2, 181, 182
  cooperi, 15, 16, 19, 30, 31, 49, 51, 52, 53, 54, 64, **181,** 182; Pls. 52, 53
  equistriata 181, 182
Pterinea? sp., 53, 56
Ptychopleurella, 69
"*Ptychopleurella*", 43
  sp., 70
punctulifera, Strophonella, 19, 20, 23, 24, 25, 27, 30, 35, 48, 50, 53, 56, 59, 64, 98, 101, 113, 115
pygmaea, Anoplia, 137, 138
pyriforma, Sieberella, 2, 16, 19, 40, 48, 53, 60, 91, 94, 95
pyxidata, Metaplasia, 215

quadracostatus, Leiorhynchus, 149, 150
quadrangularis, Pholidops, 60

Quadrikentron, 133
Quadrithyris, 209
  cf. *minuens,* 31, 34, 44, 46, 63, **209,** 210; Pl. 67

Radiastrea arachne, 61
rana, Phacops, 53, 56, 61
raricosta, Megakozlowskiella, 20, 21, 23, 56, 60, 202, 204
Rensselaeria, 57, 223, 225, 226
  sp., 19, 53, 56, 223, **225,** 226; Pl. 74
  cayuga, 226
  ovulum, 226
  subglobosa, 51
Rensselaerina, 223, 224
  sp., 30, 40, 49, 54, 55, 64, 223, **224,** 225; Pl. 74
  medioplicata, 224
resupinatus, Schizophoria, 86
Reticulariopsis, 183, 184, 208
  sp., 56
  sp. A, 34, 44, 63, **208**; Pl. 67
  sp. B, 19, 53, 60, **208,** 209; Pl. 67
reticularis, Atrypa, 31, 34, 44, 63, 153, 154, 156
Retiolites geinitzianus, 38
Retzispirifer, 193
Rhipidomella, 62, 84, 85
  sp., 21, 59, 83, **86**; Pl. 6
  musculosa, 85
Rhipidomelloides
  musculosus, 85
Rhizophyllum cf. *enorme,* 44
rhomboidalis, "*Eomartiniopsis*", 209
Rhynchonella ancillans, 144
  carolina, 141
  formosa, 141
  occidens, 147
  proserpina, 148
Rhytistrophia beckii, 127
robertsensis, Meristella, 19, 20, 21, 22, 23, 25, 30, 49, 53, 56, 64, 175, 176
robustus, Quadrithyris, 209
roemeri, Palaeomanon, 61
roemeri, Sieberella, 95
roeni, Spinoplasia, 2, 30, 35, 38, 40, 49, 50, 64, 213, 214, 215
romingeri, Sieberella, 95
rostrata, Cyrtina, 51, 183
rugosa, Leptaena, 102

Saccarchites sp., 45
salairica, Elythyna, 210

INDEX OF GENERA AND SPECIES 419

Salopina, 89, 90
  sp., 20, 56, 60
  aff. crassiformis, 33
  cf. crassiformis, 33, 90
Sanguinolites? combensis, 61
  gracilis, 61
sarcinulatus, Chonetes, 135
Schizodus (Cytherodon) orbicularis, 61
Schizophoria 43, 46, 74, 86, 87, 89, 90
  sp., 33, 39, 40, 48, 63
  allani, 74
  bisinuata, 87
  derjawinin, 87
  ferganensis, 87
  fragilis, 43, 87
  multistriata, 51, 87
  nevadaensis, 23, 24, 26, 27, 28, 36, 38, 59, 60, 87, 88, 89; Pl. 9
  oriskana, 87
  parafragilis, 2, 31, 34, 43, 44, 63, 86, 87, 89; Pl. 8
Schizoramma, 72
Schuchertella, 106, 107
"Schuchertella", 107
  sp., 20, 22, 25, 28, 33, 35, 40
  sp. A, 31, 34, 44, 63, **107**, 108, 109; Pl. 18
  sp. B, 19, 30, 48, 53, 64, **108**, 109; Pl. 18
  sp. C, 20, 21, 23, 56, 60, 108, **109**; Pl. 18
  nevadaensis, 24, 59, 60, 108, 109, **110**; Pl. 19
Scoliostoma cf. americana, 61
scutiformis, Isorthis, 74
secus, Spathognathodus, 45, 47
semifasciata, Strophonella, 112
senecta, Hirnantia, 89
seretensis, Mesodouvillina, 116, 117
Shaleria sp., 115
  (Telaeoshaleria) yukiangensis, 130, 131
Sicorhyncha, 142
Sieberella, 91, 94, 96
  kakvensis, 95
  lubimovi, 95
  cf. problematica, 34, 38, 44, 63, **96**; Pl. 13
  pseudogaleata, 97
  pyriforma, 2, 16, 19, 53, 60, 91, **94**, 95; Pl. 12
  cf. pyriforma, 40, 48
  roemeri, 95

romingeri, 95
sieberi, 95, 96
vagranica, 95
weberi, 95
sibirica, Spirigerina, 46
sieberi, Sieberella, 95, 96
simplex, Panderodus, 60
simpsoni, Atrypina, 2, 33, 34, 44, 63, 159, 160
singularis, Costellirostra, 51
Skenidioides, 69, 112
  sp., 33, 39
Spathognathodus sp., 44, 45, 46, 47, 60
  exiguus, 28, 63
  cf. frankenwaldensis, 60
  inclinatus, 45
  johnsoni, 44, 47
  cf. primus, 60
  steinhornensis, 28, 44, 63
  sulcatus, 45, 47, 53
  transitans, 44, 47, 51
speciosus, Acrospirifer, 55
Sphaerirhynchia, 141
Spenophragmus, 90
Spinatrypa sp., 31, 44, 63
Spinella, 183, 205, 206, 207
  sp., 25, 26, 27, 60, 205
  buchanensis, 205, 206
  talenti, 2, 24, 26, 60, **205**, 206, 207; Pl. 64
Spinocyrtia, 198
Spinoplasia, 213, 214
  gaspensis, 213
  roeni, 2, 30, 35, 38, 40, 49, 50, 64, **213**, 214, 215; Pl. 69
Spinulicosta, 62, 138, 139, 140
  sp., 21, 24, 28, 60, **140**; Pl. 35
spinulicostae, Spinulicosta, 138
"Spirifer"
  aculeatus, 62
  arenosus planicostatus, 194
  audaculus, 199
  carinatus, 198
  concinnus, 51, 194
  cumberlandiae, 193, 194
  gaspensis, 184
  incertus, 206
  kobehana, 16, 190
  laeviplicatus, 184
  medialis, 199
  minuens, 209
  modestus, 184, 215
  modestus nitidulus, 215

cf. *modestus*, 213
multistriatus, 179
murchisoni, 5, 189
nerei, 200
paucicostatus, 214
perforatus, 179
perlamellosus, 201
pinyonensis, 207
primaeva, 188
pyxidatus, 214
raricosta, 190
robustus, 209
ventricosa, 177
(*Crispella*) *laeviplicatus*, 184
(*Reticularia*) *dereimsi*, 208
(*Trigonotreta*) *pinonensis*, 207
Spirifera murchisoni, 189
parryana, 205, 206
pinonensis, 207
raricosta, 5, 204
(*Martinia*) *undifera* 211
Spiriferina, 188
Spirigerina, 153, 157, 158
bicostata, 158
marginaliformis, 158
supramarginalis, 34, 44, 46, 63, 153, 157, 158; Pl. 42
Spirinella, 184, 208, 215
Spongophyllum? sp., 44
squamosa, *Leptocoelina*, 2, 30, 35, 38, 49, 50, 64, 170, 171, 173
Stegerhynchus sp., 54
steinhornensis Spathognathodus, 28, 44, 63
Stenochisma althi, 142
stephani, *Cymostrophia*, 120
Straelenia sp., 180
Streptorhynchus, 111
lens, 106, 107
Strepulites sp., 49, 61
Striatopora sp., 44
cf. *gwenensis*, 49
Stropheodonta, 101, 123
sp., 24, 25, 60, 124
beckii, 127
consobrina, 124
filicosta, 2, 19, 20, 22, 30, 48, 53, 56, 64, 101, **123**; Pl. 22
iddingsi, 122
magnacosta, 2, 19, 30, 49, 53, 56, 64, 101, **124**, 125; Pl. 21
cf. *magnacosta*, 35
micropleura, 124

patersoni, 120
varistriata arata, 118
verneuili, 124
(*Brachyprion*) *subinterstrialis*, 117
(*Brachyprion*) *varistriata*, 117
Strophochonetes, 132, 133
cingulatus, 10, 133; Pl. 30
"*Strophochonetes*" sp., 30, 35, 49, 53, 64, 134; Pl. 30
filistriata, 5, 21, 23, 24, 25, 26, 27, 60, **133**; Pl. 31
cf. *filistriata*, 25, 133
Strophodonta calvini, 131
demissa, 122
punctulifera, 113
varistriata, 117
varistriata arata, 118
Strophomena aesopea, 111
bohemica, 112
demissa, 123
leda, 115
punctulifera, 113
semifasciata, 112
stephani, 120
varistriata, 117
(*Strophodonta*) *beckii*, 127
(*Strophodonta*) *concava*, 120
(*Strophodonta*) *magnifica*, 125
(*Strophodonta*) *nacrea*, 129
Strophonella, 101, 112, 115
sp., 22, 27, 33
bohemica, 112
cf. *bohemica*, 31, 34, 44, 63, **112**, 113; Pl. 20
punctulifera, 50
cf. *punctulifera*, 19, 20, 23, 24, 25, 27, 30, 35, 48, 53, 56, 59, 64, 98, 101, **113**, 155; Pls. 20, 21
Styliolina sp., 28, 61
subaculeata, *Productella*, 138
subcarinata, *Levenea*, 50, 74, 78
subattenuatum, *Loxonema*? 61
subextensa, *Chonetes*, 135
subglobosa, *Rensselaeria*, 51
subinterstrialis, *Mesodouvillina*, 116
sublepida, "*Atrypina*," 160
subsphaerica, *Nucleospira*, 2, 20, 56, 60, 177, 178, 179
suessana, *Beachia*, 51
sulcatus, *Spathognathodus*, 45, 47, 53
superba, *Nadiastrophia*, 131
supramarginalis, *Spirigerina*, 34, 44, 46, 63, 153, 157, 158

Index of Genera and Species

szajnochai, *Isorthis*, 74
symmetricus, *Cyrtograptus*, 38
symmetricus, *Icriodus*, 60
Syringaxon, 35
  sp., 38, 39, 49
  acuminatum, 49
Syringopora sp., 26, 60

taeniolata, "*Levenea*", 74
talenti, *Spinella*, 2, 24, 26, 60, 205, 206, 207
Teicherticeras nevadense, 61
Teichostrophia, 130
Tentaculites sp., 49, 61
Terebratula corvina, 148
  crispa, 184
  lepida, 160
  marginalis, 157
  michelini, 85
  ovoides, 225
  proserpina, 148
Terebratulites sarcinulatus, 135
tetragona, *Isorthis*, 74
texana, *Leptocoelia nunezi*, 171
textilis, *Howellella*, 27, 36, 60, 185, 186, 187, 188, 193
Thamnopora ("*Pachypora*") sp., 44
Theodossia, 193
Thliborhynchia, 142
Thlipsura aff. *furcoides*, 49
Toquimaella, 160
  kayi, 31, 34, 38, 44, 46, 47, 63, 153, **160**; Pl. 42
transitans, *Megastrophia*, 2, 19, 20, 22, 27, 30, 35, 49, 53, 56, 60, 101, 120, 122, 124
transitans, *Spathognathodus*, 44, 47, 51
Trematospira, 179, 181
  sp., 180
  cooperi, 181
  equistriata, 181
  infrequens, 169
  multistriata, 179, 180, 181
  perforata, 10, 19, 20, 22, 30, 38, 49, 50, 51, 53, 54, 64, **179**, 180; Pl. 52
  cf. perforata, 31, 40
  perforata atlantica, 79
Treposella ? sp., 45
triangulatus, *Monograptus*, 38
tribulis, *Howellella*, 10, 186
Trichonodella excavata, 45
Tricornina sp., 45
trifoliatum, *Pleurodictyum*, 49

Trigonirhynchia, 147
  occidens, 5, 21, 24, 26, 28, 36, 38, 60, 140, 141, **147**; Pls. 38, 39
tuberculatus, *Leonaspis*, 49
Tubulibairdia sp., 45, 49, 61
Tyersella, 74
  perelegans, 50
  typica, 74
typica, *Tyersella*, 74
typicalis, *Gypidula*, 96

ultimus, *Monograptus*, 29
umbonata, *Ambocoelia*, 212
Uncinulina fallaciosa, 147
undifera, "*Martinia*", 211, 212
unicostatus, *Panderodus*, 60
Urella, 193
uskensis, *Isorthis*, 74

Vagrania, 46, 160
vagranica, *Sieberella*?, 95
varia, *Cyrtina*, 19, 20, 22, 30, 40, 49, 53, 64, 183, 219, 220
varica, *Dicaelosia*, 33, 34
varistriata, *Mesodouvillina*, 34, 44, 63, 101, 117
varistriatus, *Aesopomum*, 2, 34, 44, 63, 111
ventricosa, *Leptaena*, 105
ventricosa, *Nucleospira*, 177
ventroplicatus, "*Brachyspirifer*", 199
verneuili, *Anastrophia*, 51, 92
verneuili, *Parachonetes*, 137
verneuili, *Stropheodonta*, 124
virginia, *Coelospira*, 33, 34, 164
vomerinus, *Monograptus*, 38

Waldheimia podolica, 224
Warrenella, 184
weberi, *Sieberella*, 95
Welleriopsis ? *occidentalis*, 45
  cf. *occidentalis*, 45, 49
windmillensis, *Protocortezorthis*, 34, 44, 45, 63, 73, 78, 79
woschmidti, *Icriodus*, 47

yukiangensis, "*Telaeoshaleria*", 130, 131
yukonensis, *Monograptus*, 46, 47, 48, 52, 79, 160

Zlichorhynchus, 142
Zygobeyrichia sp., 61

DATE DUE

NB 2-5-71			

GAYLORD · PRINTED IN U.S.A.